Lakshman Chandra De, Promila Pathak, A.N. Rao, P.K. Rajeevan
Commercial Orchids

Lakshman Chandra De, Promila Pathak, A.N. Rao, P.K. Rajeevan

Commercial Orchids

Managing Editor: Magdalena Golachowska

Language Editor: Emefa Monu

DE GRUYTER OPEN

Published by De Gruyter Open Ltd, Warsaw/Berlin
Part of Walter de Gruyter GmbH, Berlin/Munich/Boston

ISBN (paperback): 978-3-11-042638-0
ISBN (hardcover): 978-3-11-042638-0
e-ISBN: 978-3-11-042640-3

Bibliographic information published by the Deutsche Nationalbibliothek The Deutsche Nationalbibliothek lists this publication in the Deutsche Nationalbibliografie; detailed bibliographic data are available in the Internet at http://dnb.dnb.de.

Managing Editor: Magdalena Golachowska
Language Editor: Emefa Monu

www.degruyteropen.com

Cover illustration: © Lakshman Chandra De

Contents

1 Introduction

Orchidaceae are cosmopolitan, occurring in almost every habitat apart from deserts and glaciers, although they are mostly found in the tropics, particularly Asia, South America and Central America. They are distributed above the Arctic Circle, in southern Patagonia and even in Antarctica.
The following list gives a rough overview of their distribution:
Tropical America: 300 to 350 genera
Tropical Asia: 250 to 300 genera
Tropical Africa: 125 to 150 genera
Oceania: 50 to 70 genera
Europe and temperate Asia: 40 to 60 genera
North America: 20 to 30 genera

Orchids comprise the largest family of flowering plants with 25,000 to 35,000 species belonging to 600-800 genera and cover 10% of the flowering plants. They are prized for their incredible diversity in size, shape and colour and attractiveness of their flowers and high keeping qualities even up to 10 weeks. Most orchids originated in the tropical humid forests of Central and South America, India, Sri Lanka, Burma, South China, Thailand, Malaysia, Philippines, New Guinea and Australia. Brazilian *Cattleya*, Mexican *Laelia* and Indian *Cymbidium*, *Vanda* and *Dendrobium* have played a major role in developing present day beautiful hybrid orchids, which number more than 200,000. In international trade, among the top ten cut flowers, orchids rank as sixth and among orchids *Cymbidium* claims the first position, accounting for 3% of the total cut flower production in floricultural crops. Orchids are found in nearly every environment in the world starting from tropical and subtropical to alpine zones, both epiphytically and terrestrially.

Epiphytic orchids like *Thunia, Coelogyne, Cattleya, Laelia, Dendrobium, Calanthe, Bulbophyllum, Aerides, Phalaenopsis, Aranda* and *Aranthera* with thick leaves and succulent stems have Crassulacean Acid Metabolism (CAM) and are drought tolerant with higher water use efficiency. Orchids are organically viable and easily grown in locally available media including fir bark, coconut husk, sphagnum moss, tree fern fibres, coco peat, saw dust and perlite, and frequently a mixture of two or three of these materials. Several local species of *Vanda, Cymbidium, Ascocentrum, Bulbophyllum, Coelogyne, Renanthera, Rhyncostylis, Paphiopedilum, Calanthe, Dendrobium* etc. are in great demand in international market for breeding materials. Among the cut flower crops, *Cymbidium, Dendrobium, Phalaenopsis, Odontoglossum, Oncidium, Cattleya, Paphiopedilum, Vanda, Aeridovanda, Aranda, Mokara, Arachnis, Vascostylis, Renanthera, Rhyncicentrum, Rhyncovanda* etc. are important. Important orchid genera used as potted plants in the international market are *Phalaenopsis, Oncidium, Miltonia, Cymbidium, Paphiopedilum, Dendrobium, Cattleya, Ascocenda, Vanda, Brassia, Miltonia* and *Epidendrum*. The origin and distribution of these and other types of orchid species are given in Tab. 1.1. Tribal people of the North eastern hill region of India use wild orchids for a variety of folk medicine, as orchids are rich in alkaloids, flavonoids, glycosides, carbohydrates and other phytochemicals. Leaves, tubers and pseudobulbs of several species are used for edible purposes. Vanilla, a major spice crop and source of vanillin, comes from *Vanilla planifolia*. *Anoectochilus* leaves are used as vegetables in Indonesia and Malayasia. Pseudobulbs of *Cymbidium maladimum* and *Dendrobium speciosum* and tubers of *Microtis uniflora* and *Caladenia carnea* are eaten. Orchids can be considered as an item for value addition like 'Cilindra' (a gift of a glass flute containing a flowering mini Cymbidium) and 'Stylish Setting' (a festive packaging for special occasions such as Birthdays).

In India, orchids are comprised of 158 genera and 1331 species which grow at up to an elevation of 5000 m. Indian terrestrials are commonly located in humus rich moist earth under tree shades in North Western India. Western Ghats harbour the small flowered orchids. Epiphytic orchids are common in North eastern India which grow at up to an elevation of 2000 m from sea level. Indian orchid species with high ornamental values and used as breeding materials are *Aerides multiflorum, Aerides odoratum, Arundina graminifolia, Arachnis, Bulbophyllum, Calanthe masuca, Coelogyne elata, Coelogyne flavida, C. corymbosa; Cymbidium aloifolium, Cym. lowianum, Cym. devonianum, Cym. hookerianum, Cym. lancifolium,; Dendrobium aphyllum, Den. nobile, Den. chrysanthum, Den. farmeri, Den. densiflorum, Den. moschatum, Den. fimbriatum, Den. jenkinsii; Paphiopedilum venustum, P. spicerianum, P. hirsutissimum, P. insigne, Phaius wallichii, Pleione praecox, Renanthera imschootiana, Rhyncostylis retusa, Thunia alba, Vanda cristata, Vanda coerulea* and *Vanda coerulescens*.

In India, some of native genera like *Cymbidium, Renanthera, Paphiopedilum, Vanda, Arachnis* and *Dendrobium* are cultivated on a large scale for cut flower production. The Cymbidium is mainly grown in NEH Region, Sikkim, Darjeeling hills, Arunachal Pradesh and Assam. Tropical orchids are cultivated in Kerala and some parts of Tamil Nadu. We should choose those species that flower during winter and spring months to export to temperate regions from December to May. The orchids have taken a significant position in cut flower industry due to its attractiveness, long shelf life, high productivity, right season of bloom, easy in packing and transportation.
Ten orchid species under Indian Wildlife Protection Act, 1972 (Schedule VI, Sec.2)

- *Paphiopedilum charlesworthii* (Rolf.) Pfitz.
- *Paphiopedilum druryi* (Bedd.)
- *Paphiopedilum fairrieanum* (Lindl.)
- *P. hirsutissimum* (Lindl.) Stein
- *Paphiopedilum insigne* (Wall ex Lindl.)
- *Paphiopedilum spicerianum* (Rchb.f.ex Mast & Moor)
- *Paphiopedilum villosum* (Lindl.)
- *Paphiopedilum wardii* (Summerh)
- *Vanda coerulea Griff.* Ex Lindl)
- *Renanthera imschootiana* (Rolfe)

Tab. 1.1: Origin and habit of different types of orchid species

Category	Genera	Species	Description	Habitat
European open habitat orchids	*Ophrys, Orchis, Serapias, Aceras, Gymnedia*	*Ophrys apifera* *Ophrys fusca*	A new shoot grows from each underground tuber during the autumn to form rosette of leaves in winter or early spring	Suited for hot and dry in summer, and cooler and damper in autumn and winter. 12 cm plastic pots filled up with old and fresh compost (leaf mold and bark) are useful.
European Woodland Orchids	*Epipactis, Cephalenthera, Listera, Cypripedium*	*Cypripedium calceolus*	Summer flowering, die down in winter and do not form tubers	A compost mixture with extra leaf mold and more frequent watering in summer.
African Terrestrials	*Disa, Stenoglottis, Eulophia*	*Disa corunata,* *D. kewensis, D. uniflora,* *D. cardinalis* *Stenoglottis fimbriata,* *S. longifolia* *S. woodii* *Eulophia petersii*	*Disa* produces tall racemes of many bright coloured flowers *S. fimbriata* bears basal rosette of spotted leaves. *S. longifolia* produces a tuft of fresh green strap shaped leaves with white or lilac flowers with pink spots and 70 cm tall spikes. *S. woodii* have white or pink flowers in short spikes. *E. petersii* have hardy fleshy leaves, flowers loosely spaced along a spike and white lips. *E. guineensis* bears large flowers with pale pink lips. *E. streptopetala* bear pseudobulbs which produce soft broad leaves and tall flower spikes. Flowers are bright yellow with brown spotted greenish sepals.	Disas grow near streams or in damp grass lands at a minimum night temperature of 11°C or in a potting mixture of equal parts of peat and perlite. *Stenoglottis* grows terrestrially or on mossy rocks.
African Epiphytes	*Bulbophyllum, Polystachya, Ansellia, Micrococelia, Aerangis, Angraecum, Cyrtorchis, Mystacidium, Rangaeris, Tridactyle, Taeniophyllum, Polyrhizza*	*Bulbophyllumpurpureorachis, B. falcatum,* *B. buntingii, B. barbigerum* *Polystachya virginea* *P. lawrenceana, P. bella* *Ansellia africana,* *Aerangis verdickii,* *A. mystacidi, Angraecum infundibulare,* *A.distichum, A. erectum*	*B. purpureorachis* bear leathery leaves. Flower spikes are dark purple knife blades that are spirally twisted and individual flowers are purple. *B. falcatum* has a flattened yellow green rachis with two rows of tiny red and yellow flowers along its midrib. *B. buntingii* has 12 cm tall spikes, leathery leaves and cream coloured fragrant flowers. *B. barbigerum* with red hairy lips. *P. virginea* is 20 cm tall bearing helmet shaped white flowers. *P. lawrenceana* forms a small hummock of fresh green strap shaped leaves over which hang grayish green and pink flowers. *P. bella* has erect stems with 20 gold or orange nodding flowers. *A. africana* bears highly scented spotted flowers borne in panicles from the apex of the mature canes. *A. verdickii* is a vigorous plant bearing 12 flowers on a raceme.	All tropical bulbophyllums grow on cork bark in a humid semi-shaded location where the temperature range is 15-27°C. *Polystachyas* grow on cork oak bark or in a small pot with epiphytic compost. A minimum night temperature of 15°C is needed.

Tab. 1.1: Origin and habit of different types of orchid species (continued)

Category	Genera	Species	Description	Habitat
Madagascar orchids	*Angraecum, Eulophiella, Cymbidiella, Aeranthes*	*Angraecum sesquipedale* *A. eburneum* *A. compactum* *Eulophiella roempleriana* *Cymbidiella flabellata* *Aeranthes caudata* *Aeranthes grandiflora* *Aeranthes ramosa*	*Angraecum sesquipedale* has leathery leaves and silvery blooms. *A. eburneum* are 2 m tall with a clump of green strap shaped leaves and 1 m long spikes bearing 20 flowers of whitish green colour. *A. compactum* has fleshy leaves and star shaped white flowers. *E. roempleriana* a rhizomatous clump forming large plant bearing 25 pink colored flowers. *C. flabellata* has grassy foliages and yellowish green flowers with purplish black spots. *A. caudata* has pale green flowers with long graceful petals and sepals. *A. grandiflora* with large white translucent flowers having pale green tips to the petals. *Aeranthes ramosa* bears pale green flowers.	Angraecums grow on coarse compost with high humidity. *Aeranthes* are grown in a shady humid location with a winter night temperature of 16°C in hanging baskets.
Australian Temperate Terrestrial	*Pterostylis, Diuris, Caladenia, Corybas*	*Pterostylis angusta, P. nutans, P. baptistii, P. vittata, Diuris palustris*	*Pterostylis* species are tuber propagated and characterized by the presence of a hood which is formed by the fusion of the dorsal sepals and lateral petals of the flower. The hood is usually green, streaked with white, red or brown.	Australian temperate terrestrials grow in a symbiotic relationship with a mycorrhizal fungus which obtain organic matter from the compost. *Diuris palustris* grows in marshy areas.
Australian terrestrial orchids	*Phaius, Calanthe, Malaxis*	*Phaius tankervilleae* *Phaius australis var. bernaysii* *Calanthe triplicata* *Malaxis latifolia*	*P. tankervilleae* bears 2 m tall spikes bearing white flowers on the outside and is gingery brown, except for the lip which is pale on the outside and rose-pink within. *Phaius australis var. bernaysii* has yellow and white flowers. *Calanthe triplicata* bears deep green ribbed leaves with protruded white flowering spikes. *Malaxis latifolia* has handsome foliage and a raceme of tiny pale green flowers which turn purple in a few days	Tropical terrestrials grow in hot, wet summer and rest during the winter. They require humus rich rain forest soils.

continued **Tab. 1.1:** Origin and habit of different types of orchid species

Category	Genera	Species	Description	Habitat
Australian Epiphytes	*Cymbidium, Bulbophyllum, Dendrobium, Sarcochilus*	*Cymbidium madidum* *C. canaliculatum* *C. suave* *Bulbophyllum baileyi* *Dendrobium bigibbum* *D. discolor* *D. tetretifolium* *D. tetragonum* *D. speciosum* *D. delicatum* *Sarcochilus ceciliae, Sarcochilus falcatus, S. hartmannii, Phalaenopsis amabilis var papuana*	*C. madidum* has glossy green stout leaves and fleshy, yellow brown to green long lasting flowers borne on pendulous spikes. *C. canaliculatum* with thick, fleshy, grey green leaves and brown and red flowers through purple to green. *C. suave* with a soft grassy foliages and a longer slender stem and greenish, scented flowers. *B. baileyi* is the largest species with solitary flowers (4 cm diameter) that are white or yellow with purple spots. *D. bigibbum* (floral emblem of Queensland) has 45 cm tall pseudobulbs with a tuft of leaves at the top. *D. discolor* forms multistemmed, less vigorous and 30 cm long arching racemes. The petals and sepals are brown and yellow, twisted and curled. *D. teretifolium* bears small white flowers each winter. *D. tetragonum* has square pseudobulbs with a tuft of dark green leaves at the end of each one. The flowers are yellow and brown or green and white. *D. speciosum* has long spikes of densely packed cream coloured flowers. *D. delicatum* has pink or white flowers. *S. ceciliae* with compact foliages and racemes of pink flowers. *S. falcatus* has pure white flowers with with yellow and orange striped lips. *P. amabilis* var *papuana* has pendulous leaves and 5 to 10 white flowers borne on arching inflorescence.	Being a large plant *C. madidum* grows in a 30 cm pot under shady location at a night temperature of 16°C. *C. canaliculatum* requires sufficient sunlight with a night temperature of 18°C. *C. suave* grow well in hollow logs in a cooler glass house. Bulbophyllums grow on slabs of cork oak bark in a shady location with a night temperature of 15°C. *D. tetretifolium* and *D. tetragonum* grows in shadier, cooler green houses with a winter night temperature of 16°C. *D. speciosum* grow well in a cooler glass house with a winter night temperature of 14°C. *P.amabilis* var *papuana* grow in a shady and very humid location with a winter night temperature of 18°C.

continued **Tab. 1.1:** Origin and habit of different types of orchid species

Category	Genera	Species	Description	Habitat
Asian terrestrials	*Pleione, Paphiopedilum, Acanthephippium, Ludisia, Tainia, Anectochilus, Macodes*	*Pleione praecox,* *P. maculata, P. lagenaria,* *P. humilis* *Ludisia discolor* *Macodes cominsii* *Tainia hookeriana* *Acanthephippium sylhetense* *Calanthe vestita* *C. rubens* *C. arisanense C. discolor* *P. rothschildianum* *P. delentii* *Paphiopedilum fairreanum* *P. insigne* *P. micranthum* *P. wardii* *P. philippinense* *P. concolor* *P. insigne* *P. fairrieanum*	*P. praecox* bears slightly scented pink flowers with a darker frilled lip. *P. maculata* has smaller pink or white flowers whereas *P. lagenaria* has a deep pink colour with a mauve flush down the centre of each sepal and purple streaks on the lip. *P. humilis* has pale pink sepals and petals and a white frilled lip with reddish brown streaks and a conical pointed pseudobulb. *P. formosana* has rosy pink sepals and petals and a white lip with raised yellow markings and orange blotches. *L. discolor* has redish green leaves overlaid with bronze and velvety white blooms. *M. cominsii* has a fleshy heart shaped leaf with green purple veins and markings. *T. hookeriana* has petiolate leaves and 1 m tall flower spikes carrying 25 flowers marked with brown lines and a white lip. *A. sylhetense* has bold plicate leaves arising from stout conical pseudobulbs and waxy scented, cup shaped and creamy flowers with purple markings. *C. vestita* has vigorous leaves and 1.5 m long flower spikes bearing white flowers with a crimson lip. *C. rubens* bears dark pink flowers with a pinkish lip. *C. arisanense* has white flowers with a hint of mauve. *C. discolor* is purple with a pale pink lip. Paphiopedilums are recognized by their basal fans of leaves from the centre from which flower scape emerges. *P. rothschildianum* has long, strap shaped, uniform deep green leaves with creamy maroon dorsal sepals whereas *P. wardii* and *P. delentii* have shorter, dark green mottled, light green or silver leaves and slipper shaped lip and staminode column. *P. micranthum* has dark green leaves, delightful flowers, short, rounded petals and pale green sepals tinged with pink and veined with maroon. The lip is very large, spherical, and white with rosy pink blush. *P. philippinense* has several flowers in each inflorescence with pale purple and green petals, white dorsal sepals with purple markings and a yellowish green lip with purple veins. *P. concolor* has rosette of mottled leaves, attractive creamy flowers and rounded petals and sepals marked with purple spots. *P. insigne* has pale narrow leaves, a broad pale brown lip and pale brown petals with wavy edges. The dorsal sepal is white and green and heavily marked with purple blotches. *P. fairrieanum* has a white petals boldly marked with maroon veins and hairs; the petals turn backwards at their tips. The dorsal sepal is white with a network of maroon veins and the lip is pale green.	Pleiones grow in the mats of moss and fallen leaf litter that accumulate in cracks among rocks and on mountain slopes. They prefer a bark based compost. Autumn flowering species require warmer conditions while spring flowering species are kept in a cold greenhouse. *Anectochilus, Ludisia* and *Macodes* need warm temperature with a winter night temperature of 16°C, shade from direct sunlight and a high relative humidity. *T. hookeriana* grows in a semishaded position with a winter night temperature of 18°C and a high relative humidity. Deciduous calanthes grow in a sunny position with a winter night temperature of 18°C and a high relative humidity. Evergreen calanthes require a semishaded location with a minimum night temperature of 13°C and a relative humidity of 75%. They are heavy feeders and prefer leaf mould. *P. fairreanum* and *P. insigne* bloom at a winter night temperature of 13°C and need a potting mixture of bark, peat and charcoal.

continued **Tab. 1.1:** Origin and habit of different types of orchid species

Category	Genera	Species	Description	Habitat
Asian Epiphytes	*Taeniophyllum, Microcoelia, Phalaenopsis, Vanda, Aerides, Dendrobium, Cymbidium*	*Paphiopedilum parishii, P. lowii* *Taeniophyllum* *Phalaenopsis amabilis* *P. stuartiana* *P. schilleriana* *P. mannii* *P. gigantea* *Vanda coerulea* *V. tessellata* *V. tricolor* *V. lamellata* *Aerides falcata* *A. houlletiana* *Dendrobium discolor* *D. lasianthera* *D. antennatum* *D. noblie* *D. farmeri* *D. primulinum* *D. alaticaulinum* *D. subuliferum* *D. simplex* *D. cuthbertsonii* *Cymbidium ensifolium* *C. eburneum* *C. lowianum* *C. finlaysonianum* *C. aloifolium* *C. insigne* *C. traycyanum* *C. faberi* *C. devonianum* *C. dayanum* *C. elegans*	*Taeniophyllum* has diminutive pin headed flowers. *Microcoelia* has small brown scales and photosynthesizes in the roots. *Phalaenopsis amabilis* has 1 m long inflorescence with 20 flowers of whitish yellow lips and red markings. *P. stuartiana* and *P. schilleriana* have narrow deep green leaves marbles with silver. *P. stuartiana* has white flowers marked with yellow on the lip and peppered with maroon spots. *P. schilleriana* has pale pink flowers with yellow markings and red spots on the lip. *P. mannii* has plain green leaves and a short inflorescence with star shaped six pale yellow flowers with brown bars across the segments. *P. gigantea* are waxy, grey green coloured hanging like a bunch of grapes underneath the foliage. Flowers are pale yellow or pink and ground obscured by heavy brown blotches. *V. coerulea* has white or blue flowers with bold network of royal purple markings. *V. tessellata* has pale green flowers tessellated with pale brown markings. *V. tricolor* var. *suavis* has white flowers marked with magenta blotches. *V. lamellata* has slender flowers of pale lemon colours with red streaking on the lateral sepals. Aerides bear waxy flowers clustering short and pendulous rachis. *A. falcata* has white flowers marked with striking magenta spots. *A. fieldingii* has purple flowers with white markings. *A. houlletiana* has brown yellow flowers with orange markings at the tip of each segment. *Dendrobium discolor* produces 2 m tall canes and 30 cm long inflorescences. The flowers are cream, gold or brownish and petals with wavy margins. *D. lasianthera* produces taller inflorescences with 20 bronze flowers with purple markings on the lip. *D. antennatum* has densely crowded pseudobulbs with light green lanceolate leaves, 30 cm long arching inflorescences bearing 12 white flowers and pink markings in the lip. *D. noblie* has 60 cm tall pseudobulbs and deciduous leaves. The older canes remain leafless at any time of the year. The flowers are fragrant, long lasting, white with magenta tips to each tepal and a deep purple blotch in the throat. *D. farmeri* has 4 cm diameter pale pink flowers with a soft yellow lip. *D. primulinum* has a pendulous habit and flowers on leafless canes. Flowers are white tinged with pink and a pale yellow lip. *D. subuliferum* has fine grassy foliage with white flowers. *D. alaticaulinum* has intense orange colour at the tips, *D. simplex* are white dotted with purple on the outside and pale green inside. *D. cuthbertsonii* has pink. Orange and scarlet flowers.	Taeniophyllums and Microclelias grow upon the twigs of rain forest trees. *Phalaenopsis* requires epiphytic composts, a shady warm green house with a minimum night temperature of 16°C and frequent watering and misting. *Vandas and Aerides* need a lot of sunlight to induce flowering, a coarse epiphytic compost, and a winter night temperature of 18°C. *D. discolor, D. lasianthera and D. antennatum* grow well in green houses with a lot of warmth and light, and a winter night minimum temperature of 18°C and a high relative humidity. They require clay pots filled with broken crocks and coarse epiphytic composts. *D. noblie* *D. farmeri* and *D. primulinum* grow in a light airy green house with a night minimum temperature of 15°C and in small clay pots of medium epiphytic composts. *D. alaticaulinum* *D. subuliferum* *D. simplex* and *D. cuthbertsonii*

Tab. 1.1: Origin and habit of different types of orchid species *continued*

Category	Genera	Species	Description	Habitat
			Cymbidium ensifolium has Inflorescence bearing 8-10 small green flowers with streaked red and white lip. *C. eburneum* has fragrant, white to ivory colour flowers with deep yellow lip. *C. lowianum* has tall arching flower spikes of long lasting qualities with 30 apple green flowers. *C. finlaysonianum* has clumps of hard, leathery, dark green, strap shaped foliage and 1.5 m long pendulous flower spikes carrying yellowy green flowers with purple markings on the lip. *C. aloifolium* has shorter leaves and a 60 cm long flower spike bearing pale green flowers with brown markings down the middle of the segment and red brown lip. *C. insigne* has an erect flower spike, 1.5 m tall with 12 flowers at the top. Flowers are white suffused with palest pink and lip marked with deeper pink blotches. *C. tracyanum* has an arching flower spike, 1.5 m long densely crowded with large pale browny yellow flowers having deep red markings on the petals and lip. *C. faberi* has slender grassy foliage with 30 cm tall spikes bearing pale green flowers with red markings and a deep green lip with crimson streaks. *C. devonianum* has rounded leaves and pendulus flower spikes bearing pale yellow flowers thickly streaked with purple. *C. dayanum* holds the flowers on short spikes close to the base of the leaves. The flowers are star shaped and pale cream in colour with bold red line along the centre of each petal and sepal and red lip with yellow markings. *C. elegans* bears many flowered racemes of pale yellow, funnel shaped flowers.	require cool house with a minimum temperature of 14°C, 80% relative humidity and a good air circulation. *Cymbidium ensifolium* *C. eburneum* and *C. lowianum* require cool, light and airy conditions. *C. aloifolium* and *C. finlaysonianum* need a warm humid climate, quite light with a night minimum temperature of 18°C and a coarse epiphytic compost. *C. insigne* *C. traycyanum* *C. faberi* *C. devonianum* *C. dayanum* and *C. elegans* grow well under cool climate with a minimum temperature of 11°C and good air movement.

continued **Tab. 1.1:** Origin and habit of different types of orchid species

Category	Genera	Species	Description	Habitat
American Terrestrials	*Peristeria*, *Phragmipedium*	*Peristeria elata* *Phragmipedium schlimii* *P. longifolium* *P. caudatum*	*Peristeria elata*, the national flower of Panama has robust plants with flat round pseudobulb and the deciduous plicate leaves. Flowers are white, cup shaped and scented. *Phragmipedium schlimii* has grassy leaves and 30 cm tall spikes bearing white flowers with a deep pink lip. *P. longifolium* has glossy dark green foliage and pale apricot flowers with green vein. *P. caudatum* has greenish yellow long petalled flowers with dark green veins and a red hairy margin to the petals.	*Peristeria elata* grow in a semi-shaded location with a winter night minimum temperature of 18°C and in shallow clay pans filled with broken crocks and leaf mould. *Phragmipedium schlimii* and *P. caudatum* grow in a semi-shaded position in a cool intermediate glasshouse with a night minimum temperature of 14°C and plenty of ventilation. *P. longifolium* grow in warm humid conditions with a minimum night temperature of 14°C and medium grade epiphytic compost.

continued **Tab. 1.1:** Origin and habit of different types of orchid species

Category	Genera	Species	Description	Habitat
American Epiphytes	*Cattleya, Laelia, Stanhopea, Gongora, Oncidium, OdotoglossumLemboglossum, Osmoglossum, Rossioglossum, Miltonia, Epidendrum, Encyclia, Mesdavallia, Dracula, Cypripedium*	*Cattleya bowringiana* *C. skinneri* *C. mossiae* *C. percivaliana* *C. maxima* *C. warscewiczii* *C. tricolor* *Laelia rubescens* *L. gouldiana* *L. anceps* *Stanhopea ecornuta* *S. tigrina* *Gongora galeata* *G. quinquenervis* *Oncidium parviflorum* *O. cheirophorum* *O. splendidum* *O. cocciferum* *O. papilio* *Lemboglossum cervantessi* *L. bictoniense* *L. maculatum* *Rossioglossum splendens* *Epidendrum difforme* *E. ciliolare* *Encyclia fragrans* *E. cochleata* *E. citrine* *Dracula chimaera* *Dracula bella* *Masdevallia coccinea* *M. angulata* *M. barlaeana* *M. macrura*	*Cattleya bowringiana* has an arching spikes bearing 10 flowers, rosy mauve with darker markings on the lip and a white throat. *C. skinneri*, a summer flowering species. *C. mossiae*, a spring flowering species has a large pale pink flower with a splash of orange and deep pink veining on the lip. *C. percivaliana* has narrower petals with deep pink lip. *C. maxima* has weak stems with intricate pattern of deep pink veins along the lip. *C. warscewiczii* has pale pink flowers with a darker pink lip. *C. tricolor* has long, narrow and open petals, golden colour with a creamy lip outside and bright gold with red markings inside. *Laelia rubescens* bears cluster of upto seven flowers produced at the each end of each one. The flowers are white through pink and lavender to mauve. *L. anceps* is characterized by one leaved, angled pseudobulbs and one metre long flower spikes. Flowers are one to many, rosy pink, rich pink inside the velvety lip and with deeper coloured veins and a yellow callus. *L. gouldiana* has pseudobulbs bearing two to three leaves and rounded flowers. This species is a natural hybrid between *L. anceps* and *L. autumnalis*. Petals are rosy pink with a dark pink strike at the down the centre, while the lip is dark pink with darker veins and a central yellow spot. *Stanhopea ecornuta* has creamy yellow flowers with purple spots. *S. wardii* has cream coloured waxy flowers with red spots. *S. tigrina* bears cream coloured flowers obscured by the large red blotches all over the flower. *Gongora galeata* is pale brown colour with vivid brown lip. *G. quinquenervis* has scented yellow flowers covered in a profusion of red spots. *Oncidium parviflorum* has 1 m long spike bearing pretty small flowers. *O. cheirophorum* is equipped with 6 cm long leaves and 20 cm long arching flower spikes bearing tiny, frilly and yellow scented flowers. *O. splendidum* has thick hard leaves and a tall branching flower spikes carrying red and yellow flowers. *O. cocciferum*, a small plant with tall flower spikes bearing small reddish brown flowers with yellow tips to each sepal and petal. *O. papilio* has a slender stem with attractive red and yellow lip and sepals. *Lemboglossum cervantessi* has white and pink flowers with concentric circle of maroon bars around the column. *L. bictoniense* has a broad pink lip with green ad red mottled sepals.	Cattleyas require a great deal of sunlight and medium to coarse compost. Brazilian and Mexican laelias grow at lower temperature under bright sun light and in shallow, slatted wooden baskets. Stanhopeas and gongoras are grown in a semi-shaded location with a winter minimum night temperature of 16°C and high relative humidity. A medium grade epiphyte compost is ideal for them. Oncidiums require high humidity and high light levels. Odontoglossum alliances require a cool and airy location and a winter night minimum temperature of 14°C and grown as potted orchids using a medium epiphytic compost. Masdevallia and Dracula require a night temperature of 14°C and a fine epiphytic compost and liberal watering.

continued **Tab. 1.1:** Origin and habit of different types of orchid species

Category	Genera	Species	Description	Habitat
			L. maculatum have red petals and white petals spotted with red. *Rossioglossum splendens* has flowers of a blazing orange colour with a yellow lip marked with red spots. *Epidendrum difforme* are 30 cm tall, fleshy and translucent and bears pale green flowers. *E. ciliolare* are arranged with lime green star flowers and white lip with a shaggy fringed margin. *Encyclia fragrans* is scented, with whitish green flowers and violet striped lip. *E. cochleata* has pale green, slender and reflexed sepals and petals and shiny dark maroon lip stands over the column like a hood. *E. citrina* is distinguished with its pendent, grey green pseudobulbs and foliage. Flowers are yellow and buttercup sizes. *Dracula chimaera* has white sepals blotched with white and red lip. The sepals are dotted with short white hairs. *Dracula bella* has yellow sepals with deep red spots and long trailing tails at the tip of the sepal. Tip is white. *Masdevallia coccinea* with white, yellow and magenta flowers. *M. angulata* has tubular olive green flowers with deep red blotches. *M. barlaeana* is vivid magenta coloured. *M. macrura* has orange flowers with red spots and long twisty yellow tails. *Cypripedium reginae* forms splendid clumps of stem with attractive pink and white flowers. *C. acaule* has a large pink lip and smaller brownish sepals and petals. *C. calceolus* has bright yellow lip and yellow, green and brownish sepals and petals.	

Tab. 1.2: Present status of orchid distribution in India (Hajra and De, 2010)

Region	States	Genera	Species
Eastern Himalayas and North Eastern India	Arunachal Pradesh, Assam, Manipur, Meghalaya, Mizoram, Nagaland, Sikkim and Tripura	159	870
North-Western Himalayas	J& K, Himachal Pradesh, Uttarakhand	75	288
Peninsular India	M.P., Orissa, Andhra Pradesh, Gujrat, Central India, Eastern Ghats and Western Ghats	89	379
Andaman and Nicobar Islands	319 Islands and islets in the Bay of Bengal	53	115

2 Global Orchid Industry

Floriculture has emerged as an immense entrepreneurial opportunity for small and marginal farmers and a way forward to earn foreign exchange. Floriculture business deals with cut flowers, pot plants, cut foliages, seeds, bulbs, tubers, rooted cuttings and dried flowers or leaves. Major floricultural crops traded internationally include roses, carnations, chrysanthemums, gerberas, gladiolus, gypsophilas, liatris, nerines, orchids, achillea, anthuriums, tulips and lilies. With the increase demand for flowers, floriculture has become one of the important commercial trades in Indian Agriculture. Therefore, commercial floriculture has emerged as a high–tech activity taking place under controlled environments inside greenhouses. Indian floriculture is now viewed as a high growth industry, particularly as the liberalization of industrial and trade policies have paved the way for development of export oriented floriculture. The floral industry is a highly dynamic business. The characteristics of varieties, the origin of production, production technologies, markets and retailing systems as well as individual products are all undergoing continuous change, while challenging the adaptive capacity of the actors involved. In a slowly but steadily growing world market, new developing country exporters are increasing market share at the expense of existing producers. These producers try to remain ahead by increasing productivity and through diversification and innovation. Kenya, Ecuador and Zimbabwe, the rapidly rising exporters of the last decade, have become established suppliers to their ambitious new competitors including China, India, The Republic of Korea, Malaysia, Malawi, Mexico, Palestine, Peru, South Africa and Zambia. India is the world's fastest growing destination for tourism. Flower consumption growth rises by up to 30% per annum and numerous festivals, along with increasing modernization and per capita income, make India a floral super power of the future. Our country is emerging as the world's fastest growing flower and gardening market.

About 190 thousand ha area has been brought under flower cultivation during 2011-12. During 2010-2011, the production of flowers was estimated to be 870.4 metric tons of loose flowers and 43417.5 million of cut flowers. The government of India has considered floriculture as a sunrise industry and accorded it 100% export oriented status. The country has reached the ability to export 30926023 metric tons of floriculture products to the world for the worth of Rs. 365.32 crores in 2011-12. USA, Germany, UK, Japan, Canada and UAE were the major consumers of Indian floriculture. While the Indian floriculture industry is likely to reach the Rs. 8000 crore mark by 2015, it currently stands at about Rs. 3700 crores with the share of 0.65% in the international floriculture sector which is likely to reach 0.89% by 2015.

2.1 Global Orchid Trade

Imports and exports of floriculture products from all round the world are increasing year by year because industry professionals in India are always seeking new products, techniques and devices that are effective and economic. As a result, floriculture trade is flourishing and research centres and species nurseries have turned into a promising business. In addition, India has the advantage of favourable agroclimate, low cost labour, arable land, skilled manpower, proximity to emerging floriculture markets such as Japan, Australia and the Middle East, and conducive growing weather during the peak time of worldwide demand in November - March in the northern plains as well as mild climate in the southern part of India throughout the year for tropical flowers. The recent export situation of orchids focusing on India is given below (Tab. 2.1)

Tab. 2.1: Export of orchids from India (Rs. in lacs & quantity in MT)

Country	2010-2011		2011-12		2012-13	
	Quantity	Value	Quantity	Value	Quantity	Value
Bahrain	0.00	0.00	0.00	0.00	0.20	1.43
Maldives	0.04	0.18	0.13	0. 56	0.27	0.96
Kenya	0.00	0.00	0.04	0.05	0.06	0.09
Sri Lanka	0.00	0.00	0.00	0.00	0.02	0.05
Quatar	0.00	0.00	0.01	0.02	0.00	0.00
Singapore	4.77	3.08	0.00	0.00	0.00	0.00
UAE	0.04	0.06	0.20	0.35	0.00	0.00

Source: DGCIS Annual report

The orchid has taken a significant position in the cut flower industry due to its attractiveness, long shelf life, high productivity, right season of bloom, ease of packing and transportation. Orchids account for a large share of global floriculture trade both as cut flowers and as potted plants and are estimated to comprise around 10% of the international fresh cut flower trade. The average trade value of fresh cut orchids and buds trade during 2007-2012 was US$ 483 million. In 2012, there were more than 40 and 60 exporting and importing orchid countries, respectively around the world, and the total size of the global trade was US$ 504 million (Tab. 2.2).

Tab. 2.2: Value of fresh cut orchids and buds global trade (2007-2012) (Unit: Million US$)

Year	2007	2008	2009	2010	2011	2012
Import	233,734,023	252,647,645	232,568,129	251,445,523	265,702,077	267,196,847
Export	230,470,421	238,702,950	217, 781, 745	227,389,789	244,996,271	237,543,797
Total	464,204,444	491,350,595	450,349,874	478,835,312	510,698,348	504,740,644

Source: Department of Foreign Trade, Thailand (2013)

The Netherlands is the top orchid exporting country in the world (39.67% of world orchid market) followed by Thailand (28.41%), Taiwan (10%), Singapore (10%) and New Zealand (6%). Importing countries are mainly Japan (30%), UK (12%), Italy (10%), France (7%) and the USA (6%). The total orchid cut flower trade of the world mostly consists of 85% Dendrobium species and 15% Phalaenopsis and Cymbidium species, and Asia is the main source of orchid to enter the world market (Cheamuangphan *et al*, 2013)

Major markets for orchids in Asia are occupied by Japan and Singapore. The total imports of orchids by Japan accounted for US$ 57.4 mn in 2008 making it the largest importer of orchids in the world. The main sources for these imports include Thailand, Taiwan, New Zealand and Malayasia which together account for as much as 96.5% of the total orchids imports by Japan in 2008. Imports by Singapore of fresh orchids amounted to US$ 6.5 mn in 2007 with Malayasia, Thailand and Taiwan being the main sources of imports for the country. In contrast, imports of fresh orchids by Singapore from India were only US$ 1379.3 representing a share of 0.02% of the country's total imports of the product in 2007 (Tab. 2.3). This clearly indicates that there are vast possibilities for increasing India's exports to Singapore particularly considering the proximity of the country and India's East Policy.

Tab. 2.3: Imports of orchids by Singapore (2007)

Country	Value (US$)	Share (%)
Malayasia	5422069.0	83.03
Thailand	520699.7	7.97
Taiwan	307596.2	4.71
Mauritious	149655.2	2.29
Indonesia	117241.4	1.90
Netherlands	8275.9	0.13
New Zealand	2758.6	0.04
India	1379.3	0.02
China	699.7	0.01
Total imports	6530344.8	100.00

Source: International Enterprise Singapore

2.2 Trend in Growth

2.2.1 Cymbidium

Cymbidiums are among the most popular winter and spring blooming semi-terrestrial orchids originating from tropical and subtropical Asia; covering North Eastern India, China, Japan, Malayasia, the Philippines, the Borneo islands and North Australia, usually grown in cooler climates at high elevations. Cymbidiums are highly valued for genetic resources, cut flowers, hanging baskets, potted plants and herbal medicines. Cymbidium has been considered the top commercial orchid in Europe for many years. They fetch the highest price in the international markets of which major destination include Asian markets of Singapore and Japan or the Dutch market. Cymbidiums imported from the Netherlands fetched as much as US$ 11.18 per stem in Singapore and those imported by Japan from New Zealand fetched US$ 3.33 per stem. As far as the Dutch Auction market is concerned, the cymbidiums fetched the highest value, averaging Euro cents 331 per stem during 2003-2007 (Tab. 2.4).

Tab. 2.4: Average annual prices at Netherlands auction (2003-2007) (Euro cents /stem)

Orchids	2003	2005	2007	Average price
Phalaenopsis	38	46	37	40
Cymbidium (Big bud)	330	334	329	331
Cymbidium (Small bud)	138	148	140	142
Paphiopedilum	58	52	63	58

Source: CBI Market Survey, The Cut Flowers and Foliage Market in the EU

In India, the Arunachal hills, Sikkim and Darjeeling hills with cool summer nights and monsoonal summer rain are ideal for cymbidium cultivation. The growth of orchid exports from the north eastern hill region, especially Sikkim, would provide opportunities for employment as well as the development of supporting industries like packaging, cold storage and transportation. East Sikkim has been declared an Agri Export Zone exclusively for production of cymbidium orchids. In Sikkim, more than 250 hybrids of cymbidium orchids are commercially cultivated in and around 25 ha of land and about 5 lakhs spikes are produced annually.

2.2.2 Dendrobium

Dendrobiums are popular flowering potted plants and cut flowers around the world due to their floriferousness, wide range in flower color, size and shape, year round availability and lengthy vase life. Hawaii, California and Florida are major potted Dendrobium growing regions in the United States. The wholesale value of sales for this commodity in Hawaii has been established for several decades and sales increased from US $ 2.4 million in 1991 to US$ 5.6 million in 2000 (Tab. 2.5).

In the Netherlands, production of potted orchids is now 40 to 50 million units with Dendrobium increasing in popularity. Imports from Thailand, the worlds largest exporter of tropical cut orchids and second largest supplier to the EU, accounted for 22% of supplies to the EU. Thailand holds a particularly strong position in Dendrobium orchids.

Tab. 2.5: Quantity of consumption and wholesale value of sales of potted Dendrobium in Hawaii (Johnson, 1999).

Year	Number of pots sold	Value (US $)	Number of growers
1985	0.2 millions	1.1 million	88
1991	0.4 millions	2.4 million	47
2000	1.0 million	5.6 million	69

Tab. 2.6: Orchid price in Singapore (US$)

Product details	Origin	Price /stem
Orchid Cymbidium	Malaysia	1.97
Orchid Cymbidium	Netherlands	11.18
Orchid Cymbidium	Taiwan	5.26
Orchid Dendrobium XL	Thailand	0.46
Orchid Dendrobium L	Thailand	0.39
Orchid Dendrobium M	Thailand	0.33
Orchid Dendrobium S	Thailand	0.26
Orchid Oncidium XL	Malaysia	0.72
Orchid Oncidium L	Malaysia	0.59
Orchid Oncidium M	Malaysia	0.39

Source: Market News Service-Week 14, 2009, ITC

2.2.3 Phalaenopsis

Phalaenopsis is the second most valuable and popular flowering potted plant and cut flower around the world due to their easy cultural practices, diversity in flower colour, size and shape, year round availability, delicacy and longer vase life. It is commercially grown in Germany, Japan, The Netherlands, Taiwan and United States. In the United States, 75% of all orchids purchased are phalaenopsis and about 13,500,000 phalaenopsis were sold in 2005 in United States. The export value of phalaenopsis from Taiwan to the United States increased from $8 million in 2005 to $13 million in 2006. Worldwide turnover of Taiwanese phalaenopsis increased from $27.5 million to $35.4 million from 2005 to 2006.

2.2.4 Other Tropical Orchids

Vanda is widely distributed throughout Australasia from China through the Philippines, Indonesia, Malaysia, New Guinea and Australia, Myanmar, Thailand, India and Sri Lanka. In the world tropical orchid trade, *Dendrobium* is the most dominant crop in addition to *Mokara, Oncidium, Aranthera, Aranda, Vanda, Arachnis, Renanthera, Ascocenda, Phalaenopsis, Cattleya* and *Paphiopedilum* which are being grown as cut flowers and potted plants. Thailand is the largest world exporter of tropical orchids. China is the largest consumer of orchid cut flowers (7,493 tons from Thailand) followed by Japan, USA, Italy, India, Taiwan, Vietnam and the Netherlands at 4,407, 2892, 2395, 1830, 983, 793 and 689 metric tons, respectively. Other significant orchid genera being exported from Thailand were *Mokara, Aranthes, Aranda, Oncidium, Vanda, Arachnis* and *Ascocenda* with market shares of 3.69, 0.52, 0.48, 0.44, 0.13, 0.01 and 0.01% of total export value respectively (Tab. 2.7).

Tab. 2.7: Orchid export from Thailand in 2009 (% share of total export value)

Cut orchid		Orchid plant	
Orchid genera (20)	Value shared (%)	Orchid genera (211)	Value shared (%)
Dendrobium	94.73	*Dendrobium*	51.4
Mokara	3.69	*Phalaenopsis*	25.5
Aranthera	0.52	*Vanda*	8.9
Aranda	0.48	*Mokara*	3.7
Oncidium	0.44	*Oncidium*	3.1
Vanda	0.13	*Cattleya*	2.7
Arachnis	0.01	*Ascocenda*	1.2
Ascocenda	0.01	*Epidendrum*	0.6
		Cymbidium	0.3
		Rhyncostylis	0.3
		Spathoglottis	0.3
		Paphiopedilum	0.2
		Others	1.8

Considering the import value of orchid plants, Japan was also the largest importer of Thai orchid plants at US$ 2.4 million worth followed by the Netherlands, USA, Germany, Rep. of Korea and Vietnam with values of 1.6, 1.4, 1.0, 0.9 and US$ 0.9 million respectively. In 2009, Thailand exported 211 genera of orchid plants. *Dendrobium* and *Phalaenopsis* plants were the most important orchid genera for exports with market shares of 51.4 and 25.5%, respective. Other important orchid genera exported from Thailand were *Vanda, Mokara, Oncidium, Cattleya* and *Ascocenda* with market shares of 8.9, 3.7, 3.1, 2.7 and 1.2% of total export value, respectively (Lekawatana, 2010).

3 Biodiversity, Conservation and Bio-piracy of Genetic Resources in India

Biodiversity, especially species diversity, indicates the number of species of plants and animals present in a region. Maintaining a wide diversity of species in each ecosystem is necessary to preserve the web of life that sustains all living things.

Orchids, believed to have evolved in this region, form a very noticeable feature of the vegetation here. There are about 25,000 species of orchids estimated to occur in the world. In India, about 1350 species belonging to 186 genera represent approximately 5.98% of the world orchid flora and 6.83% of the flowering plants in India. The Eastern Himalayas and North Eastern; North West Himalayas; Peninsular India; and Andaman & Nicobar Islands are the major orchid regions of India.

3.1 Region Wise Distribution of Orchids

3.1.1 Eastern Himalayas and North-Eastern India

This region includes the Darjeeling district of West Bengal and other North-eastern states, i.e., Arunachal Pradesh, Assam, Manipur, Meghalaya, Mizoram, Nagaland, Sikkim and Tripura. This region is relatively warmer with high humidity and heavier precipitation in comparison to western Himalayan region and endowed with 870 species in 159 genera constituting 72.8% of the total orchid species in the country. A state wise analysis indicates that Arunachal Pradesh has the highest number of orchid species (622) followed by Sikkim (543 species) and Meghalaya with 389 species. Among the other states of this region, Assam accounts for 290, Nagaland for 246, Mizoram for 234, Manipur for 215 and Tripura for 57 species (Hajra and De, 2010).

3.1.1.1 Endemic Orchids

Orchids are distributed from tropical to alpine zones in forest trees, secondary vegetations, river banks, bamboo and palm thickets, forest floor, grassy slopes and rocky areas and are considered an element in Farming System Research (Chowdhery, 1998, 2001). It is to be noted that there are some orchid species which are endemic not only to this region, but also to the home states in which they are distributed in Sikkim and Arunachal Pradesh Himalayas, the Naga and Manipur hills, the Lusai-Mizo hills and Khasi –Jaintia hills (Nayar, 1996).

These are:

- *Dendrobium spatella*, *Dendrobium parciflorum* and *Luisia macrotis* from Assam.
- *Vanda coerulea* and *Dendrobium palpebrae* from Arunachal Pradesh.
- *Renanthera imschootiana* and *Cymbidium tigrinum* from Nagaland.

- *Anoectochilus crispus, Cymbidium eburneum, Habenaria khasiana, Liparis deliculata, Paphiopedilum venustum, Taeniophyllum khasianum and Tainia khasiana* from Meghalaya.
- *Renanthera imschootiana* from Tripura.
- *Dendrobium palpebrae* from Mizoram.
- *Ascocentrum ampullaceum var. auranticum, Epidendrum radicans* and *Vanda stangeana* from Manipur.
- *Calanthe whiteana, Cymbidium whiteae* and *Vanda pumila* from Sikkim (Nayar, 1996)

North East India has the highest flora of monotypic orchid genera (Tab. 3.1) (Tandon *et al*, 2007).

Tab. 3.1: Some monotypic orchid genera of North East India

Anthogonium Wall. Ex Lindl.	*Arundina* Bl.
Acrochaene Lindl.	*Bulleyia* Schltr.
Cremastra Lindl.	*Cleisocentron* Bruhl
Dickasonia L.O. Williams	*Diglyphosa* bl.
Eriodes Rolfe	*Herpysma* Lindl.
Jejosephia A.N. Rao & Mani	*Mischobulbum* schltr.
Myrmechis (Lindl.) Bl.	*Neogyne* reichb.f.
Ornithochilus (Lindl.) Wall. Ex Benth.	*Risleya* King & Pantl.
Renanthera Lour.	*Tipularia* Nutt.

North East India is reported to harbour a large number of valuable threatened orchids (Tab. 3.2).

Tab. 3.2: Some threatened orchids of North East India (Tandon et al, 2007)

Acanthephippium sylhetense	*Bulleyia yunnanensis*
Cymbidium eburneum	*Cym. devonianum*
Cym. tigrinum	*Dendrobium chrysotoxum*
Den. densiflorum	*Den. falconeri*
Eria fragrans	*Paphiopedilum fairreanum*
P. insigne	*P. venustum*
P. villosum	*Renanthera imschootiana*
Vanda coerulea	*Vanilla pilfera*
Vanda pumila	*Epigeneium rotundatum*

A state wise distribution of orchid species of North Eastern states of India is listed in Tab. 3.3.

Tab. 3.3: Distribution of major orchids in North East States (Chowdhery, 2001, 2009; Nayar and Sastry, 1987, 1988, 1990, 1997-1998, 1999)

State	Orchid species
Arunachal Pradesh	*Cymbidium ensifolium, C. grandiflorum, Dendrobium aphylla, D. chrysanthum, D. gibsonii, D. nobile, Paphiopedilum fairrieanum, P. venustum, P. spicerianum, Calanthe masuca, Rhyncostylis retusa*
Assam	*Arundina graminifolia, Eulophia mannii, Goodyera procera, Calanthe angusta, Rhyncostylis retusa, Aerides multiflora, Aerides odorata, Acampe papillosa, Cymbidium aloifolium, Dendrobium aphyllum, D. acinaciforme*
Manipur	*Ascocentrum ampullaceum, Paphiopedilum spicerianum, Vanda amsiana, Vanda stangeana, Vanda coerulea*
Meghalaya	*Paphiopedilum insigne, P. venustum, R. retusa, Coelogyne corymbosa, Phaius tankervilliae, Dendrobium devonianum, Cymbidium elegans, Vanda coerulea*
Mizoram	*Vanda coerulea, Renanthera imschootiana, Paphiopedilum hirsutissimum, P. villosum*
Nagaland	*Goodyera viridiflora, Liparis caespitosa, Luisia trichorrhiza, Malaxis latifolia, Oberonia pyrulifera, Spiranthes sinensis*
Sikkim	Alpine zone*(2500-3000 m): Orchis, Habenaria, Tipularia, Satyrium, Cypripedium* Temperate & Sub-temperate zone*(1850-3500 m): Goodyera, Calanthe, Pleione humilis, Cymbidium hookerianum, C. devonianum, C. longifolium, Coelogyne cristata, Dendrobium hookerianum* Subtropical zone*(850-1250 m): Anoectichilus, Calanthe, Phaius, Eulophia, Paphiopedilum fairrieanum, P. venustum, Dendrobium, Bulbophyllum, Coelogyne, Arachnis.* Tropical Zone *(250-850 m): Phalaenopsis, Dendrobium, Aerides, Vanda, Arundina graminifolia*
Tripura	*Dendrobium (*14 spp.), *Vanda teres, V. coerulea, Renanthera imschootiana*

3.1.2 North-Western Himalayas

This region covers the states of Jammu & Kashmir, Himachal Pradesh, and Uttarakhand. The Western Himalayan region is large and blessed with higher elevations, cooler drier climate and a wide mountain mass. The pre-dominant forest flora are pine, deodar, fir etc. In all, 288 species under 75 genera are found to occur in this region, constituting 24.1% of the total Indian orchids.

Four species of ground orchids found mainly in this region, *viz.*, *Platanthera edgeworthii, Habenaria intermedia, Crepidium acuminatum* and *Malaxis muscifera* form important constituents of *Astavarga* (a group of eight herbs used in preparation of *Chyavanprash*).

Among the epiphytes, *Dendrobium* Sw. and *Bulbophyllum* Thouars. are the largest genera comprising of 16 and 12 species, respectively whereas, among the terrestrial orchids, *Habenaria* Willd. is the largest group with 13 species. 34 species of Orchids are represented by single species in western Himalaya. Eleven species are found to be endemic to this region: namely *Dendrobium normale, Eria occidentalis, Flickingeria hesperis, Gastrochilus garhwalensis, Herminium kumaunensis, Neottia mackinnonii, Neottia nandadeviensis, Peristylus duthiei, Peristylus kumaunensis, Peristylus fallax* and *Ponerorchis renzii* (Jalal *et al*, 2009a, 2009b, 2010a, 2010 b).

3.1.3 Peninsular India

This region comprises Madhya Pradesh, parts of Orissa, Andhra Pradesh, Gujrat, the extra peninsular region of Central India and Gangetic plains along with Eastern and Western Ghats. Western Ghats harbours dense forests with high humidity and rainfall and the vegetation is rich in scrub jungles, moist and dry deciduous forests, tropical evergreen forests and montane grasslands containing huge numbers of orchid species. Eastern Ghats consists of broken hills. Orchid diversity in Eastern Ghats is poor as compared to that in Western Ghats; the region is enriched with 379 species belonging to 89 genera constituting 31.72% of the total Indian orchid flora.

Predominant terrestrial orchid species available in Western Ghats are *Acanthephippium bicolor* Lindl., *Aenhanhreya rotundifolia* Blatt., *Anoectochillus elatus* Lindl., *Calanthe masuca* Lindl., *Calanthe triplicata* Willemet, *Eulophia epidendrea* Koenig., *Epipogium roseum* D.Don, *Geodorum densiflorum* Lamk., *Habenaria longicorniculata* Grah., *H. longicornu* Lindl., *H. multicaudata* Sedgw., *H. roxburghii* Nicolson, *Liparis atropurpurea* Lindl., *Malaxis versicolor* Lindl., *Nervilia aragona* Gaud., *Paphiopedilum druryii* Bedd., *Tainia bicornis* Lindl., *Vanilla walkeriae* Wight., *Zeuxine gracilis* Breda. and *Z. longilabris* Lindl (Sharief, 2011). Some rare and endemic epiphytic species from Western Ghats are *Aerides ringens* Fischer, *Bulbophyllum aureum* Hook.f., *B. fuscopurpureum* Wight, *B. fimbriatum* Lindl., *B. fischeri* Seid, *B. tremulum* Wight., *B. neilgherrense* Wight., *Coelogyne breviscapa* Lindl., *C. nervosa* A. Rich., *Cottonia peduncularis* Lindl., *Cymbidium bicolor* Lindl., *Dendrobium aqueum* Lindl., *D. crepidatum* Lindl., *D. herbaceum* Lindl., *D. microbulbon* A. Rich, *D. ovatum* Lindl., *Diplocentrum recurvum* Lindl., *Eria nana* A. Rich, *E. pseudoclavicaulis* Blatt., *Flickingeria nodosa* Daiz., *Liparis elliptica* Wight., *L. viridiflora* Bl., *Luisia zeylanica* Lindl., *Oberonia brunoniana* Wight., *Papilionanthe subulata* J. Koen., *Pholidota imbricata* Hook., *Rhyncostylis retusa* Bl., *Sirhookeria lanceolata* Wight., *Taeniophyllum alwisii* Lindl., *Trias stocksii* Benth., *Vanda testacea* Lindl. and *Xenikophyton smeeanum* Reich.

Orchids distributed in Eastern Ghats are given in Tab. 3.4.

Tab. 3.4: Orchid flora of Eastern Ghats of India

Name of species	Habit	Flowering time	Remarks
Acampe carinata	Epiphyte	October–January	Flowers light yellow with maroon blotches on sub-umbellate raceme
A. ochracea	Epiphyte	December–Feb	Flowers light yellow with maroon markings on lax panicle
A. praemorsa	Epiphyte		Flowers small creamy yellow with maroon transverse bands clustered on compact corymbose cyme
A . rigida	Epiphyte	August–September	Flowers pale yellow on sub-corymbose raceme
Acanthephippium bicolor	Terrestrial	May	Flowers yellow specked with red
Aerides maculosa	Epiphyte	June	Flowers mild violet on axillary spreading raceme
A .multiflora	Epiphyte	June-July	Flowers purple on long axillary pendulus floral axis
A . odorata	Epiphyte	June-july	Flowers white, lemon scented on extra axillary, deflexed raceme
A . ringens	Epiphyte	June -August	Flowers white on axillary, sub-erect second raceme
Bulbophyllum crassipes	Tropical epiphyte	Octo–Nov.	Flowers small, yellow maroon, densely spirally arranged in catkin like raceme
B. umbellatum	Epiphyte	April–May	Flowers pale yellow with purple spots on umbellate raceme
Calanthe sylvatica	Terrestrial	July	Flowers large, purple coloured
Cleisostoma appendiculatum	Epiphyte	August-October	Flowers pale yellow on recurved raceme
Coelogyne breviscopa	Epiphyte	April	Flowers white raceme with scarious bract, lip 3-lobed
Cottonia peduncularis	Tropical epiphyte	March–June	Flowers dark purple on long and branched raceme with a bee like lip.
Cymbidium aloifolium	Epiphyte	April-June	Flowers pale yellow with purple centre on long pendulus raceme
C. bicolor	Epiphyte	Feb–April	Flowers pale yellow at base and maroon towards top borne on lax raceme
Dendrobium aphyllum	Tropical epiphyte	Feb–April	Flowers pale purple, translucent with sub-panduriform, fenely dendate lip
D. aqueum	Epiphyte	Sept–Nov	Flowers showy white. Lip faintly yellowish
D. bicameratum	Epiphyte	August	Flowers pale yellow, crowded on sub-terminal capitate raceme.

continued Tab. 3.4: Orchid flora of Eastern Ghats of India

Name of species	Habit	Flowering time	Remarks
D. cathcartii	Tropical epiphyte	March–April	Flowers yellow-ochre on leaf opposed fascicles of 2-3 flowers.
D. crepidatum	Tropical epiphyte	Feb–April	Flowers white with a large yellow spot on the lip, 1-3 on very short bracteate peduncle from the swollen nodes of the leafless stems.
D. fimbriatum	Epiphyte	April	Flowers golden yellow. Lip orbicular, fimbriate.
D. formosum	Tropical epiphyte	May-June	Flowers white with bright yellow patch on lip, on subterminal racme.
D. herbaceum	Epiphyte	Feb–March	Flowers greenish white on subterminal condensed racemes.
D. macrostachyum	Tropical epiphyte	May-July	Flowers lemon green, 1-4 in cymes.
D. moschatum	Epiphyte	May–June	Flowers large, spreading, peach coloured, musk scented on a spreading raceme from a node near to apex.
D. regium	Tropical epiphyte	April–June	Flowers magenta coloured with yellow throat on sub-terminal raceme.
Diplocentrum recurvum	Epiphyte	April–August	Flowers small in long racemes, pink to rose or pink or brownish tinged with pink. Lip sessile.
Eria bambusifolia	Epiphyte	Dec–Jan	Flowers pale brown on axillary lax raceme
E . lasiopetala	Tropical epiphyte	March-April	Flowers greenish yellow, lip light maroon, on axillary racemes.
E. pauciflora	Epiphyte	Aug–Sept	1-2 white flowered inflorescence
E. reticosa	Epiphyte		Single flowered
Eulophia epidendraea	Terrestrial	Dec–Jan	Scape laxly many flowered. Flowers green white purple.
E. graminea	Terrestrial	Jan–May	Inflorescence 1-3 per plant, lax flowered produced from the nodes. Flowers inodorous, leaf-green with maroon veins.
E. pulchera	Terrestrial	May–July	Flowers in few-many flowered raceme, greenish purple.
E. spectabilis	Tropical terrestrial	May -June	Flowers greenish white or purple, lax racemes; lip crenate and crisped.
Flickingeria macraei	Epiphyte	July–Sept	Flowers cream with a pale yellow lip, solitary on top of the pseudobulb.

continued **Tab. 3.4:** Orchid flora of Eastern Ghats of India

Name of species	Habit	Flowering time	Remarks
Gastrochilus acaulis	Epiphyte	March–April	Flowers pale yellow, lip white with violet blotch, on axillary corymb; hypochile cup-shaped
Gastrochilus inconspicuum	Tropical epiphyte	June–Sept	Flowers greenish purple on a condensed spike, hypochile cup-shaped
Geodorum densiflorum	Terrestrial	July–Aug	Flowers pinkish white on decurved racemes, lip sub-panduriform.
G. recurvum	Tropical terrestrial	May–June	Flowers white, inodorous on long slender decurved corymbose raceme.
Goodyera procera	Terrestrial	March–May	Flowers minute, white; lip saccate at base.
Habenaria commelinifolia	Terrestrial	Aug–oct	Flowers greenish white, inodorous on long dense spike with a long curved spur.
H. digitata	Terrestrial	July	Flowers greenish white with sickle-shaped flower petals.
H. furcifera	Terrestrial	Aug–Sept	Flowers green on long and stout peduncle
H. gibsoni var. foetida	Terrestrial	July–Aug	Flowers green, foul scented, on lax raceme
H. longicorniculata	Terrestrial	August	Flowers greenish white faintly scented on long peduncled raceme.
H. longicornu	Terrestrial	Aug–Sept	Flowers white, about 8 on lax raceme
H. marginata		Sept–oct	Flowers yellow on many flowered lax raceme
H. ovalifolia	Terrestrial	Aug-Oct	Flowers green. Side lobes of lip linear-lanceolate, midlobe inflexed, ovate-oblong
H. panigrahiana	Terrestrial	Oct–Dec	Flowers green on lax racemes, sweet scented at night
H. rariflora	Terrestrial	Aug–Sept	Flowers white with bipartite petals
H. reniformis	Terrestrial	July–Sept	Flowers green on lax spike
H. roxburghii	Terrestrial	August	Flowers white, mildly fragrant on densely many flowered spike
H. stenopetala	Terrestrial	Oct–Dec	Flowers grayish green with brown lip on densely many flowered raceme.
H. virens	Terrestrial	July–Oct	Flowers white, lip scabrid-pilose, 3-furcate
Kingidium deliciosum	Epiphyte	May–Sept	Flowers pale yellow and purple
Liparis nervosa	Terrestrial	May–Aug	Flowers purple or green tinged with pink on long spikes.

continued Tab. 3.4: Orchid flora of Eastern Ghats of India

Name of species	Habit	Flowering time	Remarks
L. paradoxa	Terrestrial	July	Flowers small, pale yellow to maroon in terminal raceme
L. viridiflora	Epiphyte	Nov–Dec	Flowers pale yellow on slightly recurved, dense flowered raceme
Luisia abrahami	Epiphyte	March–May	Flowers pale green flushed with pale purple
L. birchea	Epiphyte	Throughout the year	Flowers stray coloured, thick. Lip sessile, dark purple.
L. trichorhiza	Epiphyte	March–July	Flowers maroon on extra axillary, condensed raceme
L. zeylanica	Epiphyte	Feb–March	Flowers green and dark purple on extra-axillary condensed raceme
Micropera pallida	Epiphyte	June–Sept	Flowers light yellow on leaf opposed decurved raceme
Nervilia aragoana	Terrestrial	May–July	Flowers many in lax, slightly drooping, nodding, light green in colour
N. crociformis	Terrestrial	May-July	Flowers greenish white, single on top of the peduncle
N. infundibulifolia	Terrestrial	June	Flowers maroon green on one solitary flowered, pale maroon peduncle
N. plicata	Terrestrial	May–June	Flowers two on top of the peduncle, dull brownish green and white
Oberonia brunoniana	Epiphyte	Sept–Oct and Feb-March	Flowers large, brown
O. ensiformis	Epiphyte	Oct–Nov	Flowers minute, greenish-yellow on sigmoid curved spiciform raceme
O. falconeri	Epiphyte	Sept–Oct	Flowers minute; in whorls, greenish yellow on densely flowered curved raceme
O. mucronata	Epiphyte	Sept-Nov	Flowers are minute, greenish yellow at decurved, spiciform raceme
O. proudlockii	Epiphyte	February	Flowers yellowish, partly sunk in pits of the rachis
Odisha cleistantha	Terrestrial	October	Flowers verdant green, inodorous
Papilionanthe cylindrica	Epiphyte	Feb–June	Flowers showy, fragrant, white. Lip 3-lobed yellow at the tip, pink with white spot below.
Pecteilis gigantea	Terrestrial	August–Sept	Flowers are large, snow white, fragrant in terminal raceme
Pelatantheria insectifera	Epiphyte	Oct–Dec	Flowers greenish yellow with deep brown streaks lip purple on extra-axillary corymb

continued **Tab. 3.4:** Orchid flora of Eastern Ghats of India

Name of species	Habit	Flowering time	Remarks
Peristylus constrictus	Terrestrial	June- July	Flowers white, fragrant, dense flowered on long terminal spikes
P. goodyeroides	Terrestrial	July	Flowers greenish white, sub-secund on dense flowered spike
P. lawii	Terrestrial	June–July	Flowers minute, white on slender, terete peduncle
P. plantagineus	Terrestrial	Sept	Flowers greenish white on many flowered spike
Phaius tankervilleae	Terrestrial	Feb–March	Flowers large, downwardly faced, mildly scented, Silvery on the outside having yellow throats.
Pholidota imbricata	Epiphyte	July–Sept	Flowers small, cream or pale yellow on drooping racemes arising from the base of the petiole
Polystachya concreta	Epiphyte	July	Flowers pale greenish yellow, on terminal raceme
Pomatocalpa decipiens	Tropical epiphyte	Jan–march	Flowers yellow with reddish blotches on extra-axillary spikes
Robiquetia josephiana	Epiphyte	Aug–Dec	Inflorescence pendulous of few flowered raceme. Flowers inodorous, white, small
Seidenfia versicolor	Terrestrial or Lithophyte	July–Aug	Flowers minute, light yellow or maroon
Smitinandia micrantha	Epiphyte	April -June	Flowers minute, waxy white, faintly scented, small on densely flowered raceme
Spathoglotis plicata	Terrestrial	Throughout the year	Flowers racemed, reddish violet, pink or rarely white
Staurochilus ramosus	Epiphyte	April–May	Flowers pale yellow, mildly scented on paniculate raceme
Taprobanea spathulata	Epiphyte	March–Septem-ber	Flowers large, golden yellow in raceme
Thunia bracteata	Epiphyte	August	Flowers nodding, large pinkish white on compressed racemes
Tropidia angulosa	Terrestrial	July	Inflorescence terminal, very short, slender with a few crowded small white flowers
Vanda tessellata	Epiphyte	March–May & Sept-Dec	Flowers greenish yellow with brown-ish tessellations, on axillary, zigzag racme, lip purple
Vanda testacea	Epiphyte	April	Flowers light yellow on axillary raceme
Zeuxine longilabris	Terrestrial	Jan–Feb	Flowers small, brownish and white

3.1.4 Andaman & Nicobar Islands

The A & N islands are a group of about 319 islands and islets in the Bay of Bengal. The South East monsoon governs the climate of these islands. Heavy mist over the forests in morning, high rainfall from May to November and constant sea currents keep the forest moist through out the year. This type of climate provides congenial habitats for luxuriant growth of unique orchid flora in the area. This region is home to 115 species belonging to 53 genera. Genera like *Grosourdya*, *Plocoglottis* and *Vrydagzynea* are confined only to this region (Singh *et al*, 2001).

Some orchid genera of Andaman and Nicobar islands are listed in Tab. 3.5.

Tab. 3.5: Orchid genera of Andaman and Nicobar islands

Name of species	Habit	Categories
Bulbophyllum protractum	Epiphyte	Rare
Dendrobium tenuicaule	Epiphyte	Endangered
Habenaria andamanica	Terrestrial	Rare
Malaxis andamanica	Terrestrial	Endangered
Malleola andamanica	Epiphyte	Rare
Phalaenopsis speciosa	Epiphyte	Rare
Taeniophyllum scaberulum	Epiphyte	Vulnerable
Zeuxine andamanica	Terrestrial	Rare

3.2 Conservation of Orchid Genetic Resources

There are three prominent methods of conservation of genetic resources of orchid species, namely:

- Legislative measures
- *In -situ* conservation in Sanctuaries /Reserves
- *Ex-situ* conservation in Orchidaria /Botanic gardens by cultivation (Hedge, 2012).

3.2.1 Legislative Measures of Conservation

Vanda coerulea (Blue Vanda) and *Renanthera imschootiana* (Red Vanda) are now included in schedule VI of the Wild Life Protection Act 1972 of Government of India as amended in 1992. As a result, all orchids are protected under the Wild Life Protection Act.

The International Union for Conservation of Nature (IUCN) has a Species Survival Commission (SSC) with a well defined preservation program for the Convention on

International Trade in Endangered Species of Wild Flora and Fauna (CITES). Under this provision, orchids are treated as protected species. In India, three genera and eleven species are being treated as protected under Schedule-VI of Wild Life Protection Act of 1972 and simultaneously under CITES Appendix-I. All other species of orchid in India have been included in Appendix–II of CITES. The following species have been kept under CITES Appendix-I and Schedule–VI of Wild Life Protection Act of Govt. of India.

Paphiopedilum charlesworthii, P. fairrieanum, P. hirsuitissimum, P. spicerianum, P. insigne, P. venustum, P. wardii, P druryii, Renanthera imschootiana and *Vanda coerulea*. However, excepting *P. druryii* which is reportedly from the state of Kerala, all other species of Paphiopedilum belong to North East India.

As per law, no wild orchids can be traded with and so, plants listed above can not be allowed for export. However, under CITES rules and regulations, whenever CITES Appendix-I species are cultivated, these can be allowed for export subject to the condition that proper permits for possessing and growing these scheduled plants are obtained from the concerned State, provided the nursery is registered under Wild Life Preservation Office, Government of India.

Appendix-I includes threatened and extinct species. In fact, no trade of wild plants is allowed. Trade is allowed in cultivated and artificially propagated plants subject to licensing.

Appendix-II covers species which may be threatened unless trade is strictly regulated. Orchidaceae is listed on Appendix-II. Trade in wild and propagated specimens is allowed, subject to licensing. Further, all cultured orchid seedlings in flasks or those aseptically cultured from seeds and tissues are now exempt from CITES control w.e.f. 16th April, 1993.

These legislations have helped to check illegal collection of orchids from the wild, as no importing countries accept plants of wild origin. In this regard, the Proceedings of the Seminar on CITES Implementation for Plants (1997), illustrates the method of recognizing and procedure for dealing with the wild plants at the port.

3.2.2 *In Situ* Conservation

This term refers to the maintenance of the germplasm in its natural habitat, allowing continual adaptation to the environment without any human interference.

3.2.2.1 Biosphere Reserve

These are versatile protected areas established to preserve the genetic diversity in the representative ecosystem, and are internationally recognized. The proposal for development of biosphere reserves was initiated by UNESCO in 1971 under the 'Man & Biosphere' (MAB) programme. The first biosphere reserve in the world was established in 1979. To date, 564 biosphere reserves have been developed in 109 countries across the world. India has 17 biosphere reserves, namely Achanakamar-Amarkantak, Agasthymalai, Cold Desert, Dihang-Dibang, Dibru Saikhowa, Great Nicobar, Gulf of Mannar, Kachchh, Khangchendzonga, Manas, Nanda Devi, Nilgiri, Nokrek, Pachmarhi, Seshachalam Hills, Simlipal and Sunderbans for conservation of endemic, endangered and vulnerable orchid species.

3.2.2.2 National Parks

These are areas of adequate natural biological and geomorphological interest owned by a sovereign state having one or several ecosystems where conservation of wild life (both flora and fauna) is practiced along with educative and recreative interest; and are designated, created and protected by legislation. Hailey National Park, presently known as Jim Corbett National Park was the first developed National Park in India in 1936. Presently, there are 98 National Parks in India. Two examples are the Simlipal National Park of Orissa where 96 different species of orchids are found, and the Buxa Tiger Reserve of West Bengal with 150 different species of orchids.

3.2.2.3 Sacred Groves

A sacred grove is a special type of area where all forms of life, particularly the sacred tree species related to any particular culture, are protected by a particular human community, race or tribe in the name of their respective deity. Himachal Pradesh, Karnataka, Kerala, Maharashtra, Andhra Pradesh, West Bengal and Chhattishgarh are very prominent states for sacred grooves. About 13928 sacred grooves presently exist in India. This may be an important area for *in situ* conservation of orchids of that particular locality.

3.2.2.4 Gene Sanctuary

A gene sanctuary is a protected area where a broad spectrum of genetic variability is conserved to act as a reserve for future use and crop improvement. At present, India has 480 wildlife gene sanctuaries. The Sessa Orchid Sanctuary of Arunachal Pradesh with 100 Sq. kilometer area conserves about 200 species of orchids. Similar types of sanctuaries have also been created in Sikkim at Deorali and Singtam.

3.2.2.5 Individual Trees

The epiphytic orchid species are conserved on tree species in their natural habitat. In their natural habitat, they attach themselves to the bark of trees, or the surface of other plants (Tab. 3.6). Their thick, white roots are specially adapted to absorb moisture and dissolved nutrients. Because these tropical orchids usually grow high in the trees, rather than on the forest floor, they are accustomed to good air circulation and plenty of light.

Tab. 3.6: Host – tree and orchid species specificity (De et al, 2011)

Common trees	Botanical name	Orchid species
Angeri	*Lyonia ovalifolia*	*Bulbophyllum viridiflorum, Chiloschista usneoides, Coelogyne corymbosa, Coel. cristata, Coel. ovalis, Cym. elegans, Cym. iridioides, Den. aphyllum, Den. chryseum, Den. denudans, Den. eriiflorum, Den. heterocarpum, Den. longicornu, Gastrochilus calceolaris, Kingidium taenialis, Oberonia acaulis, Pleione hookeriana, Vanda cristata*
Ban Litchi	*Benthamidia capitata*	*Den. aphyllum, Den. longicornu*
Banjh	*Quercus leucotricha*	*Bulbophyllum viridiflorum, Coelogyne corymbosa, Coel. cristata, Cym. elegans, Cym. iridioides, Den. aphyllum, Den. chryseum, Den. denudans, Den. eriiflorum, Den. heterocarpum, Den. longicornu, Gastrochilus calceolaris, Kingidium taenialis, Oberonia acaulis, Pleione hookeriana, Vanda cristata, Cleisostema spp.*
Chiuri	*Diploknema butyracea*	*Aerides multiflora, Aerides odorata, Coelogyne flaccida, Coel. ovalis, Den. aphyllum, Den. heterocarpum, Den. longicornu, Gastrochilus calceolaris, Oberonia acaulis, Rhyncostylis retusa, Vanda cristata, Oberonia spp.*
Chutro	*Berberis asiatica*	*Chiloschista usneoides, Den. denudans, Den. eriiflorum, Gastrochilus calceolaris, Kingidium taenialis*
Kaphal	*Myrica esculenta*	*Coelogyne corymbosa, Coelogyne flaccida, Coel. ovalis, Cym. iridioides, Den. aphyllum, Den. denudans, Den. eriiflorum, Den. longicornu, Gastrochilus calceolaris, Oberonia acaulis, Vanda cristata*
Katush	*Castanopsis indica*	*Bulbophyllum viridiflorum, Chiloschista usneoides, Coelogyne corymbosa, Coel. cristata, Coel. ovalis, Cym. iridioides, Den. aphyllum, Den. chryseum, Den. denudans, Den. eriiflorum, Den. heterocarpum, Den. longicornu, Gastrochilus calceolaris, Kingidium taenialis, Oberonia acaulis, Pleione hookeriana, Vanda cristata*
Kaulo	*Persea odoratissima*	*Aerides odorata, Chiloschista usneoides, Coelogyne corymbosa, Coel. cristata, Coelogyne flaccida, Coel. ovalis, Den. aphyllum, Den. denudans, Den. eriiflorum, Den. heterocarpum, Den. longicornu, Oberonia acaulis, Pholidota spp., Vanda cristata*
Khote Salla	*Pinus roxburghii*	*Den. aphyllum, Den. denudans, Den. longicornu, Rhyncostylis retusa*

continued **Tab. 3.6:** Host – tree and orchid species specificity (De et al, 2011)

Common trees	Botanical name	Orchid species
Lali Gurans	*Rhododendron arboreum*	*Bulbophyllum viridiflorum, Chiloschista usneoides, Coel. cristata, Coelogyne flaccida, Coel. ovalis, Cym. elegans, Cym. iridioides, Den. aphyllum, Den. chryseum, Den. denudans, Den. eriiflorum, Den. heterocarpum, Den. longicornu, Gastrochilus calceolaris, Kingidium taenialis, Oberonia acaulis, Pleione hookeriana, Vanda cristata*
Lek Chutro	*Berberis cristata*	*Chiloschista usneoides, Den. aphyllum, Den. denudans, Den. eriiflorum, Den. longicornu, Kingidium taenialis, Vanda cristata*
Mauwa	*Engelhardia spicata*	*Aerides multiflora, Aerides odorata, Coelogyne flaccida, Coel. ovalis, Den. aphyllum, Den. eriiflorum, Den. heterocarpum, Den. longicornu, Den. nobile, Gastrochilus calceolaris, Oberonia acaulis, Rhyncostylis retusa, Vanda teres, Epidendrum radicans, Pholidota spp.*
Mayal	*Pyrus pashia*	*Chiloschista usneoides, Den. aphyllum, Den. longicornu, Kingidium taenialis, Vanda cristata*
Sal	*Shorea robusta*	*Aerides multiflora, Aerides odorata, Coelogyne flaccida, Den. aphyllum, Den. eriiflorum, Den. heterocarpum, Gastrochilus calceolaris, Oberonia acaulis, Rhyncostylis retusa*
Pashi	*Mioromeles rhamnoides*	*Rhyncostylis retusa, Epidendrum radicans, Den. heterocarpum*
Utis	*Alnus nepalensis*	*Den. nobile, Coelogyne flaccida, Epidendrum, Vanda cristata, Thunia alba,*
Wild Cherry	*Prunus cerasus*	*Aerides multiflora, Epidendrum radicans, Coelogyne flaccida, Vanda tessellata, Rhyncostylis retusa, Vanda cristata*

3.2.3 *Ex situ* Conservation

This refers the preservation of germplasm outside the natural habitat. In India, the Botanical Survey of India maintains three National Orchidaria and Experimental Gardens; one each at Yercaud (Tamil Nadu), Howrah (West Bengal), and Shillong (Meghalya) where representative species of the region are cultivated. Similarly, Arunachal Pradesh State Forest Research Institute is maintaining large number of orchid species at Orchid Research Centre, Tipi, Itanagar, Sessa, Dirrang, Jenging and Roing as a measure of *ex situ* conservation of orchids. In Karnataka, three *ex situ* conservation Centers have been established: one in Kodagu, another in Kudremukh and the third in Dhandeli (Rao and

Sridhar, 2007). A natural Orchidarium for the conservation of orchid germplasm has also been planned in Bangalore within Lal Bagh Botanical Garden.

3.2.3.1 Field Gene Banks

In this area, germplasm are collected from natural habitats or from other sources including commercial houses and nurseries and maintained in field or protected structures. The Centre for Orchid Gene Conservation of the Eastern Himalayan region at Hengbung of Senapati district of Manipur, the country's first orchid gene bank, has already been established to conserve orchids as well as to facilitate research work.

In the field gene banks of TBGRI, Trivandrum, 600 different species and 150 hybrids of orchids are maintained. Orchids of 90 different genera and a number of hybrids of commercial orchids are also maintained at NRC for Orchids, Pakyong, Sikkim.

3.2.3.2 Botanical Gardens

These are protected areas where living plant specimen are conserved in fields or in protected structures provide significant information regarding mode of perpetuation, reproductive biology, taxonomical characters and propagation technique. At present, there are 13 botanical gardens in India maintaining a number of orchid species. About 43 species of orchids are collected and displayed in the orchid house of Lloyd Botanical garden, Darjeeling, West Bengal.

3.2.3.3 Herbal Gardens

In these areas, medicinal plant genetic resources are reared in a protected area to maintain them generation after generation. The Government of India sanctioned funds for development of herbal gardens in 16 SAU's and research institutions in different agro-climatic regions of the country to conserve and maintain regional medicinal plants and endangered species (Gupta, 1993). In India, a network among the herbal gardens has already developed at DMAPR where 83 such gardens are registered, along with details of the species they house.

3.2.3.4 Orchid Seed Gene Bank

Million of seeds are produced in a single capsule of orchid. However, they lack the functional endosperm and require specific mycorrhizal association for germination under natural conditions and, consequently, the percentage of germination is low. Many orchids have been germinated through an asymbiotic technique, where the rate of germination can reach as high as 90%. The seeds of orchids are orthodox in nature and provide a great scope for long term storage through low temperature.

3.2.3.5 *In vitro* Conservation

This technique can be used for the revitalization of orchid germplasm affected by virus and virus-like diseases through apical meristem culture. As a matter of fact, orchids were the first plants to be tissue cultured (Pritchard, 1989). There is a need for studies on genetic stability to avoid the soma-clonal variants, and slow growth cultures for longer storage duration to avoid frequent transfers.

3.2.3.6 Cryo-preservation

Cryo-preservation means long term storage or conservation of plant parts and reproductive materials at a very low temperature under laboratory conditions either in liquid (-196°C) or vapour phase (-150°C) nitrogen. Tissues/ explants of orchids can be cryo-preserved in liquid nitrogen cylinders as a long term storage procedure after proper treatment of cryo-protectants and Plant Vitrification Solutions. In our country, NBPGR has created a cryobank facility where 2.5 lakhs of germplasm lines can be stored (Singh, 2005).

3.3 Bio-piracy

Bio-piracy defines the gaining of exclusive monopoly rights over the biological material of one country by individuals, institutions or companies of other countries that ultimately leads to the denial of the rights of the country of origin. Several objectives of bio-piracy are (Akurugoda, 2013):

- Introduction of new plant varieties
- Introduction of new living organisms
- Production of pharmaceuticals
- Privatization of traditional knowledge

Further, bio-piracy is also known as the stealing of knowledge from traditional and indigenous communities or individuals. The term also means to suggest a breach of a contractual agreement on the access and use of traditional knowledge to the detriment of the provider, and bio-prospecting without the consent of the local communities. Accordingly, it can be suggested that bio-piracy is a twofold phenomenon, which includes traditional knowledge bio-piracy and genetic resources bio-piracy. As such 'bio-piracy' has been labelled as a term to describe the ways that corporations from the developed world claim ownership of, take a free ride on, or otherwise take unfair advantage of, the genetic resources and traditional knowledge and technologies of developing countries.

Bio-piracy has many negative effects on biodiversity such as extinction of endemic genotypes, genetic erosion of biodiversity and privatization of bio-treasures of the country. Further, this practice impairs the economy of the country – bio-piracy is an extremely lucrative business and as a result most racketeers tend to exploit the bioresources of developing countries and obtain patents for these uses. The emergence of monopolies over seeds and medicines through patents is becoming a major threat to farmers, livelihoods and public health.

3.3.1 Legal Regime Pertaining to Biodiversity and Bio-piracy

3.3.1.1 International Law

Before 1994, the legalities of obtaining samples of plant, microbes and animals were straightforward. Anyone including researches, scientists and tourists could simply reach a field site, collect samples and take them home because living species were regarded as the common heritage of mankind. On that basis, as common resources, private companies and individuals could take and use the resources without stating valid justifications or giving compensation. But, after the implementation of the Convention on Biodiversity (CBD), principles have been set in order to strengthen the national protection against bio- piracy.

3.3.1.2 Convention on Biological Diversity (1994)

Under Article 038 of the CBD, a sovereign nation has national rights over biological resources. This principle allows developing countries to get better benefits from their biological resources, and the traditional knowledge pertaining to them. Further, according to Article 01 of the CDB the main objectives of the convention are: conservation of bio diversity, the sustainable use of its components and the equitable sharing of the benefits arising out of the utilization of genetic resources. Under Article 8(j), the CBD needs each state party to respect, preserve and maintain knowledge, innovations and practices of indigenous and local communities embodying traditional lifestyles relevant for the conservation and sustainable use of biological diversity. The CBD commits member countries to conserve and develop biological resources for sustainable use. Sustainable use of biological resources includes finding new drugs, crops and industrial products, while conserving the resources for future generation. Since the CBD recognizes the sovereign rights over biological resources it establishes the concept of bio-prospecting under article 15(5) which can be read with 15(7) and 19 of the CBD. According to article 15(5), access to genetic resources shall be subject to prior informed consent of the contracting party providing such resources, unless otherwise determined by that party.

Further, article 15(7) provides that each contracting party shall take legislative, administrative or policy measures, as appropriate, and in accordance with Articles 16 and 19 and, where necessary, through the financial mechanism established by Articles 20 and 21 with the aim of sharing in a fair and equitable way, the results of research and development and the benefits arising from the commercial and other utilization of genetic resources with the Contracting Party providing such resources. Such sharing shall be upon mutually agreed terms. Further, under CBD, it is expected that developing countries should receive a share in the benefits from biotechnology. Hence, state parties must take all practicable measures to promote and advance priority access, on a fair and equitable basis, to the results and benefits arising from biotechnologies based upon genetic resources. It can be argued that under the CBD, prior informed consent is the standard for ensuring a fair and equitable access and benefit sharing agreement.

3.3.1.3 Cartagena Protocol (2003)

The *Cartagena Protocol on Biosafety to the Convention on Biological Diversity* is an international agreement which aims to ensure the safe handling, transport and use of living modified organisms (LMOs) resulting from modern biotechnology that may cause adverse effects on biological diversity, also taking into account human health hazards. It was adopted on 29 January 2000 and entered into force on 11 September 2003. The Protocol contains reference to a precautionary approach and reaffirms the precaution language in Principle 15 of the Rio Declaration on Environment and Development.

3.3.1.4 Nagoya Protocol (2010)

The *Nagoya Protocol on Access to Genetic Resources and the Fair and Equitable Sharing of Benefits Arising from their Utilization to the Convention on Biological Diversity* is an international agreement which aims at sharing the benefits arising from the utilization of genetic resources in a fair and equitable way, by appropriate access to genetic resources and by appropriate transfer of relevant technologies. It takes into account all rights over those resources and technologies, and by appropriate funding, thereby it contributes on the conservation of biological diversity and the sustainable use of its components.

3.3.1.5 CITES 1972

The Convention on International Trade in Endangered Species of Wild Fauna and Flora is an international agreement between governments. It ensures that international trade in specimens of wild animals and plants does not threaten their survival. Bio-piracy is understood at two levels:

3.3.2 Traditional Knowledge (TK) Bio-piracy

3.3.2.1 Collection and Use

Different types of TK bio-piracy: the un-authorised use of common TK; The unauthorised use of TK only found among one indigenous group; the unauthorised use of TK acquired by deception or failure to fully disclosure the commercial motive behind the acquisition; the unauthorised use of TK acquired on the basis of a transaction deemed to be exploitative; the unauthorised use of TK acquired on the basis of a conviction that all such transactions are inherently exploitative ("all bio-prospecting is bio-piracy"); and the commercial use of TK on the basis of a literature search.

3.3.2.2 Patenting

Paten claim under TK: the patent claims TK in the form in which it was acquired; the patent covers a refinement of the TK; patent covers an invention based on TK and other modern/traditional knowledge.

3.3.3 Genetic Resource Bio-piracy

3.3.3.1 Collection and Use

The unauthorised extraction and use of widespread resources; the unauthorized extraction and use of resources that can be found in one location; the unauthorised extraction and export of resources in breach of ABS regulations of the relevant country; the unauthorized extraction and export of resources in countries lacking ABS regulations; the authorized extraction of resources on the basis of a transaction deemed to be exploitative; and the authorized extraction of resources on the basis of a conviction that all such transactions are inherently exploitative.

3.3.3.2 Patenting

The patent claims the resource itself; the patent claims a purified version of the resource; and the patent covers a derivative of the resource and/or is based on more than one resources.

3.3.4 Measures of Bio-piracy in Some Orchids

The species of *Paphiopedilum*, a genus first described by E. Pfitzer in 1886 (Chowdhery, 1998), are commonly known as 'Lady's or Venus's Slipper' orchids. These plants are characterized by their luxuriant and multi-coloured flowers with shoe-shaped labellum or synsepalum, a structure unique to orchids that is formed by the fusion of two lateral sepals. The flowers have long vase life, and they remain on the plant for 60-90 days (Rao, 2006). The beautiful and long-lasting flowers of *Paphiopedilum* species are among the most expensive and horticulturally important orchid flowers (Rao, 2006). The genus is native to south-east Asia, northern India, southern China, Myanmar, Thailand and New Guinea, with 80 species distributed worldwide

(Chung *et al*, 2006). Out of the nine species of *Paphiopedilum* occurring in India, one (*P. druryii*) is endemic to South India, and the remaining eight species are found in different parts of north-east India, viz. Sikkim, Meghalaya, Manipur and Arunachal Pradesh (Chowdhery, 1998). All species of *Paphiopedilum* are highly endangered and are currently listed in Appendix I of CITES (http://www.cites.org; Sun *et al*, 2011). Consequently, the international trade of these naturally occurring wild species is strictly prohibited (http://www.cites.org). Many hybrids of *Paphiopedilum* with complex parentage exist in the international orchid markets (Sun *et al*, 2011). It is fairly easy to identify these species at the flowering stage, but very difficult in vegetative stage or as fragments. Moreover, it is difficult to differentiate their inter-species hybrids from the endangered *Paphiopedilum* species morphologically, especially in their vegetative stage. Hence, the illicit trade of the endangered species cannot be easily checked.

The efficacy of DNA barcoding with matK is considered as the signature sequence for the identification of closely related endangered species of Indian Paphiopedilums and also in elucidating the parentage of their inter-specific hybrids. The uniqueness of the generated matK sequences of the investigated species as evidenced by the BLAST results further confirms their species discrimination capability. This approach could be used for generating DNA barcoding for all other endangered *Paphiopedilum* species. These barcodes, once developed, could become potent tools in the hands of enforcement agencies entrusted with the responsibility of checking illicit trade. An effective check on the collection of these plants from wild, in turn, would help in their conservation *in situ*.

4 Morphological and Molecular Characterization of Valuable species

Plant patent is usually granted for newly invented asexually propagated plants which provides protection for 20 years. In 1930, the Plant Patent Act was enacted for the protection of vegetatively propagated plants, covering fruits and ornamental plants, and the U.S. was the only country to issue plant patents. During 1961, the Association of Plant Breeders for the protection of plant varieties (ASSINSEL) led to an 'International Convention for the Protection of New Varieties of Plants at the Paris to address the protection of rights of Plant Breeders and farmers and to encourage the development of new plant varieties. This in turn led to the development of the International Union for the Protection of New Varieties of Plants, UPOV, an inter-governmental organization with headquarter in Geneva, Switzerland, which was constituted at the Paris Convention of 1961, to set the basic rules and principles for plant variety protection. It also provides the general principles for the examination of plant varieties and sets specific guidelines for some 160 genera and species. Currently, DUS (Distinctiveness, Uniformity and Stability) testing is used as an effective system of plant variety protection, with the aim of encouraging the development of new varieties of plants and for the benefit of society. It protects rights for breeders to exploit or develop new plant varieties, to allow access to foreign varieties with wider gene pools, to promote intensive breeding activities and to prevent unauthorized varieties genetic degradation (Gautam *et al*, 2012)..

4.1 Cymbidium, Dendrobium and Vanda

4.1.1 Subject

The Test Guidelines (DUS) apply to all vegetatively propagated varieties of *Cymbidium* Sw., *Dendrobium* Sw. and *Vanda* Jones ex R.Br. of the family Orchidaceae (PVJ, October, 2011).

4.1.2 Plant Material Required

- The Protection of Plant Varieties and Farmers' Rights Authority (PPV & FRA) shall decide when, where and in what quantity and quality the plant material are required for testing of a variety denomination for registration under the Protection of Plant Varieties and Farmers' Rights (PPV & FR) Act, 2001. Applicants submitting such plant material from a country other than India shall make sure that all customs and quarantine requirements stipulated under relevant national legislations and regulations are complied with.
- For all varieties, only 20 full grown plants (10 for each Centre) previously not flowered shall be required for DUS testing.

- The plant material supplied should be visibly healthy, not lacking in vigour or affected by any pests or diseases or mechanical damage.
- Plant material shall not have undergone any chemical or bio-physical treatment unless the competent authority allow or request such treatment. If it has been treated, details of the treatment must be given.

4.1.3 Conduct of Tests

- The test should normally be conducted for two growing seasons.
- Tests should normally be conducted at two places. If any essential characteristics of the variety are not expressed during visual observations at these places, the variety shall be considered for further examination at another appropriate test site or under special test protocol at the expressed request of the applicant.
- The tests should be carried out in the greenhouse under conditions ensuring normal growth for the plant concerned. The size of the pots/ containers should be such that plants or parts of plants may be removed for measurement and counting without prejudice to the observations, which must be conducted up to the end of the growing period. Each test should include a total of 10 plants.
- Tests should be done only in pots of prescribed size. The pots should be kept at a standard spacing recommended for each type or at the spacing specified by the applicant.
- Additional tests for special purposes may be established.
- Normally growth regulators should not be used.

4.1.4 Methods and Observations

- The characteristics described in the Table of characteristics (Tab. 4.1, 4.2, 4.3) shall be used for the testing of varieties for their DUS.
- For the assessment of Distinctiveness and Stability, all observations shall be made on 10 plants or parts taken from each of 10 plants.
- For the assessment of Uniformity, a population standard of 1% and an acceptance probability of at least 95% shall be applied. In the case of a sample size of 10 plants, the maximum number of off-types allowed would be 1.
- All observations of the pseudobulbs shall be made on the flowering pseudobulb.
- All observations on the leaf shall be made on the longest leaf of a flowering pseudobulb.
- All observations on the inflorescence and the flower shall be made at the time when 50% of the flowers on the inflorescence have opened, and the observation should be made on the most recently fully opened flower on the inflorescence before fading of colour.

- All observations on the length and width of the flower and parts of the flower shall be made on the spread out positions.
- All observations on the colour of sepal, petal, lip and column shall be made both on the inner and outer sides.
- For the assessment of colour characteristics, the Royal Horticultural Society (RHS) colour chart shall be used.

4.1.5 Grouping of Varieties

The candidate hybrids for DUS testing were divided into groups to facilitate the assessment of Distintiveness. Characteristics which are known from experience not to vary, or to vary only slightly within a species, and which in their various states are fairly evenly distributed across all species in the collection are suitable for grouping purpose (Henke, 2008). Out of 66 characteristics, the following characteristics were used for grouping of *Cymbidium* hybrids (Tab. 4.1):

(a) Pseudobulb shape, (b) Leaf shape, (c) Inflorescence type, (d) Inflorescence length, (cm), (e) Inflorescence: number of flowers, (f) Flower width (cm), (g) Flower duration (days), (h) Flower predominant colour, (i) Lip ornamentation and (j) Flowering season

4.1.6 Characteristics and Symbols

- To assess Distinctiveness, Uniformity and Stability, the characteristics and their states as given in the Table of characteristics shall be used.
- Notes (1-9) (Table 4.1) shall be used to describe the state of each character for the purpose of electronic data processing.
- *Legend*: **(*)** = Characteristics that shall be observed during every growing season for all varieties and hybrids shall always be included in the description of the varieties and hybrids, except when the state of expression of any of these characters is rendered impossible by a preceding phenological characteristic or by the environment conditions of the testing region. Under such exceptional situation, adequate explanation shall be provided. **(+)** = See explanations on the Table of characteristics.
- Characteristics denoted with symbols QL, QN and PQ in the first column of the Table of characteristics shall be indicated as:

QL: Qualitative characteristic **QN:** Quantitative characteristic **PQ:** Pseudo-qualitative characteristic

- **(a)-(e):** see 4.1.7 for explanation
- Type of assessment of characteristics indicated in column six of the Table of characteristics are as follows:

MG: Measurement by a single observation of a group of plants or parts of plants
MS: Measurement of a number of individual plants or parts of plants
VG: Visual assessment by a single observation of a group of plants or parts of plants
VS: Visual assessment by observations of individual plants or parts of plants

Tab. 4.1: Characteristics for DUS Test guidelines in *Cymbidium*

Sl. No.	Characteristics	States	Notes	Example Varieties /hybrids	Type of Assessment
1. * (+) PQ (a)	Pseudobulb shape	Narrow	1		VG
		Round	3	Cym. 'Lucky Rainbow	
		Ovoid	5	Cym. 'Soul Hunt ', Cym. 'Fire Storm', Cym. Bob Marlin 'Lucky', Cym. 'Winter Beach Sea Green', Cym. 'Madrid Forest King', Cym. 'Show Girl', Cym. 'Sleeping Nymph'	
		Conical	7	Cymb. 'Luna Pink'	
2. * (+) PQ (a)	Leaf shape	Linear	1	Cym. 'Yankilla', Cym. 'Soul Hunt', Cym. 'Fire Storm', Cym. Bob Marlin 'Lucky', Cym. 'Winter Beach Sea Green', Cym. 'Madrid Forest King', Cym. 'W.W. Wondrous', Cym. 'Sleeping Nymph'	VG
		Linear-oblong	3	Cym. 'Lucky Rain Bow', Cym. 'Sainte Lapine'	
		Lanceolate	5	Cym. 'Stanley Fouraker White Magic'	
3. PQ (b)	Inflorescence orientation	Erect	1	Cym. 'H.C. Aurora', Cym. 'Ammesbury', Cym. 'Red Imperial Red Tower', Cym. 'Hawtescens', Cym 'Jungfrau Snow Queen', Cym. 'Sleeping Nymph'	VG
		Horizontal/ Arching	3	Cym. 'Yankilla', Cym. 'Fire Storm', Cym. Bob Marlin 'Lucky', Cym. 'Winter Beach Sea Green', Cym. 'Madrid Forest King', Cym. 'W.W. Wondrous'	
		Drooping/ Pendulus	5		
4. * QN (b)	Inflorescence length of Standard type	Short (<75 cm)	3	Cym. 'Soul Hunt', Cym. 'Stanley Fouraker White Magic' Cym. 'Yankilla', Cym. 'Ammesbury', Cym. 'Fire Storm', Cym. Bob Marlin 'Lucky', Cym. 'W.W. Wondrous', Cym. 'Sleeping Nymph'	MS
		Medium (75-90 cm)	5	Cym. 'Hawtescens', Cym. 'Winter Beach Sea Green'	
		Long (>90 cm)	7	Cym. 'Red Star', Cym. 'Madrid Forest King'	

continued **Tab. 4.1:** Characteristics for DUS Test guidelines in *Cymbidium*

Sl. No.	Characteristics	States	Notes	Example Varieties /hybrids	Type of Assessment
5. * QN (b)	Number of flowers/ inflorescence	<12	1	Cym. 'Fire Storm', Cym. Bob Marlin 'Lucky', Cym. 'Winter Beach Sea Green', Cym. 'W.W. Won-drous', Cym. 'Sleeping Nymph'	VS
		12-20	3	Cym. 'Madrid Forest King', Cym. 'Red Star'	
		>20	5		
6. * QN (c)	Flower width	Narrow (<4 cm)	3		MS
		Medium (cm-8 cm)	5	Cym. 'Soul Hunt', Cym. 'Yankilla', Cym. 'Show Girl', Cym. 'Ammesbury', Cym. 'Velvet Green', Cym. 'Takarjoke', Cym. 'Sleeping Nymph'	
		Large (>8 cm)	7	Cym. 'H.C. Aurora' Cym. Tracey Reddaway', Cym. 'Fire Storm', Cym. Bob Marlin 'Lucky', Cym. 'Winter Beach Sea Green', Cym. 'Madrid Forest King', Cym. 'W.W. Wondrous'	
7. * QN (b)	Flower duration on the plant	Short (<20 days)	3		VG
		Medium (20-40 days)	5	Cym. 'Red Star'	
		Long (>40 days)	7	Cym. 'H.C. Aurora', Cym. 'Yankilla', Cym. 'Amesbury', Cym. 'Fire Storm', Cym. Bob Marlin 'Lucky', Cym. 'Winter Beach Sea Green', Cym. 'Madrid Forest King', Cym. 'W.W. Wondrous', Cym. 'Sleeping Nymph'	
8. QL (b)	Flower Colour in general appearance	White	1	Cym. 'Takarjoke', Cym. 'W.W. Wondrous	VG
		Pink	2		
		Yellow	3	Cym. 'Angelica December Gold', Cym. 'Hawtescens'	
		Green	4	Cym. 'Yankilla', Cym. 'Winter Beach Sea Green', Cym. 'Madrid Forest King'	
		Red	5	Cym. 'Soul Hunt', Cym. 'Miss Sanders', Cym. 'Fire Storm', Cym. Bob Marlin 'Lucky'	
		Brown	6	Cym. 'Tracey Reddaway'	
		Maroon	7	Cym. 'Show Girl', Cym. 'Ammesbury'	

continued Tab. 4.1: Characteristics for DUS Test guidelines in *Cymbidium*

Sl. No.	Characteristics	States	Notes	Example Varieties /hybrids	Type of Assessment
9. QL (d)	Inside Lip colour pattern	Absent	1		VG
		Spotted	3	Cym. 'Jungfrau Dos Pueblos'(red purple), Cym. 'Takarjoke'(red purple), Cym. 'Fire Storm'(red purple), Cym. Bob Marlin 'Lucky'(red purple), Cym. 'W.W. Wondrous'(red purple), Cym. 'Winter Beach Sea Green'(red purple), Cym. 'Madrid Forest King'(red purple)	
		Blotched	5	Cym. 'Pine Clash Moon Venus'(red purple), Cym. 'H.C. Aurora', (red purple), Cym. 'Sun Gold' (yellow orange), Cym. 'Jungfrau Dos Pueblos'(red purple), Cym. Bob Marlin 'Lucky', Cym. 'Madrid Forest King'(red purple), Cym. 'Sleeping Nymph'(red purple)	
		Streaked/ Striped	7	Cym. 'Winter Beach Sea Green'(red purple), Cym. 'Red Star' (red purple)	
10. QL (d)	Outside lip colour pattern	Absent	1	Cym. 'Jungfrau Dos Pueblos', Cym. 'Madrid Forest King', Cym. 'Sleeping Nymph'	VG
		Spotted	3	Cym. 'H.C. Aurora' (red purple), Cym. 'Takarjoke'(red purple), Cym. 'Fire Storm'(red purple), Cym. Bob Marlin 'Lucky'(red purple)	
		Blotched	5	Cym. 'Fire Storm' (red purple), Cym. Bob Marlin 'Lucky'(red purple), Cym. 'Sun Gold'(yellow orange)	
		Streaked/ Striped	7	Cym. 'Fire Storm' (red purple)	
11. (+) PQ	Flowering season	Winter	1	Cym. 'Fire Storm', Cym. Bob Marlin 'Lucky', Cym. 'Red Star', Cym. 'Soul Hunt', Cym. 'Yankilla', Cym. 'Sleeping Nymph'	VG
		Spring	3	Cym. 'Ammesbury', Cym. 'H.C. Aurora', Cym. 'Red Imperial Red Tower', Cym. 'Hawtescens', Cym. 'Fire Storm', Cym. 'Bob Marlin Lucky', Cym. 'Winter Beach Sea Green', Cym. 'W.W. Wondrous'	
		Summer	5		
		Rainy	7		

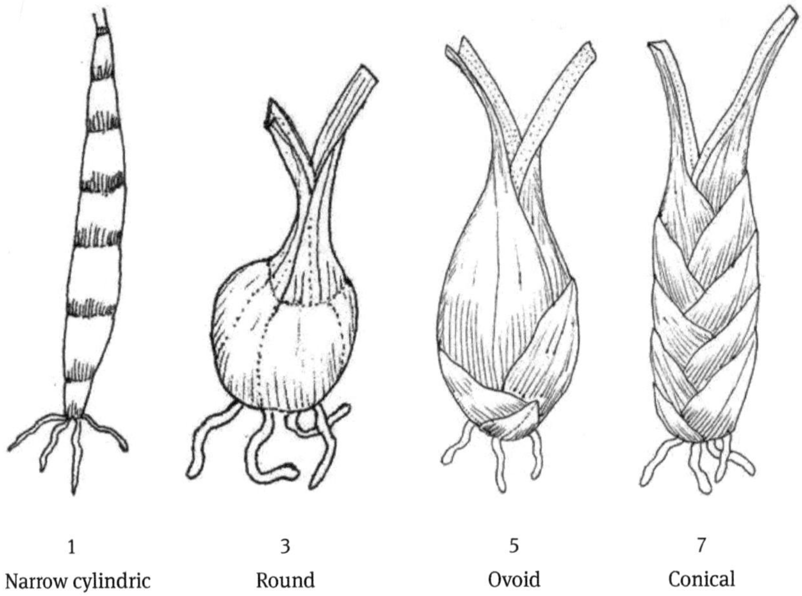

Fig. 4.1: Pseudobulb shape in Cymbidium

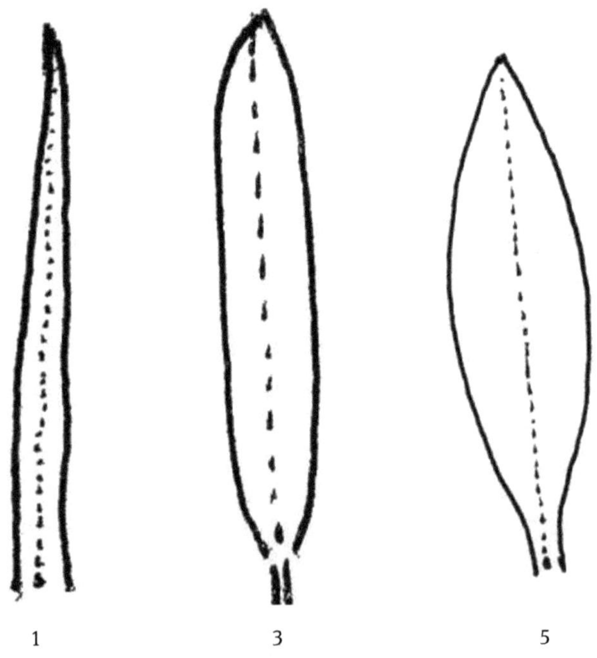

Fig. 4.2: Leaf shape in Cymbidium

In *Dendrobium*, out of 62 descriptors, the following characteristics were used for grouping of *Dendrobium* hybrids (Tab. 4.2):

Tab. 4.2: Characteristics for DUS test guide lines in *Dendrobium*

(a) Nature of stem, (b) Inter-node length (cm), (c) Internode number, (d) Inflorescence length (cm), (e) Inflorescence number/cane, (f) Flower width (cm) (g) Lip colour, (h) Lip ornamentation and (i) Flowering season.

Sl. No.	Characteristic	States	Note	Example Varieties /hybrids	Type of Assess-ment
1. * QN (+) (a)	Nature of shoot at flowering	Cane (woody) type	1		VG
		Cane cylindric	3		
		Cane clavate fleshy	5	Den. 'Bangkok Blue', Den. 'Big White Jumbo', ' Den. 'Thongchai Gold', Den. 'Erika', Den. 'Triple Pink',	
		Bulbous round	7	Den. 'Julie'	
2. * QN (a)	Internode number (flowering shoot)	Few (<5)	3	--	VS
		Medium (5-10)	5	Den. 'Big White-4N', Den. 'Thongchai Gold', Den. 'Bangkok Blue', Den. 'Julie', Den. 'Erika', Den. 'Earsakul', Den. 'Dang Saard', Den.'Madam Pink', Den. 'Triple Pink'	
		Many (>10)	7	Den. 'Abraham', Den. 'Emma White', Den. 'Big White Jumbo', Den. 'Lervia', Den. 'Kating Dang'	
3. QN (a)	Internode diameter (widest part of shoot)	Small (<2 cm)	3	Den. 'Abraham,', Den. 'Big White-4N', Den. 'Emma White', Den. 'Thongchai Gold', Den. 'Bangkok Blue', Den. 'Big White Jumbo', Den. 'Erika', Den. 'Lervia', Den. 'Earsakul', Den. 'Dang Saard', Den. 'Kating Dang', Den.'Madam Pink', Den. 'Triple Pink'	MS
		Medium (2-4 cm)	5	Den. 'Julie'	
		Large (>4 cm)	7	--	
4. * QN (b)	Inflorescence number/plant	Few (<3)	3	Den. 'Abraham', Den. 'Big White Jumbo', Den. 'Erika', Den. 'Lervia', Den. 'Dang Saard', Den. 'Kating Dang', Den. 'Triple Pink' Den. 'Big White-4N', Den. 'Madam Pink'	VS
		Medium(3-5)	5	Den. 'Bangkok Blue', Den. 'Julie'	
		Many (>5)	7	Den. 'Earsakul', Den.'Emma White' Den. 'Thongchai Gold'	

continued Tab. 4.2: Characteristics for DUS test guide lines in *Dendrobium*

Sl. No.	Characteristic	States	Note	Example Varieties /hybrids	Type of Assess-ment
5. * QN (b)	Inflorescence Length	Short (<20 cm)	3	Den. 'Thongchai Gold', Den. 'Big White Jumbo'	MS
		Medium (20-40 cm)	5	Den. 'Emma White', Den. 'Abraham', Den. 'Big White-4N', Den. 'Emma White', Den. 'Bangkok Blue', Den. 'Julie', Den. 'Erika', Den. 'Lervia', Den. 'Ear-sakul', Den. 'Dang Saard', Den. 'Kating Dang', Den. 'Triple Pink'	
		Long (41-60 cm)	7	Den.'Madam Pink'	
		Very long (>60 cm)	7		
6. * QN (c)	Flower width (Tip distance of two lateral petals)	Narrow (<3.0 cm)	3		MS
		Medium (3.0-6.0 cm)	5	Den. 'Abraham', Den. 'Emma White', Den. 'Julie', Den. 'Big White Jumbo', Den. 'Erika', Den. 'Dang Saard	
		Broad (>6.0 cm)	7	Den. 'Abraham', Den. 'Big White-4N', Den. 'Thongchai Gold', Den. 'Bangkok Blue', Den. 'Lervia', Den. 'Earsakul', Den. 'Kating Dang', Den. 'Madam Pink', Den. 'Triple Pink'	
7. * QL (d)	Lip predomin-ant colour (as per RHS chart)	Green	1		VS
		White	2	Den. 'Big White-4N', Den. 'Emma White', Den. 'Big White Jumbo'	
		Yellow	3	Den. 'Thongchai Gold'	
		Pink	4		
		Red	5	Den. 'Kating Dang'	
		Purple	6	Den. 'Abraham', Den. 'Julie', Den. 'Lervia', Den. 'Earsakul', Den. 'Dang Saard', Den. 'Madam Pink'	
		Blue	7	Den. 'Thongchai Blue'	
		Violet	8	Den. 'Bangkok Blue', Den. 'Erika', Den. 'Triple Pink'	
8. * QL (d)	Lip Colour pattern (Colour as per RHS chart)	Uniform	1	Den. 'Big White Jumbo'	VG
		Mixed	3		
		Spotted	5		
		Striped/shaded	7	Den. 'Abraham', Den. 'Big White-4N, Den. 'Emma White', Den. 'Thongchai Gold', Den. 'Julie', Den. 'Triple Pink'	
		Netted	9	Den. 'Bangkok Blue', Den. 'Lervia', Den. 'Earsakul', Den. 'Dang Saard', Den. 'Kating Dang', Den.'Madam Pink'	

continued **Tab. 4.2:** Characteristics for DUS test guide lines in *Dendrobium*

Sl. No.	Characteristic	States	Note	Example Varieties /hybrids	Type of Assessment
9. * PQ	Flowering season	Winter	1	Den. 'Kating Dang', Den. 'Erika'	VG
		Spring	3	Den. 'Abraham', Den. 'Big White-4N', Den. 'Thongchai Gold', Den. 'Bangkok Blue'	
		Summer	5	Den. 'Julie', Den. 'Big White Jumbo', Den. 'Erika', Den. 'Lervia', Den.'Dang Saard'	
		Rainy	7	Den. 'Abraham', Den. 'Kating Dang', Den. 'Madam Pink', Den. 'Triple Pink'	
		Year round	9	Den. 'Earsakul', Den. 'Emma White',	

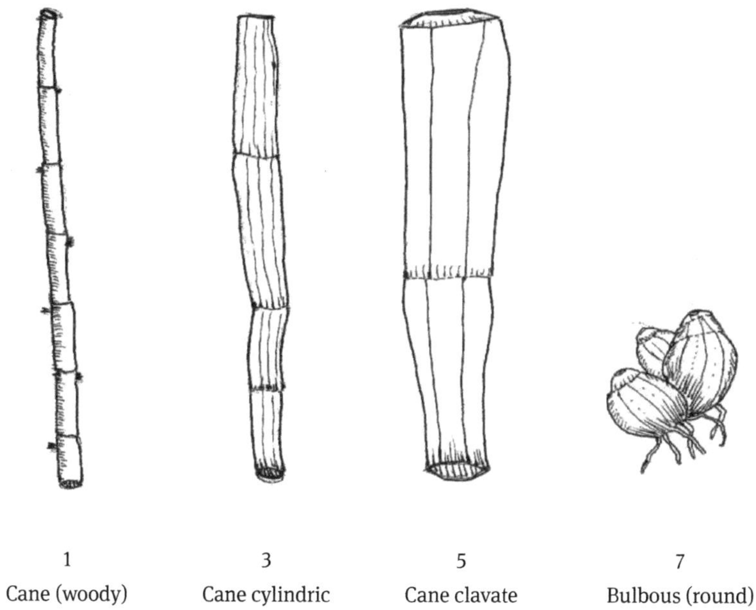

Fig. 4.3: Nature of stem in Dendrobium

In *Vanda*, out of 66 descriptors, the following characteristics were used for grouping of *Vanda* hybrids (Tab. 4.3):

Tab. 4.3: Characteristics for DUS test guidelines in *Vanda*

(a) Plant type, (b) Inter-node length (cm), (c) Leaf type, (d) Spike length (cm), (e) Flower number, (f) Flower colour, (g) Sepal ornamentation, (h) Petal ornamentation, (i) Lip colour, (j) Lip ornamentation, (k) Spur type and (l) Flowering season.

S. No.	Characteristics	States	Notes	Example Varieties /hybrids	Type of Assess-ment
1. QN (a)	Internode length	Short (0.5-0.1 cm)	3	Vanda 'KS.SD', V. 'Pures Wax', V. 'Pakchong Blue'	MS
		Medium (1.0-2.0 cm)	5	V. 'Prao Sky Blue', V. 'RBSD Black', V. 'PAT D', V. 'Sansai Blue', V. 'Motes Indigo', V. 'Roberts Delight Blue'	
		Long (>2 cm)	7	V. 'John Clubb' V. 'Miss Joaquim'	
2. * PQ (+)	Leaf type	Terete (Round)	1	V. 'John Clubb' V. 'Miss Joaquim'	VG
		Semi-terete	3		
		Channelled	5		
		Strap (Flat)	7	V. 'KS.SD', V. 'Prao Sky Blue', V. 'Pures Wax', V. 'RBSD Black', V. 'PAT D', V. 'Sansai Blue', V. 'Motes Indigo', V. 'Pakchong Blue', V. 'Roberts Delight Blue	
3. * QN (b)	Inflorescence length	Short (<15 cm)	3		MS
		Medium (15-30 cm)	5	V. 'KS.SD', V. 'Prao Sky Blue', V. 'Pures Wax', V. 'RBSD Black', V. 'PAT D', V. 'Sansai Blue'	
		Long (>30 cm)	7	V. 'Motes Indigo', V. 'Pakchong Blue', V. 'Roberts Delight Blue'	
4. * QN (b)	No. of flowers / inflorescence	Few (<5)	3	V. 'KS.SD', V. 'Prao Sky Blue', V. 'PAT D'	VG
		Medium (5-10)	5	V. 'Pures Wax', V. 'RBSD Black', V. 'Sansai Blue', V. 'Motes Indigo, V. 'Pakchong Blue', V. 'Roberts Delight Blue'	
		Many (>10)	7		
5. * QL (b)	Flower predom-inant colour	Red	1	V. 'RBSD Red'	VS
		Blue	3	V. „Roberts Delight Blue', V. 'Prao Sky Blue', V. 'Sansai Blue', V. 'Pakchong Blue'	
		Violet	5	V. 'KS.SD', V. 'Pures Wax', V. 'Motes Indigo'	
		Purple	7	V. 'RBSD Black', V. 'PAT D'	

continued Tab. 4.3: Characteristics for DUS test guidelines in *Vanda*

S. No.	Characteristics	States	Notes	Example Varieties /hybrids	Type of Assess-ment
6. * QL (d)	Sepal colour pattern	Uniform	1	-	VG
		Spotted	3	V. 'KS.SD', V. 'Pures Wax', V. 'RBSD Black' V. 'PAT D'	
		Blotched	5	V. 'KS.SD', V. 'Roberts Delight Blue'	
		Streaked /Striped	7	-	
		Tessellated	9	V. 'KS.SD', V. 'Prao Sky Blue', V. 'RBSD Black', V. 'PAT D', V. 'Sansai Blue', V. 'Motes Indigo', V. 'Pakchong Blue'	
7. * QL (d)	Petal colour pattern (as per RHS colour chart)	Uniform	1	-	VG
		Spotted	3	V. 'KS.SD', V. 'Pures Wax', V. 'RBSD Black', V. 'PAT D', V. 'Roberts Delight Blue'	
		Blotched	5		
		Streaked/Striped	7		
		Tessellated	9	V. 'Prao Sky Blue' V. 'KS.SD', V. 'Sansai Blue', V. 'Motes Indigo', V. 'Pakchong Blue'	
8. QL (d)	Lip colour	Single	1	V. 'KS.SD', (violet) V. 'Prao Sky Blue', (violet) V. 'Pures Wax', (violet) V. 'Motes Indigo'(violet)	VG
		Double	3	V. 'RBSD Black' (purple & blue) V. 'PAT D', (purple & blue) V. 'Sansai Blue', (purple & violet) V. 'Pakchong Blue'(violet & blue)	
		Triple or more	5	V. 'Roberts Delight Blue' (purple, yellow & blue)	
9. * QL (d)	Lip colour pattern	Uniform	1		VG
		Spotted	2	V. 'RBSD Black', V. 'Motes Indigo'	
		Blotched	3		
		Streaked/Striped	4	V. 'KS.SD', V. 'Prao Sky Blue', V. 'Pures Wax', V. 'RBSD Black', V. 'PAT D', V. 'Sansai Blue', V. 'Pakchong Blue', V. 'Roberts Delight Blue'	
		Tessellated	5		

continued **Tab. 4.3:** Characteristics for DUS test guidelines in *Vanda*

S. No.	Characteristics	States	Notes	Example Varieties /hybrids	Type of Assess-ment
10. * PQ (+) ©	Spur type	Saccate	1	-	VG
		Conical	3	V. 'KS.SD', V. 'Prao Sky Blue', V. 'Pures Wax', V. 'RBSD Black', V. 'PAT D', V. 'Sansai Blue', V. 'Motes Indigo', V. 'Pakchong Blue', V. 'Roberts Delight Blue'	
		Cylindric	5	-	
		Tubular	7	-	
11. * PQ	Flowering season	Winter	1	V. 'Pakchong Blue'	VG
		Spring	3	V. 'KS.SD', V. 'Prao Sky Blue', V. 'Pures Wax'	
		Summer	5	V. 'Sansai Blue', V. 'Prao Sky Blue	
		Rainy	7	V. 'Motes Indigo', V. 'PAT D', V. 'RBSD Black', V. 'Roberts Delight Blue	

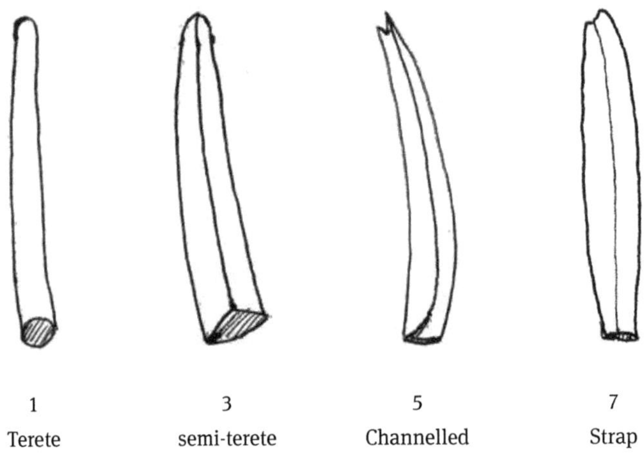

Fig. 4.4: Leaf type in Vanda

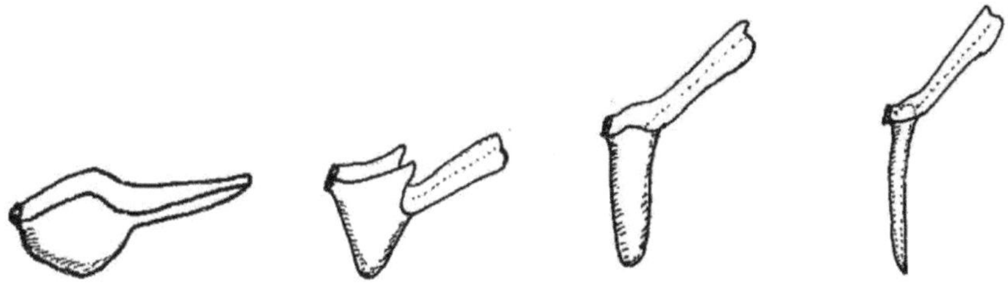

1	3	5	7
Saccate	Conical	Cylindric	Tubular

Fig. 4.5: Spur type in Vanda

4.1.7 Explanations on the Table of Characteristics

Guidelines for recording the observations of vegetative and flowering characteristics
Characteristics indicated with (a), (b), (c), (d) and (e) in the first column of the Table of characteristics should be examined as indicated below:

a) Observations on the leaf, shoot and internode length shall be made on the flowering shoot.
b) Observations on the inflorescence and the flower shall be made at the time when 50% of the flowers on the inflorescence have opened and the most recently fully opened flower on the inflorescence before the color starts to fade.
c) Observations on the length and width of the flower and parts of the flower shall be made on the spread out positions.
d) Observations on the color of the sepal, the petal and the lip shall be made on inner side at apex, mid and base portion.
e) Observations on the colour of column shall be made on inner side at apex, mid and basal region.

4.2 Phalaenopsis (PVJ, Nov, 2012)

4.2.1 Subject

These test guidelines apply to all vegetatively propagated varieties of *Phalaenopsis* Blume and alliance of the family Orchidaceae.

4.2.2 Plant Material Required

- The Protection of Plant Varieties and Farmers' Rights Authority (PPV & FRA) shall decide when, where and in what quantity and quality the plant material are required for testing of a variety denomination for registration under the Protection of Plant Varieties and Farmers' Rights (PPV & FR) Act, 2001. Applicants submitting such plant material from a country other than India shall make sure that all customs and quarantine requirements stipulated under relevant national legislations and regulations are complied with.
- For all varieties, 20 flowering size plants (10 for each Centre) shall be required for DUS testing.
- The plant material supplied should be visibly healthy, not lacking in vigour nor affected by any pests, diseases or mechanical damage.
- Plant material shall not have undergone any chemical or bio-physical treatment unless the competent authority allows or requests such treatment. If it has been treated, details of the treatment must be given.

4.2.3 Conduct of Test

- The minimum duration of a test should normally be two similar flowering seasons.
- The test shall normally be conducted at two places. If any essential characteristic of the variety is not expressed for visual observations at these places, the variety shall be considered for further examination at another appropriate test site or under special test protocol on request of the applicant.
- The test should be carried out under greenhouse conditions ensuring satisfactory growth for the expression of the relevant characteristics of the variety and for the conduct of the examination.
- The design of the test should be such that the plants or parts of the plants may be removed for measurement and counting without prejudice to the observations, which must be made up to the end of the flowering period. Each test should be designed to result in a total of at least 10 plants.
- Unless otherwise indicated, all observations determined by measuring or counting should be made on 10 plants or parts taken from each of 10 plants.
- Additional tests for special purposes may be established.
- Normally, growth regulators shall not be used.

4.2.4 Methods and Observations

- The characteristics described in the Table of Characteristics (Tab.4.4) shall be used for the testing of varieties for their DUS.

- For the assessment of Distinctiveness and Stability, all observations shall be made on 10 plants or parts taken from each of 10 plants.
- For the assessment of Uniformity, a population standard of 1% and an acceptance probability of at least 95% shall be applied. In the case of a sample size of 10 plants, the maximum number of off-types allowed would be 1.
- All observations on the shoot shall be made on the flowering shoot.
- All observations on the leaf shall be made on the longest leaf of a flowering shoot.
- All observations on the inflorescence and the flower shall be made at the time when 50% of the flowers on the inflorescence have opened and on the most recently fully opened flower on the inflorescence before fading of colour.
- All observations on the length and width of the flower and parts of the flower shall be made in the spread out position.
- All observations on the colour of sepal, petal, lip and column shall be made on the inner side.
- For the assessment of colour characteristics, the Royal Horticultural Society (RHS) colour chart shall be used.

4.2.5 Grouping of Varieties

- The selection of varieties of common knowledge to be grown in the trial with the candidate varieties and the way in which these varieties are divided into groups to facilitate the assessment of distinctiveness is aided by the use of grouping characteristics.
- Grouping characteristics are those in which the documented states of expression, even where produced at different locations, can be used, either individually or in combination with other such characteristics (Tab. 4.4) (a) to select varieties of common knowledge that can be excluded from the growing trial used for examination of distinctiveness; and (b) to organize the growing trial so that similar varieties are grouped together.
- The following have been agreed upon as useful grouping characteristics:
  - Plant: size (Characteristic 1)
  - Flower size: width in front view (Characteristic 15)
  - Petal: colour pattern (Characteristic 38)
  - Petal: predominant colour (Characteristic 39)
  - Lip: colour pattern of apical lobe (Characteristic 50)
  - Lip: predominant colour of apical lobe (Characteristic 51)

4.2.6 Characteristics and Symbols

- To assess Distinctiveness, Uniformity and Stability, the characteristics and their states as given in the Table of Characteristics shall be used.

- Notes 1-9 (numbers) shall be used to describe the state of each character for the purpose of electronic data processing.
- Legend

(*) Characteristics that shall be observed during every growing season for varieties and hybrids and shall always be included in the description of the varieties and hybrids, except when the state of expression of any of these characters is rendered impossible by a preceding phenological characteristic or by the environment conditions of the testing region. Under such exceptional situation, adequate explanation shall be provided.

(+) See explanations on the Table of Characteristics

- Characteristics denoted with symbols QL, QN and PQ in the first column of the Table of Characteristics shall be indicated as:

QL: Qualitative characteristic

QN: Quantitative characteristic

PQ: Pseudo-qualitative characteristic

- **(a)-(e)** see 4.2.7.1 for explanation
- Type of assessment of characteristics indicated in column six of the Table of Characteristics are as follows:

MG: Measurement by a single observation of a group of plants or parts of plants

MS: Measurement of a number of individual plants or parts of plants

VG: Visual assessment by a single observation of a group of plants or parts of plants

VS: Visual assessment by observations of individual plants or parts of plant

Tab. 4.4: Table of Characteristics of DUS Test Guidelines in *Phalaenopsis*

Sl. No	Characteristics	States	Notes	Example Varieties /Hybrids	Type of Assessment
1. (*) QN	Plant: size (cm)	Very small (<3.0)	1	--	MS
		Small (3.0-4.5)	3	Dtps. 'Younghome Orange Lip', Dtps. 'Acker's Sweetie', Dtps. 'Big Red Robe', Dtps. 'Mount Lip Taisuco', P. 'Amabilis', Dtps. 'Tying Shin Zebra', P. 'Luchia Pink', P. 'Jin Cheng Sun', P. 'Sin Yuan Golden Beauty', P. 'Little Gem Stripe', P. 'Surf Song', Dtps. 'Lian Her Happy', P. 'Timothy Christopher', Dtps. 'Happy UFO', P. 'Big White Pink Stripe', P. Medium Pink, P. Leodora, P. Magic Kiss, P. Goldie, P. Roxanne, P. Ivory Pearl, P. Lucyna, P. Brother Girl, Dtps. Ox Prince Thunder, P. Carlotta, Dtps. Leopard Princess M611, Dtps. Gan Lin Fairy 'GL', P. Shu Long Spotted Deer, P. Ho's Little Caroline, Dtps. Sogo Soft, Dtps. Gan Lin Beauty Hsiung.	
		Medium (4.6-6.0)	5	Dtps. Plum Rose x Ox Black Jack, P. Memoria Francis Hunter, Dtps. Chin Ann Diamond 'Alisun', Dtps. Gan Lin Diamond, Dtps. Shulong Sun Beauty, P. Chian Xen Pearl 'Cx#1', P. Miki Wata Nabe, Dtps. Chian Xen Magpie	
		Large (>6.0-7.5)	7	Dtps. Shu Long Pearl, P. Taida Salu Red, Dtps.Shih Hua Long First Love	
		Extra large (>7.5)	9	P. Kaleidoscope, P. Strawberry, Dtps. Hsin Yang Fortune	

continued Tab. 4.4: Table of Characteristics of DUS Test Guidelines in *Phalaenopsis*

Sl. No	Characteristics	States	Notes	Example Varieties /Hybrids	Type of Assessment
2. (*) QN (a)	Leaf: length (cm)	Short (<10)	3	P. Big White Pink Stripe	MS
		Medium (10-15)	5	Dtps. Mount Lip Taisuco, P. Amabilis, P. Strawberry, Dtps. Shih Hua Long First Love, P. Taida Salu, Dtps. Tying Shin Zebra, P. Medium Pink, P. Magic Kiss, Goldie, Roxanne, P. Lucyna, P. Brother Girl, P. Memoria Francis Hunter, P. Carlotta, Dtps. Leopard Princess M611, P. Ho's Little Caroline, Dtps. Shulong Sun Beauty.	
		Long (>15-20)	7	Dtps. Acker's Sweetie, Dtps. Big Red Robe, P. Miki Wata Nabe, P. Champion Porter, Dtps. Hsin Yang Fortune, Dtps. Taida Salu Red, P. Kaleidoscope, P. Luchia Pink, P. Jin Cheng Sun, P. Sin Yuan Golden Beauty, Dtps. Chian Xen Magpie, P. Little Gem Stripe, P. Surf Song, Dtps. Lian Her Happy, P. Timothy Christopher Dtps. Happy UFO, Leodora, P. Ivory Pearl, Dtps. Ox Prince Thunder, Dtps. Gan Lin Fairy 'GL', Dtps. Gan Lin Diamond, P. Chian Xen Pearl 'Cx#1', Dtps. Gan Lin Beauty Hsiung, P. Shu Long Spotted Deer, Dtps. Shu Long Pearl	
		Extra long (>20)		Dtps. Plum Rose x Ox Black Jack, Dtps. Younghome Orange Lip, Dtps. Sogo Soft, Dtps. Chin Ann Diamond 'Alisun'	
3. (*) QN (a)	Leaf: width (cm)	Narrow (<3.0)	1	--	MS
		Medium (3.0-6.0)	3	P. Amabilis, P. Sin Yuan Golden Beauty, Dtps. Chian Xen Magpie, P. Big White Pink Stripe, P. Medium Pink, Roxanne, P. Lucyna, P. Brother Girl, P. Memoria Francis Hunter, Dtps. Ox Prince Thunder, P. Carlotta	
		Broad (>6.0)	5	Dtps. Plum Rose x Ox Black Jack, Dtps. Younghome Orange Lip, Dtps. Acker's Sweetie, Dtps. Hsin Yang Fortune, Dtps. Big Red Robe, Dtps. Mount Lip Taisuco, P. Miki Wata Nabe, P. Champion Porter, P. Strawberry, Dtps.Shih Hua Long First Love, P. Taida Salu, Dtps. Tying Shin Zebra, Dtps. Taida Salu Red, P. Kaleidoscope, P. Luchia Pink, P. Jin Cheng Sun, P. Little Gem Stripe, P. Surf Song, Dtps. Lian Her Happy, P. Timothy Christopher Dtps. Happy UFO, Leodora, P. Magic Kiss, Goldie, P. Ivory pearl, Dtps. Leopard Princess M611, Dtps. Gan Lin Fairy 'GL', P. Shu Long Spotted Deer, P. Ho's Little Caroline, Dtps. Sogo Soft, Dtps. Gan Lin Beauty Hsiung, Dtps. Chin Ann Diamond 'Alisun', Dtps. Gan Lin Diamond, Dtps. Shulong Sun Beauty, P. Chian Xen Pearl 'Cx#1, Dtps. Shu Long Pearl	

continued **Tab. 4.4:** Table of Characteristics of DUS Test Guidelines in *Phalaenopsis*

Sl. No	Character-istics	States	Notes	Example Varieties /Hybrids	Type of Assess-ment
4. (*) PQ (+) (a)	Leaf: shape	Oblong	1	Dtps. Plum Rose x Ox Black Jack, Dtps. Younghome Orange Lip, Dtps. Acker's Sweetie, Dtps. Hsin Yang Fortune, Dtps. Big Red Robe, Dtps. Mount Lip Taisuco, P. Miki Wata Nabe, P. Champion Porter, P. Amabilis, P. Strawberry, Dtps.Shih Hua Long First Love, P. Taida Salu, Dtps. Tying Shin Zebra, Dtps. Taida Salu Red, P. Kaleidoscope, P. Luchia Pink, P. Jin Cheng Sun, P. Sin Yuan Golden Beauty, Dtps. Chian Xen Magpie, P. Little Gem Stripe, P. Surf Song, Dtps. Lian Her Happy, P. Timothy Christopher Dtps. Happy UFO, P. Big White Pink Stripe, P. Medium Pink, P. Leodora, P. P. Magic Kiss, P. Goldie, P. Roxanne, P. Ivory pearl, P. Lucyna, P. Brother Girl, P. Memoria Francis Hunter, Dtps. Ox Prince Thunder, P. Carlotta, Dtps. Leopard Princess M611, P. Shu Long Spotted Deer, Dtps. Sogo Soft, Dtps. Gan Lin Beauty Hsiung, P. Chian Xen Pearl 'Cx#1' Dtps. Chin Ann Diamond 'Alisun', P. Ho's Little Caroline	VG
		Narrow obovate	3	Dtps. Shulong Sun Beauty, Dtps. Chin Ann Diamond 'Alisun', Dtps. Gan Lin Diamond, Dtps. Gan Lin Fairy 'GL'	
		Suborbicular	5	--	
5. (+) PQ (a)	Leaf: apex	Acute	1	P. Memoria Francis Hunter	VG
		Notched	3	Dtps.Shih Hua Long First Love	
		Obtuse	5	Dtps. Plum Rose x Ox Black Jack, Dtps. Younghome Orange Lip, Dtps. Acker's Sweetie, Dtps. Hsin Yang Fortune, Dtps. Big Red Robe, Dtps. Mount Lip Taisuco, P. Miki Wata Nabe, P. Champion Porter, P. Amabilis, P. Strawberry, P. Taida Salu, Dtps. Tying Shin Zebra, Dtps. Taida Salu Red, P. Kaleidoscope, P. Luchia Pink, P. Jin Cheng Sun, P. Sin Yuan Golden Beauty, Dtps. Chian Xen Magpie, P. Little Gem Stripe, P. Surf Song, Dtps. Lian Her Happy, P. Timothy Christopher Dtps. Happy UFO, P. Big White Pink Stripe, P. Medium Pink, P. Leodora, P. P. Magic Kiss, P. Goldie, P. Roxanne, P. Ivory pearl, P. Lucyna, P. Brother Girl, Dtps. Ox Prince Thunder, P. Carlotta, Dtps. Leopard Princess M611, Dtps. Gan Lin Fairy 'GL', P. Shu Long Spotted Deer, P. Ho's Little Caroline, Dtps. Sogo Soft, Dtps. Gan Lin Beauty Hsiung, Dtps. Chin Ann Diamond 'Alisun', Dtps. Gan Lin Diamond, Dtps. Shulong Sun Beauty, P. Chian Xen Pearl 'Cx#1', Dtps. Shu Long Pearl	
6. QL (a)	Leaf: anthocyanin colouration	Absent	1	Dtps. Younghome Orange Lip, Dtps. Acker's Sweetie, Dtps. Big Red Robe, Dtps. Mount Lip Taisuco, P. Champion Porter, P. Amabilis, P. Strawberry, P. Taida Salu, Dtps. Taida Salu Red, P. Luchia Pink, P. Jin Cheng Sun, P. Little Gem Stripe, P. Surf Song, Dtps. Lian Her Happy, P. Timothy Christopher Dtps. Happy UFO, P. Big White Pink Stripe, P. Medium Pink, Leodora, Goldie, Roxanne, P. Ivory Pearl, P. Memoria Francis Hunter, P. Carlotta, Dtps. Leopard Princess M611, Dtps. Gan Lin Fairy 'GL', Dtps. Sogo Soft, Dtps. Gan Lin Beauty Hsiung, Dtps. Chin Ann Diamond 'Alisun', Dtps. Gan Lin Diamond, Dtps. Shulong Sun Beauty, Dtps. Shu Long Pearl	VG
		Present	9	Dtps. Plum Rose x Ox Black Jack, Dtps. Hsin Yang Fortune, P. Miki Wata Nabe, Dtps.Shih Hua Long First Love, Dtps. Tying Shin Zebra, P. Kaleidoscope, P. Sin Yuan Golden Beauty, Dtps. Chian Xen Magpie, P. Magic Kiss, P. Lucyna, P. Brother Girl, Dtps. Ox Prince Thunder, P. Chian Xen Pearl 'Cx#1' P. Ho's Little Caroline, P. Shu Long Spotted Deer	

continued Tab. 4.4: Table of Characteristics of DUS Test Guidelines in *Phalaenopsis*

Sl. No	Character-istics	States	Notes	Example Varieties /Hybrids	Type of Assess-ment
7. (+) PQ (b)	Inflorescence type	Solitary	1	--	VS
		Raceme	3	Dtps. Plum Rose x Ox Black Jack, Dtps. Younghome Orange Lip, Dtps. Acker's Sweetie, Dtps. Hsin Yang Fortune, Dtps. Big Red Robe, Dtps. Mount Lip Taisuco, P. Miki Wata Nabe, P. Champion Porter, P. Amabilis, P. Strawberry, Dtps.Shih Hua Long First Love, P. Taida Salu, Dtps. Tying Shin Zebra, Dtps. Taida Salu Red, P. Kaleidoscope, P. Luchia Pink, P. Jin Cheng Sun, P. Sin Yuan Golden Beauty, Dtps. Chian Xen Magpie, P. Little Gem Stripe, P. Surf Song, Dtps. Lian Her Happy, P. Timothy Christopher Dtps. Happy UFO, P. Big White Pink Stripe, P. Medium Pink, P. Leodora, P. P. Magic Kiss, P. Goldie, P. Roxanne, P. Ivory pearl, P. Lucyna, P. Brother Girl, P. Memoria Francis Hunter, Dtps. Ox Prince Thunder, P. Carlotta, Dtps. Leopard Princess M611, Dtps. Gan Lin Fairy 'GL', P. Shu Long Spotted Deer, P. Ho's Little Caroline, Dtps. Sogo Soft, Dtps. Gan Lin Beauty Hsiung, Dtps. Chin Ann Diamond 'Alisun', Dtps. Gan Lin Diamond, Dtps. Shulong Sun Beauty, P. Chian Xen Pearl 'Cx#1', Dtps. Shu Long Pearl	
		Panicle	5	--	
8. (*) (+) QN (b)	Rachis length (cm)	Short (<10)	3	Dtps. Acker's Sweetie, Dtps. Big Red Robe, Dtps. Mount Lip Taisuco, P. Champion Porter, P. Amabilis, Dtps.Shih Hua Long First Love, Dtps. Tying Shin Zebra, P. Kaleidoscope, P. Jin Cheng Sun, P. Sin Yuan Golden Beauty, P. Little Gem Stripe, P. Surf Song, P. Timothy Christopher Dtps. Happy UFO, Leodora, Roxanne, P. Ivory pearl, P. Lucyna, P. Brother Girl, P. Memoria Francis Hunter, Dtps. Ox Prince Thunder, P. Carlotta, Dtps. Gan Lin Fairy 'GL', P. Shu Long Spotted Deer, P. Ho's Little Caroline, Dtps. Sogo Soft, Dtps. Gan Lin Beauty Hsiung, Dtps. Gan Lin Diamond, Dtps. Shulong Sun Beauty, P. Chian Xen Pearl 'Cx#1', Dtps. Shu Long Pearl	MS
		Medium (10-20)	5	Dtps. Younghome Orange Lip, Dtps. Hsin Yang Fortune, Dtps. Taida Salu Red, P. Luchia Pink, Dtps. Lian Her Happy, P. Big White Pink Stripe, P. Medium Pink, P. Magic Kiss, Goldie, Dtps. Leopard Princess M611, Dtps. Chin Ann Diamond 'Alisun', Taida Salu Red, P. Strawberry, P. Miki Wata Nabe, Dtps. Chian Xen Magpie, Dtps. Plum Rose x Ox Black Jack	
		Long (>20)	7	--	
9. (*) QN (b)	Inflorescence: number of flowers	Few (<5)	3	Dtps. Plum Rose x Ox Black Jack, Dtps. Acker's Sweetie, Dtps. Big Red Robe, Dtps. Mount Lip Taisuco, P. Miki Wata Nabe, P. Champion Porter, P. Amabilis, Dtps.Shih Hua Long First Love, P. Kaleidoscope, P. Jin Cheng Sun, P. Sin Yuan Golden Beauty, Dtps. Chian Xen Magpie, P. Little Gem Stripe, P. Surf Song, Roxanne, P. Ivory pearl, P. Lucyna, P. Brother Girl, P. Memoria Francis Hunter, Dtps. Ox Prince Thunder, P. Carlotta, Dtps. Gan Lin Fairy 'GL', P. Shu Long Spotted Deer, Dtps. Sogo Soft, Dtps. Gan Lin Beauty Hsiung, Dtps. Gan Lin Diamond, Dtps. Shulong Sun Beauty, P. Chian Xen Pearl 'Cx#1'	MS
		Medium (5-10)	5	Dtps. Hsin Yang Fortune, P. Strawberry, Dtps. Tying Shin Zebra, P. Luchia Pink, Dtps. Lian Her Happy, P. Timothy Christopher Dtps. Happy UFO, Leodora, P. Magic Kiss, Goldie, Dtps. Leopard Princess M611, P. Ho's Little Caroline, Dtps. Shu Long Pearl, Dtps. Chin Ann Diamond 'Alisun', Taida Saln. Red	
		Many (>10)	7	Dtps. Younghome Orange Lip, Dtps. Taida Salu Red, P. Big White Pink Stripe, P. Medium Pink.	

continued **Tab. 4.4:** Table of Characteristics of DUS Test Guidelines in *Phalaenopsis*

Sl. No	Character-istics	States	Notes	Example Varieties /Hybrids	Type of Assess-ment
10. (*) QN (b)	Peduncle: length (cm)	Short (<10)	3	--	MS
		Medium (10-20)	5	Dtps. Mount Lip Taisuco, P. Amabilis, P. Strawberry, Dtps. Tying Shin Zebra, P. Kaleidoscope, P. Sin Yuan Golden Beauty, P. Surf Song, P. Big White Pink Stripe, P. Medium Pink, Leodora, P. Brother Girl, P. Memoria Francis Hunter, Dtps. Leopard Princess M611, P. Ho's Little Caroline	
		Long (>20-30)	7	Dtps. Acker's Sweetie, Dtps. Hsin Yang Fortune, P. Champion Porter, Dtps.Shih Hua Long First Love, P. Taida Salu, Dtps. Dtps. Taida Salu Red, P. Jin Cheng Sun, Dtps. Chian Xen Magpie, P. Little Gem Stripe, P. Timothy Christopher Dtps. Happy UFO, P. Magic Kiss, Roxanne, P. Ivory pearl, Dtps. Ox Prince Thunder, Dtps. Gan Lin Fairy 'GL', P. Shu Long Spotted Deer, Dtps. Sogo Soft, Dtps. Shulong Sun Beauty, P. Chian Xen Pearl 'Cx#1', Taida Saln. Red	
		Extra long (>30)	9	Dtps. Plum Rose x Ox Black Jack, Dtps. Younghome Orange Lip, Dtps. Big Red Robe, P. Miki Wata Nabe, P. Luchia Pink, Dtps. Lian Her Happy, Goldie, P. Lucyna, P. Carlotta, Dtps. Shu Long Pearl, Dtps. Gan Lin Beauty Hsiung, Dtps. Gan Lin Diamond, Dtps. Chin Ann Diamond 'Alisun'	
11. QN (b)	Peduncle: thickness (cm)	Thin (<0.3)	3	--	MS
		Medium (0.3-0.6)	5	Dtps. Plum Rose x Ox Black Jack, Dtps. Younghome Orange Lip, Dtps. Acker's Sweetie, Dtps. Hsin Yang Fortune, Dtps. Big Red Robe, Dtps. Mount Lip Taisuco, P. Miki Wata Nabe, P. Champion Porter, P. Amabilis, P. Strawberry, Dtps.Shih Hua Long First Love, P. Taida Salu, Dtps. Tying Shin Zebra, Dtps. Taida Salu Red, P. Kaleidoscope, P. Luchia Pink, P. Jin Cheng Sun, P. Sin Yuan Golden Beauty, Dtps. Chian Xen Magpie, P. Little Gem Stripe, P. Surf Song, Dtps. Lian Her Happy, P. Timothy Christopher Dtps. Happy UFO, P. Big White Pink Stripe, P. Medium Pink, P. Leodora, P. P. Magic Kiss, P. Goldie, P. Roxanne, P. Ivory pearl, P. Lucyna, P. Brother Girl, P. Memoria Francis Hunter, Dtps. Ox Prince Thunder, P. Carlotta, Dtps. Leopard Princess M611, P. Shu Long Spotted Deer, P. Ho's Little Caroline, Dtps. Gan Lin Beauty Hsiung, Dtps. Gan Lin Diamond, Dtps. Shulong Sun Beauty, Dtps. Shu Long Pearl	
		Thick (>0.6)	7	Dtps. Gan Lin Fairy 'GL', P. Chian Xen Pearl 'Cx#1', Dtps. Sogo Soft, Dtps. Chin Ann Diamond 'Alisun'	
12. QL (b)	Peduncle: anthocyanin colouration	Absent	1	Dtps. Mount Lip Taisuco, P. Amabilis, P. Surf Song, P. Big White Pink Stripe, P. Leodora, P. Goldie, P. Roxanne, Dtps. Leopard Princess M611, Dtps. Gan Lin Fairy 'GL', Dtps. Gan Lin Diamond, Dtps. Sogo Soft, Dtps. Gan Lin Beauty Hsiung, Dtps. Shu Long Pearl.	VG
		Present	9	Dtps. Plum Rose x Ox Black Jack, Dtps. Younghome Orange Lip, Dtps. Acker's Sweetie, Dtps. Hsin Yang Fortune, Dtps. Big Red Robe, P. Miki Wata Nabe, P. Champion Porter, P. Strawberry, Dtps.Shih Hua Long First Love, P. Taida Salu, Dtps. Tying Shin Zebra, Dtps. Taida Salu Red, P. Kaleidoscope, P. Luchia Pink, P. Jin Cheng Sun, P. Sin Yuan Golden Beauty, Dtps. Chian Xen Magpie, P. Little Gem Stripe, Dtps. Lian Her Happy, P. Timothy Christopher, Dtps. Happy UFO, P. Medium Pink, P. Magic Kiss, P. Ivory pearl, P. Lucyna, P. Brother Girl, P. Memoria Francis Hunter, Dtps. Ox Prince Thunder P. Carlotta, P. Shu Long Spotted Deer, P. Ho's Little Caroline, Dtps. Chin Ann Diamond 'Alisun', Dtps. Shulong Sun Beauty, P. Chian Xen Pearl 'Cx#1'	

continued **Tab. 4.4:** Table of Characteristics of DUS Test Guidelines in *Phalaenopsis*

Sl. No	Character-istics	States	Notes	Example Varieties /Hybrids	Type of Assess-ment
13. QL (b)	Flower: texture of the surface of sepals and petals	Smooth	1	P. Big White Pink Stripe, P. Medium Pink, P. Magic Kiss, P. Roxanne, P. Ivory pearl, P. Lucyna, P. Brother Girl, P. Memoria Francis Hunter, Dtps. Ox Prince Thunder, P. Carlotta, Dtps. Leopard Princess M611, P. Shu Long Spotted Deer, Dtps. Sogo Soft, Dtps. Gan Lin Diamond, P. Chian Xen Pearl 'Cx#1'	VG
		Rough	9	Dtps. Plum Rose x Ox Black Jack, Dtps. Younghome Orange Lip, Dtps. Acker's Sweetie, Dtps. Hsin Yang Fortune, Dtps. Big Red Robe, Dtps. Mount Lip Taisuco, P. Miki Wata Nabe, P. Champion Porter, P. Amabilis, P. Strawberry, Dtps.Shih Hua Long First Love, P. Taida Salu, Dtps. Tying Shin Zebra, Dtps. Taida Salu Red, P. Kaleidoscope, P. Luchia Pink, P. Jin Cheng Sun, P. Sin Yuan Golden Beauty, Dtps. Chian Xen Magpie, P. Little Gem Stripe, P. Surf Song, Dtps. Lian Her Happy, P. Timothy Christopher Dtps. Happy UFO, P. Leodora, Goldie, Dtps. Gan Lin Beauty Hsiung, P. Ho's Little Caroline, Dtps. Gan Lin Fairy 'GL', Dtps. Chin Ann Diamond 'Alisun', Dtps. Shulong Sun Beauty, Dtps. Shu Long Pearl.	
14. (*) (+) QN (c)	Flower: length in front view (cm)	Short (<4)	3	--	MS
		Medium (4-8)	5	Dtps. Younghome Orange Lip, Dtps. Hsin Yang Fortune, Dtps. Mount Lip Taisuco, P. Champion Porter, P. Amabilis, P. Strawberry, Dtps.Shih Hua Long First Love, P. Taida Salu, Dtps. Tying Shin Zebra, Dtps. Taida Salu Red, P. Jin Cheng Sun, P. Sin Yuan Golden Beauty, Dtps. Chian Xen Magpie, P. Surf Song, P. Timothy Christopher Dtps. Happy UFO, P. Big White Pink Stripe, P. Medium Pink, P. Leodora, P. Magic Kiss, Goldie, P. Lucyna, P. Brother Girl, P. Carlotta, Dtps. Leopard Princess M611, P. Ho's Little Caroline, Dtps. Sogo Soft, Dtps. Chin Ann Diamond 'Alisun', Dtps. Shulong Sun Beauty, Dtps. Shu Long Pearl	
		Long (>8)	7	Dtps. Plum Rose x Ox Black Jack, Dtps. Acker's Sweetie, Dtps. Big Red Robe, P. Miki Wata Nabe, P. Kaleidoscope, P. Luchia Pink, P. Little Gem Stripe, Dtps. Lian Her Happy, P. Roxanne, P. Ivory pearl, P. Memoria Francis Hunter, Dtps. Ox Prince Thunder, Dtps. Gan Lin Fairy 'GL', P. Chian Xen Pearl 'Cx#1', P. Shu Long Spotted Deer, Dtps. Gan Lin Diamond, Dtps. Gan Lin Beauty Hsiung	
15. (*) (+) QN (c)	Flower: width in front view (cm)	Narrow (<4)	3	--	MS
		Medium (4-8)	5	Dtps. Younghome Orange Lip, Dtps. Hsin Yang Fortune, Dtps. Mount Lip Taisuco, P. Amabilis, P. Strawberry, Dtps. Shih Hua Long First Love, P. Taida Salu, Dtps. Tying Shin Zebra, Dtps. Taida Salu Red, P. Sin Yuan Golden Beauty, Dtps. Chian Xen Magpie, P. Surf Song, P. Timothy Christopher Dtps. Happy UFO, P. Medium Pink, P. Leodora, P. Magic Kiss, Goldie, P. Lucyna, P. Brother Girl, P. Carlotta, P. Ho's Little Caroline, Dtps. Shulong Sun Beauty, Dtps. Shu Long Pearl, Dtps. Leopard Princess M611	
		Broad (>8)	7	Dtps. Plum Rose x Ox Black Jack, Dtps. Acker's Sweetie, Dtps. Big Red Robe, P. Miki Wata Nabe, P. Champion Porter, P. Kaleidoscope, P. Luchia Pink, P. Jin Cheng Sun, P. Little Gem Stripe, Dtps. Lian Her Happy, P. Big White Pink Stripe, P. Roxanne, P. Ivory Pearl, P. Memoria Francis Hunter, Dtps. Ox Prince Thunder, Dtps. Gan Lin Fairy 'GL', P. Shu Long Spotted Deer, Dtps. Sogo Soft, Dtps. Gan Lin Beauty Hsiung, Dtps. Chin Ann Diamond 'Alisun', Dtps. Gan Lin Diamond, P. Chian Xen Pearl 'Cx#1'	

continued **Tab. 4.4:** Table of Characteristics of DUS Test Guidelines in *Phalaenopsis*

Sl. No	Characteristics	States	Notes	Example Varieties /Hybrids	Type of Assessment
16. (*) (+) PQ (b)	Flower: arrangement of petals	Open	1	Dtps. Plum Rose x Ox Black Jack, Dtps. Younghome Orange Lip, Dtps. Acker's Sweetie, Dtps. Hsin Yang Fortune, Dtps. Big Red Robe, Dtps. Mount Lip Taisuco, P. Miki Wata Nabe, P. Champion Porter, P. Amabilis, Dtps.Shih Hua Long First Love, P. Taida Salu, Dtps. Tying Shin Zebra, Dtps. Taida Salu Red, P. Kaleidoscope, P. Luchia Pink, P. Jin Cheng Sun, P. Sin Yuan Golden Beauty, Dtps. Chian Xen Magpie, P. Little Gem Stripe, P. Surf Song, Dtps. Lian Her Happy, P. Timothy Christopher Dtps. Happy UFO, P. Big White Pink Stripe, P. Medium Pink, P. Leodora, P. Magic Kiss, P. Goldie, P. Roxanne, P. Ivory pearl, P. Lucyna, P. Brother Girl, P. Memoria Francis Hunter P. Carlotta, Dtps. Leopard Princess M611, Dtps. Gan Lin Fairy 'GL', P. Shu Long Spotted Deer, P. Ho's Little Caroline, Dtps. Sogo Soft, Dtps. Gan Lin Beauty Hsiung, Dtps. Chin Ann Diamond 'Alisun', Dtps. Gan Lin Diamond, P. Chian Xen Pearl 'Cx#1', Dtps. Shu Long Pearl	VG
		Touching	3	P. Strawberry, Dtps. Ox Prince Thunder, Dtps. Shulong Sun Beauty	
		Overlapping	5	--	
17. QL	Flower: fragrance	Absent	1	Dtps. Plum Rose x Ox Black Jack, Dtps. Younghome Orange Lip, Dtps. Acker's Sweetie, Dtps. Hsin Yang Fortune, Dtps. Big Red Robe, Dtps. Mount Lip Taisuco, P. Miki Wata Nabe, P. Champion Porter, P. Amabilis, P. Strawberry, Dtps.Shih Hua Long First Love, P. Taida Salu, Dtps. Tying Shin Zebra, Dtps. Taida Salu Red, P. Kaleidoscope, P. Luchia Pink, P. Jin Cheng Sun, P. Sin Yuan Golden Beauty, Dtps. Chian Xen Magpie, P. Little Gem Stripe, Dtps. Lian Her Happy, P. Timothy Christopher Dtps. Happy UFO, P. Big White Pink Stripe, P. Medium Pink, P. Magic Kiss, P. Roxanne, P. Ivory pearl, P. Lucyna, P. Brother Girl, P. Memoria Francis Hunter, Dtps. Ox Prince Thunder, P. Carlotta, Dtps. Leopard Princess M611, Dtps. Gan Lin Fairy 'GL', P. Shu Long Spotted Deer, P. Ho's Little Caroline, Dtps. Sogo Soft, Dtps. Gan Lin Beauty Hsiung, Dtps. Chin Ann Diamond 'Alisun', Dtps. Gan Lin Diamond, Dtps. Shulong Sun Beauty, P. Chian Xen Pearl 'Cx#1', Dtps. Shu Long Pearl	VS
		Present	9	P. Surf Song, P. Leodora, Goldie, Dtps. Acker's Sweetie	
18. (*) QN (c)	Dorsal sepal: length (cm)	Short (<3)	3	P. Strawberry, Dtps. Tying Shin Zebra, P. Leodora, Goldie, P. Carlotta, P. Ho's Little Caroline, Dtps. Leopard Princess M611	MS
		Medium (3-4)	5	Dtps. Plum Rose x Ox Black Jack, Dtps. Younghome Orange Lip, Dtps. Hsin Yang Fortune, Dtps. Mount Lip Taisuco P. Champion Porter, P. Amabilis, Dtps. Shih hua long first love, P. Taida Salu, Dtps. Taida Salu Red, P. Kaleidoscope, P. Jin Cheng Sun, P. Sin Yuan Golden Beauty, P. Surf Song, Dtps. Lian Her Happy, P. Timothy Christopher Dtps. Happy UFO, P. Big White Pink Stripe, P. Medium Pink, P. Magic Kiss, P. Lucyna, P. Brother Girl, Dtps. Ox Prince Thunder, Dtps. Sogo Soft, Dtps. Shulong Sun Beauty, Dtps. Shu Long Pearl	
		Long (>4)	7	Dtps. Big Red Robe, P. Miki Wata Nabe, P. Luchia Pink, P. Little Gem Stripe, Dtps. Lian Her Happy, P. Roxanne, P. Ivory Pearl, Dtps. Acker's Sweetie, P. Memoria Francis Hunter, Dtps. Gan Lin Fairy 'GL', P. Shu Long Spotted Deer, Dtps. Gan Lin Beauty Hsiung, Dtps. Chin Ann Diamond 'Alisun', Dtps. Gan Lin Diamond, P. Chian Xen Pearl 'Cx#1', Dtps. Chian Xen Magpie	

continued Tab. 4.4: Table of Characteristics of DUS Test Guidelines in *Phalaenopsis*

Sl. No	Character-istics	States	Notes	Example Varieties /Hybrids	Type of Assess-ment
19. (*) QN (c)	Dorsal sepal: width (cm)	Narrow (<2)	3	Dtps. Tying Shin Zebra, P. Medium Pink, P. Leodora, P. Magic Kiss, Goldie, P. Ho's Little Caroline	MS
		Medium (2-4)	5	Dtps. Plum Rose x Ox Black Jack, Dtps. Younghome Orange Lip, Dtps. Acker's Sweetie, Dtps. Hsin Yang Fortune, Dtps. Big Red Robe, Dtps. Mount Lip Taisuco, P. Miki Wata Nabe, P. Champion Porter, P. Amabilis, P. Strawberry, Dtps.Shih Hua Long First Love, P. Taida Salu, Dtps. Taida Salu Red, P. Kaleido-scope, P. Luchia Pink, P. Jin Cheng Sun, P. Sin Yuan Golden Beauty, Dtps. Chian Xen Magpie, P. Little Gem Stripe, P. Surf Song, Dtps. Lian Her Happy, P. Timothy Christopher Dtps. Happy UFO, P. Big White Pink Stripe, P. Roxanne, P. Ivory pearl, P. Lucyna, P. Brother Girl, P. Memoria Francis Hunter, Dtps. Ox Prince Thunder P. Carlotta, Dtps. Leopard Princess M611, Dtps. Gan Lin Fairy 'GL', P. Shu Long Spotted Deer, Dtps. Sogo Soft, Dtps. Gan Lin Beauty Hsiung, Dtps. Chin Ann Diamond 'Alisun', Dtps. Gan Lin Diamond, Dtps. Shulong Sun Beauty, P. Chian Xen Pearl 'Cx#1', Dtps. Shu Long Pearl, Taida Salu Red	
		Broad (>4)	7	--	
20. (*) (+) PQ	Dorsal sepal: shape	Oblong	1	--	
		Ovate	3	Dtps. Plum Rose x Ox Black Jack, Dtps. Younghome Orange Lip, Dtps. Acker's Sweetie, Dtps. Big Red Robe, Dtps. Mount Lip Taisuco, P. Miki Wata Nabe, P. Champion Porter, P. Strawberry, Dtps.Shih Hua Long First Love, P. Taida Salu, Dtps. Taida Salu Red, P. Kaleidoscope, P. Luchia Pink, P. Jin Cheng Sun, P. Sin Yuan Golden Beauty, Dtps. Chian Xen Magpie, P. Little Gem Stripe, P. Surf Song, Dtps. Lian Her Happy, P. Timothy Christopher Dtps. Happy UFO, P. Big White Pink Stripe, P. Roxanne, P. Ivory pearl, P. Brother Girl, P. Memoria Francis Hunter, Dtps. Ox Prince Thunder, P. Car-lotta, Dtps. Leopard Princess M611, Dtps. Gan Lin Fairy 'GL', P. Shu Long Spotted Deer, P. Ho's Little Caroline, Dtps. Sogo Soft, Dtps. Gan Lin Beauty Hsiung, Dtps. Chin Ann Diamond 'Alisun', Dtps. Gan Lin Diamond, Dtps. Shulong Sun Beauty, P. Chian Xen Pearl 'Cx#1', Dtps. Shu Long Pearl	VG
		Elliptic	5	Dtps. Hsin Yang Fortune, P. Amabilis, Dtps. Tying Shin Zebra, Medium pink, P. Leodora, P. Magic Kiss, Goldie, P. Lucyna	
		Obovate	7	--	
		Round	9	--	
21. (*) PQ	Dorsal sepal: curvature of longitudinal axis	Incurved	1	Dtps. Plum Rose x Ox Black Jack, Dtps. Younghome Orange Lip, Dtps. Acker's Sweetie, Dtps. Hsin Yang Fortune, Dtps. Big Red Robe, Dtps. Mount Lip Taisuco, P. Miki Wata Nabe, P. Cham-pion Porter, P. Amabilis, P. Strawberry, Dtps.Shih Hua Long First Love, Dtps. Tying Shin Zebra, Dtps. Taida Salu Red, P. Kaleidoscope, P. Luchia Pink, P. Sin Yuan Golden Beauty, Dtps. Chian Xen Magpie, P. Little Gem Stripe, P. Surf Song, Dtps. Lian Her Happy, P. Big White Pink Stripe, P. Roxanne, P. Ivory pearl, P. Brother Girl, P. Memoria Francis Hunter, Dtps. Ox Prince Thunder, Dtps. Leopard Princess M611, Dtps. Shulong Sun Beauty, P. Shu Long Spotted Deer, Dtps. Shu Long Pearl	VG
		Straight	3	P. Medium Pink	
		Reflexed	5	P. Taida Salu, P. Jin Cheng Sun, P. Leodora, P. Magic Kiss, Goldie, P. Lucyna, P. Carlotta, Dtps. Gan Lin Fairy 'GL', P. Ho's Little Caroline, Dtps. Sogo Soft, Dtps. Gan Lin Beauty Hsiung, Dtps. Chin Ann Diamond 'Alisun', Dtps. Gan Lin Diamond, P. Chian Xen Pearl 'Cx#1'	

continued **Tab. 4.4:** Table of Characteristics of DUS Test Guidelines in *Phalaenopsis*

Sl. No	Character-istics	States	Notes	Example Varieties /Hybrids	Type of Assess-ment
22. (*) PQ	Dorsal sepal: undulation of margin	Absent	1	Dtps. Plum Rose x Ox Black Jack, Dtps. Younghome Orange Lip, Dtps. Hsin Yang Fortune, Dtps. Big Red Robe, Dtps. Mount Lip Taisuco, P. Miki Wata Nabe, P. Champion Porter, P. Amabilis, P. Strawberry, Dtps.Shih Hua Long First Love, P. Taida Salu, Dtps. Tying Shin Zebra, Dtps. Taida Salu Red, P. Kaleidoscope, P. Luchia Pink, P. Jin Cheng Sun, P. Sin Yuan Golden Beauty, Dtps. Chian Xen Magpie, P. Little Gem Stripe, P. Surf Song, Dtps. Lian Her Happy, P. Timothy Christopher Dtps. Happy UFO, P. Big White Pink Stripe, P. Medium Pink, P. P. Leodora, P. P. Magic Kiss, P. Goldie, P. P. Roxanne, P. Ivory pearl, P. Lucyna, P. Brother Girl, P. Memoria Francis Hunter, Dtps. Ox Prince Thunder P. Carlotta, Dtps. Leopard Princess M611, Dtps. Gan Lin Fairy 'GL', P. Shu Long Spotted Deer, P. Ho's Little Caroline, Dtps. Sogo Soft, Dtps. Gan Lin Beauty Hsiung, Dtps. Chin Ann Diamond 'Alisun', Dtps. Gan Lin Diamond, P. Chian Xen Pearl 'Cx#1', Dtps. Shu Long Pearl	VG
		Present	9	Dtps. Acker's Sweetie, Dtps. Shulong Sun Beauty	
23. QN (d)	Dorsal sepal: number of colours	One	1	Dtps. Younghome Orange Lip, Dtps. Hsin Yang Fortune, Dtps. Mount Lip Taisuco, P. Champion Porter, P. Amabilis, P. Luchia Pink, P. Sin Yuan Golden Beauty, P. Memoria Francis Hunter	VG
		Two	3	Dtps. Plum Rose x Ox Black Jack, Dtps. Acker's Sweetie, Dtps. Big Red Robe, P. Miki Wata Nabe, P. Champion Porter, P. Strawberry, Dtps.Shih Hua Long First Love, P. Taida Salu, Dtps. Tying Shin Zebra, Dtps. Taida Salu Red, P. Kaleidoscope, P. Jin Cheng Sun, P. Little Gem Stripe, P. Surf Song, P. Timothy Christopher Dtps. Happy UFO, P. Big White Pink Stripe, P. Medium Pink, P. P. Leodora, P. P. Magic Kiss, P. Goldie, P. P. Roxanne, P. Ivory pearl, P. Lucyna, P. Brother Girl, Dtps. Ox Prince Thunder, P. Carlotta, Dtps. Leopard Princess M611, P. Ho's Little Caroline, Dtps. Sogo Soft, Dtps. Gan Lin Beauty Hsiung, Dtps. Chin Ann Diamond 'Alisun', Dtps. Shulong Sun Beauty, P. Chian Xen Pearl 'Cx#1', Dtps. Shu Long Pearl	
		Three or more	5	Dtps. Lian Her Happy Dtps. Gan Lin Fairy 'GL', P. Shu Long Spotted Deer, Dtps. Gan Lin Diamond, Dtps. Chian Xen Magpie	
24. QL (d)	Dorsal sepal: colour pattern	Uniform	1	Dtps. Younghome Orange Lip, Dtps. Hsin Yang Fortune, Dtps. Mount Lip Taisuco, P. Amabilis, P. Luchia Pink, P. Ivory pearl, P. Memoria Francis Hunter	VS
		Shaded	2	Dtps. Acker's Sweetie, P. Miki Wata Nabe, Dtps.Shih Hua Long First Love, Dtps. Taida Salu Red, P. Jin Cheng Sun, Dtps. Lian Her Happy, P. Roxanne, Dtps. Ox Prince Thunder Dtps. Leopard Princess M611, P. Shu Long Spotted Deer, P. Ho's Little Caroline, Dtps. Chin Ann Diamond 'Alisun', Dtps. Gan Lin Diamond, P. Chian Xen Pearl 'Cx#1'	
		Edged	3	Dtps. Big Red Robe, P. Strawberry P. Carlotta	
		Striped	4	P. kaleidoscope	
		Netted	5	Dtps. Tying Shin Zebra, P. Sin Yuan Golden Beauty	
		Spotted	6	Dtps. Plum Rose x Ox Black Jack, P. Champion Porter, P. Surf Song, P. Leodora, P. Magic Kiss, Goldie, P. Lucyna, Dtps. Gan Lin Fairy 'GL', Dtps. Shulong Sun Beauty, Dtps. Sogo Soft, Dtps. Gan Lin Beauty Hsiung, Dtps. Shu Long Pearl	
		Mixed (specify)	7	P. Big White Pink Stripe (Shaded and striped) , P. Taida Salu, P. Little Gem Stripe, P. Timothy Christopher Dtps. Happy UFO, P. Brother Girl, Dtps. Chian Xen Magpie (Striped and spotted), P. Medium Pink (Edged and striped)	

continued **Tab. 4.4:** Table of Characteristics of DUS Test Guidelines in *Phalaenopsis*

Sl. No	Character-istics	States	Notes	Example Varieties /Hybrids	Type of Assess-ment
25. QL (d)	Dorsal sepal: dominant colour as per RHS colour chart	Green	1	--	VS
		White	2	P. Kaleidoscope, P. Miki Wata Nabe, P. Memoria Francis Hunter, Dtps. Ox Prince Thunder, Dtps. Chian Xen Magpie	
		Pink	3	--	
		Yellow	4	Dtps. Taida Salu Red, Dtps. Hsin Yang Fortune, Dtps.Shih Hua Long First Love	
		Red	5	--	
		Purple	6	P. Strawberry, Dtps. Plum Rose x Ox Black Jack	
		Blue	7	--	
		Violet	8	--	
26. QN (c)	Lateral sepal: length (cm)	Short (<3)	3	P. Strawberry	MS
		Medium (3-4)	5	Dtps. Taida Salu Red, P. Kaleidoscope, Dtps.Shih Hua Long First Love	
		Long (>4)	7	P. Miki Wata Nabe, Dtps. Hsin Yang Fortune, P. Memoria Francis Hunter, Dtps. Ox Prince Thunder, Dtps. Chian Xen Magpie, Dtps. Plum Rose x Ox Black Jack	
27. QN (c)	Lateral sepal: width (cm)	Narrow (<2)	3	P. Strawberry	MS
		Medium (2-4)	5	Dtps. Taida Salu Red, P. Kaleidoscope, P. Miki Wata Nabe, Dtps. Hsin Yang Fortune, Dtps.Shih Hua Long First Love, P. Memoria Francis Hunter, Dtps. Ox Prince Thunder, Dtps. Chian Xen Magpie, Dtps. Plum Rose x Ox Black Jack	
		Broad (>4)	7	--	
28. (+) PQ	Lateral sepal: shape	Oblong	1	--	VG
		Ovate	3	Dtps. Taida Saln Red, P. Strawberry, Dtps. Hsin Yang Fortune, Dtps. Shih Hua Long First Love, P. Memoria Francis Hunter, Dtps. Ox Prince Thunder, Dtps. Chian Xen Magpie, Dtps. Plum Rose x Ox Black Jack	
		Elliptic	5	P. Kaleidoscope, P. Miki Wata Nabe	
		Obovate	7	--	
		Round	9	--	
29. QL (d)	Lateral sepal: Number of colours	One	1	Dtps. Mount Lip Taisuco, P. Amabilis	MS
		Two	3	Dtps. Plum Rose x Ox Black Jack, Dtps. Younghome Orange Lip, Dtps. Acker's Sweetie, Dtps. Hsin Yang Fortune, Dtps. Big Red Robe, P. Miki Wata Nabe, P. Champion Porter, P. Strawberry, Dtps.Shih Hua Long First Love, P. Taida Salu, Dtps. Taida Salu Red, P. Kaleidoscope, P. Luchia Pink, P. Jin Cheng Sun, P. Sin Yuan Golden Beauty, P. Little Gem Stripe, P. Surf Song, Dtps. Lian Her Happy, P. Timothy Christopher Dtps. Happy UFO, P. Big White Pink Stripe, P. Medium Pink, P. P. Leodora, P. P. Magic Kiss, P. Goldie, P. P. Roxanne, P. Ivory Pearl, P. Lucyna, P. Brother Girl, P. Memoria Francis Hunter, Dtps. Ox Prince Thunder, P. Carlotta, Dtps. Leopard Princess M611, Dtps. Sogo Soft, Dtps. Gan Lin Beauty Hsiung, Dtps. Chin Ann Diamond 'Alisun', Dtps. Shulong Sun Beauty, P. Chian Xen Pearl 'Cx#1', Dtps. Shu Long Pearl	
		Three or more	5	Dtps. Tying Shin Zebra Dtps. Gan Lin Fairy 'GL', P. Shu Long Spotted Deer, P. Ho's Little Caroline, Dtps. Gan Lin Diamond Dtps. Chian Xen Magpie	

continued **Tab. 4.4:** Table of Characteristics of DUS Test Guidelines in *Phalaenopsis*

Sl. No	Characteristics	States	Notes	Example Varieties /Hybrids	Type of Assessment
30. QL (d)	Lateral sepal: colour pattern	Uniform	1	Dtps. Mount Lip Taisuco	VG
		Shaded	2	Dtps. Ox Prince Thunder, Dtps. Leopard Princess M611, P. Chian Xen Pearl 'Cx#1'	
		Edged	3	--	
		Striped	4	--	
		Netted	5	--	
		Spotted	6	P. Champion Porter, P. Surf Song, P. Magic Kiss, Goldie, P. Lucyna, Dtps. Sogo Soft, Dtps. Shu Long Pearl	
		Mixed (specify)	7	Dtps. Plum Rose x Ox Black Jack (spotted and edged), Dtps. Younghome Orange Lip (evenly colored and spotted at the base), Dtps. Acker's Sweetie (shaded and spotted), Dtps. Hsin Yang Fortune (evenly colored and spotted at the base), Dtps. Big Red Robe (spotted and edged), P. Miki Wata Nabe (shaded and spotted at the base), P. Amabilis (evenly colored and spotted at the base), P. Strawberry (edged and spotted), Dtps.Shih Hua Long First Love (one half evenly colored and other half shaded and spotted), P. Taida Salu (striped and spotted), Dtps. Tying Shin Zebra (netted and spotted), Dtps. Taida Salu Red (shaded and spotted), P. Kaleidoscope (shaded, striped and spotted), P. Luchia Pink (netted and spotted), P. Jin Cheng Sun (evenly coloured and shaded at the base), P. Sin Yuan Golden Beauty (netted and spotted), Dtps. Chian Xen Magpie (netted and spotted), P. Little Gem Stripe (striped and spotted), Dtps. Lian Her Happy (shaded and spotted), P. Timothy Christopher Dtps. Happy UFO (striped and spotted), P. Big White Pink Stripe (shaded, striped and spotted), P. Medium Pink (striped, spotted and edged), P. Leodora (shaded and spotted), P. Roxanne (shaded and spotted), P. Ivory pearl (evenly colored and spotted at the base), P. Brother Girl (striped and spotted), P. Memoria Francis Hunter (evenly colored and shaded at the base), P. Carlotta(Edged and spotted), Dtps. Gan Lin Fairy 'GL' (shaded and spotted), P. Shu Long Spotted Deer (shaded and spotted), P. Ho's Little Caroline (shaded and spotted), Dtps. Gan Lin Beauty Hsiung (shaded and spotted), Dtps. Chin Ann Diamond 'Alisun'(shaded and spotted), Dtps. Gan Lin Diamond (shaded and spotted), Dtps. Shulong Sun Beauty (shaded and spotted)	
31. QL (d)	Lateral sepal: dominant colour as per RHS colour chart	Green	1	--	VG
		White	2	P. Kaleidoscope, P. Miki Wata Nabe, P. Memoria Francis Hunter, Dtps. Ox Prince Thunder, Dtps. Chian Xen Magpie	
		Pink	3	--	
		Yellow	4	Taida Salu Red, Dtps. Hsin Yang Fortune, Dtps.Shih Hua Long First Love	
		Red	5	--	
		Purple	6	P. Strawberry, Dtps. Plum Rose x Ox Black Jack	
		Blue	7	--	
		Violet	8	--	

continued **Tab. 4.4:** Table of Characteristics of DUS Test Guidelines in *Phalaenopsis*

Sl. No	Character-istics	States	Notes	Example Varieties /Hybrids	Type of Assess-ment
32. (+) PQ	Petal: shape	Oblong	1	--	VS
		Elliptic	3	--	
		Ovate	5	Dtps. Plum Rose x Ox Black Jack, Dtps. Younghome Orange Lip, Dtps. Acker's Sweetie, Dtps. Big Red Robe, Dtps. Mount Lip Taisuco, P. Miki Wata Nabe, P. Champion Porter, P. Strawberry, P. Taida Salu, Dtps. Taida Salu Red, P. Kaleidoscope, P. Luchia Pink, P. Jin Cheng Sun, P. Sin Yuan Golden Beauty, Dtps. Chian Xen Magpie, P. Little Gem Stripe, P. Surf Song, Dtps. Lian Her Happy, P. Timothy Christopher Dtps. Happy UFO, P. Big White Pink Stripe, P. Roxanne, P. Ivory pearl, P. Brother Girl, P. Memoria Francis Hunter, Dtps. Ox Prince Thunder, P. Carlotta, Dtps. Gan Lin Fairy 'GL', P. Shu Long Spotted Deer, P. Ho's Little Caroline, Dtps. Gan Lin Beauty Hsiung, Dtps. Chin Ann Diamond 'Alisun', Dtps. Gan Lin Diamond, Dtps. Shulong Sun Beauty, P. Chian Xen Pearl 'Cx#1', Dtps. Shu Long Pearl	
		Sub-orbicular	7	Dtps. Hsin Yang Fortune, P. Amabilis, Dtps. Tying Shin Zebra, P. Medium Pink, P. Magic Kiss, Goldie, P. Leodora, P. Lucyna, Dtps. Leopard Princess M611, Dtps. Sogo Soft., Dtps.Shih Hua Long First Love	
33. (*) QN (c)	Petal: length (cm)	Short (<3.0)	3	Dtps. Younghome Orange Lip, P. Amabilis, P. Strawberry, P. Medium Pink, P. Magic Kiss, Goldie, P. Leodora, P. Carlotta, Dtps. Leopard Princess M611, P. Ho's Little Caroline	MS
		Medium (3.0-4.5)	5	Dtps. Plum Rose x Ox Black Jack, Dtps. Acker's Sweetie, Dtps. Hsin Yang Fortune, Dtps. Big Red Robe, Dtps. Mount Lip Taisuco, P. Champion Porter, Dtps.Shih Hua Long First Love, P. Taida Salu, Dtps. Taida Salu Red, P. Kaleidoscope, P. Luchia Pink, P. Jin Cheng Sun, P. Sin Yuan Golden Beauty, Dtps. Chian Xen Magpie, P. Little Gem Stripe, Dtps. Lian Her Happy, P. Timothy Christopher Dtps. Happy UFO, P. Big White Pink Stripe, P. Roxanne, P. Lucyna, P. Brother Girl, P. Memoria Francis Hunter, Dtps. Ox Prince Thunder, Dtps. Gan Lin Fairy 'GL', P. Shu Long Spotted Deer, Dtps. Gan Lin Beauty Hsiung, Dtps. Dtps. Sogo Soft, Chin Ann Diamond 'Alisun', Dtps. Gan Lin Diamond, Dtps. Shulong Sun Beauty, P. Chian Xen Pearl 'Cx#1', Dtps. Shu Long Pearl	
		Long (>4.5)	7	P. Miki Wata Nabe, Dtps. Tying Shin Zebra, P. Surf Song, P. Ivory pearl	
34. (*) QN (c)	Petal: width (cm)	Narrow (<3.0)	3	P. Medium Pink, P. Magic Kiss, Goldie, P. Leodora, Dtps. Tying Shin Zebra, P. Lucyna, P. Carlotta, Dtps. Leopard Princess M611, P. Ho's Little Caroline	MS
		Medium (3.0-4.5)	5	Dtps. Younghome Orange Lip, Dtps. Hsin Yang Fortune, Dtps. Mount Lip Taisuco, P. Champion Porter, P. Amabilis, P. Strawberry, Dtps.Shih Hua Long First Love, P. Taida Salu, Dtps. Tying Shin Zebra, Dtps. Taida Salu Red, P. Kaleido-scope, P. Jin Cheng Sun, P. Sin Yuan Golden Beauty, P. Surf Song, P. Timothy Christopher dtps. happy UFO, P. Big White Pink Stripe, Dtps. Sogo Soft, Dtps. Shu Long Pearl	
		Broad (>4.5)	7	Dtps. Plum Rose x Ox Black Jack, Dtps. Acker's Sweetie, Dtps. Big Red Robe, P. Miki Wata Nabe, P. Luchia Pink, Dtps. Chian Xen Magpie, P. Little Gem Stripe, Dtps. Lian Her Happy, P. Roxanne, P. Ivory pearl, P. Brother Girl, P. Memoria Francis Hunter, Dtps. Ox Prince Thunder, Dtps. Gan Lin Fairy 'GL', P. Shu Long Spotted Deer, Dtps. Gan Lin Beauty Hsiung, Dtps. Chin Ann Diamond 'Alisun', Dtps. Gan Lin Diamond, Dtps. Shulong Sun Beauty, P. Chian Xen Pearl 'Cx#1'	

continued **Tab. 4.4:** Table of Characteristics of DUS Test Guidelines in *Phalaenopsis*

Sl. No	Characteristics	States	Notes	Example Varieties /Hybrids	Type of Assessment
35. (*) (+) PQ	Petal: curvature of longitudinal axis	Incurved	1	Dtps. Younghome Orange Lip, Dtps. Hsin Yang Fortune Dtps. Acker's Sweetie, Dtps. Big Red Robe, Dtps. Mount Lip Taisuco, P. Miki Wata Nabe, Dtps. Tying Shin Zebra, P. Kaleidoscope, P. Little Gem Stripe, Dtps. Lian Her Happy, P. Big White Pink Stripe, P. Ivory pearl, P. Memoria Francis Hunter, Dtps. Ox Prince Thunder, Dtps. Leopard Princess M611, P. Ho's Little Caroline, Dtps. Shulong Sun Beauty, Dtps. Sogo Soft	VG
		Straight	3	Dtps. Plum Rose x Ox Black Jack, P. Champion Porter, P. Amabilis, P. Strawberry, Dtps. Shih hua long first love, P. Sin Yuan Golden Beauty, Dtps. Chian Xen Magpie, P. Luchia Pink, P. Surf Song, P. Medium Pink, Goldie, P. Roxanne, Taida Salu. Red	
		Reflexed	5	P. Taida Salu, Taida salu red, P. Jin Cheng Sun, P. Timothy Christopher Dtps. Happy UFO, P. Magic Kiss, P. Leodora, P. Lucyna, P. Brother Girl, P. Carlotta, Dtps. Gan Lin Fairy 'GL', P. Shu Long Spotted Deer, Dtps. Gan Lin Beauty Hsiung, Dtps. Chin Ann Diamond 'Alisun', Dtps. Gan Lin Diamond, P. Chian Xen Pearl 'Cx#1', Dtps. Shu Long Pearl	
36. PQ	Petal: undulation of margin	Absent	1	Dtps. Hsin Yang Fortune, Dtps. Big Red Robe, Dtps. Tying Shin Zebra, P. Kaleidoscope, P. Luchia Pink, P. Jin Cheng Sun, Dtps. Gan Lin Fairy 'GL', P. Ho's Little Caroline, Dtps. Gan Lin Diamond, Dtps. Gan Lin Beauty Hsiung, Taida Saln. Red	VG
		Present	9	Dtps. Plum Rose x Ox Black Jack, Dtps. Younghome Orange Lip, Dtps. Acker's Sweetie, Dtps. Mount Lip Taisuco, P. Miki Wata Nabe, P. Champion Porter, P. Amabilis, P. Strawberry, Dtps.Shih Hua Long First Love, P. Taida Salu, Dtps. Taida Salu Red, P. Sin Yuan Golden Beauty, Dtps. Chian Xen Magpie, P. Little Gem Stripe, P. Surf Song, Dtps. Lian Her Happy, P. Timothy Christopher Dtps. Happy UFO, P. Big White Pink Stripe, P. Medium Pink, P. P. Leodora, P. P. Magic Kiss, P. Goldie, P. P. Roxanne, P. Ivory pearl, P. Lucyna, P. Brother Girl, P. Memoria Francis Hunter, Dtps. Ox Prince Thunder, P. Carlotta, Dtps. Leopard Princess M611, P. Shu Long Spotted Deer, Dtps. Sogo Soft, Dtps. Chin Ann Diamond 'Alisun', Dtps. Shulong Sun Beauty, P. Chian Xen Pearl 'Cx#1', Dtps. Shu Long Pearl	
37. QL (d)	Petal: number of colors	One	1	Dtps. Younghome Orange Lip, Dtps. Big Red Robe, Dtps. Hsin Yang Fortune, Dtps. Mount Lip Taisuco, P. Amabilis, P. Luchia Pink, P. Sin Yuan Golden Beauty, P. Surf Song, P. Ivory pearl	VG
		Two	3	Dtps. Plum Rose x Ox Black Jack, Dtps. Acker's Sweetie, P. Miki Wata Nabe, P. Champion Porter, Dtps.Shih Hua Long First Love, P. Strawberry, P. Taida Salu, Dtps. Tying Shin Zebra, Dtps. Taida Salu Red, P. Kaleidoscope, P. Jin Cheng Sun, P. Little Gem Stripe, P. Timothy Christopher Dtps. Happy UFO, P. Big White Pink Stripe, P. Medium Pink, P. P. Leodora, P. P. Magic Kiss, P. Goldie, P. P. Roxanne, P. Lucyna, P. Brother Girl, P. Memoria Francis Hunter, Dtps. Ox Prince Thunder, P. Carlotta, Dtps. Leopard Princess M611, P. Ho's Little Caroline, Dtps. Sogo Soft, Dtps. Chin Ann Diamond 'Alisun', Dtps. Shulong Sun Beauty, P. Chian Xen Pearl 'Cx#1', Dtps. Shu Long Pearl	
		Three or more	5	Dtps. Lian Her Happy, Dtps. Gan Lin Fairy 'GL', P. Shu Long Spotted Deer, Dtps. Gan Lin Diamond, Dtps. Gan Lin Beauty Hsiung, Dtps. Chian Xen Magpie	

continued Tab. 4.4: Table of Characteristics of DUS Test Guidelines in *Phalaenopsis*

Sl. No	Characteristics	States	Notes	Example Varieties / Hybrids	Type of Assessment
38. (*) QL (d)	Petal: color pattern	Uniform	1	Dtps. Younghome Orange Lip, Dtps. Mount Lip Taisuco, Dtps. Hsin Yang Fortune, P. Amabilis, P. Ivory pearl	VG
		Shaded	2	P. Miki Wata Nabe, P. Roxanne, Dtps. Ox Prince Thunder, Dtps. Leopard Princess M611, P. Chian Xen Pearl 'Cx#1'	
		Edged	3	P. Strawberry, P. Carlotta	
		Striped	4	--	
		Netted	5	P. Luchia Pink, P. Sin Yuan Golden Beauty, Dtps. Chian Xen Magpie	
		Spotted	6	P. Champion Porter, P. Surf Song, P. Magic Kiss, Goldie, P. Lucyna, Dtps. Shulong Sun Beauty, Dtps. Sogo Soft, Dtps. Shu Long Pearl.	
		Mixed (specify)	7	Dtps. Plum Rose x Ox Black Jack (shaded and spotted), Dtps. Acker's Sweetie (evenly colored and shaded at the base), Dtps. Big Red Robe (netted and edged), Dtps.Shih Hua Long First Love (shaded and spotted at the base and evenly colored), P. Taida Salu (netted and spotted), Dtps. Lian Her Happy (shaded and spotted), Dtps. Tying Shin Zebra (netted and shaded at the base), Dtps. Taida Salu Red(shaded and spotted), P. kaleidoscope (striped and netted), P. Jin Cheng Sun (shaded and spotted), P. Little Gem Stripe (netted and spotted), P. Timothy Christopher Dtps. Happy UFO (netted and spotted), P. Big White Pink Stripe (shaded and netted), P. Medium Pink (netted and edged), P. Leodora (shaded and spotted) P. Brother Girl (netted, spotted and edged), P. Memoria Francis Hunter (evenly colored and spotted) Dtps. Gan Lin Fairy 'GL'(shaded and spotted), P. Shu Long Spotted Deer (shaded and spotted), P. Ho's Little Caroline (shaded and netted), Dtps. Gan Lin Beauty Hsiung (shaded and spotted), Dtps. ChinAnn Diamond 'Alisun' (shaded and spotted), Dtps. Gan Lin Diamond (shaded and spotted),	
39. (*) QL (d)	Petal: predominant colour as per RHS colour chart	Green	1	--	VS
		White	2	P. Kaleidoscope, P. Miki Wata Nabe, P. Memoria Francis Hunter, Dtps. Ox Prince Thunder, Dtps. Chian Xen Magpie	
		Pink	3	--	
		Yellow	4	Dtps. Taida Salu Red, Dtps. Hsin Yang Fortune, Dtps.Shih Hua Long First Love	
		Red	5	--	
		Purple	6	P. Strawberry, Dtps. Plum Rose x Ox Black Jack	
		Blue	7	--	
		Violet	8	--	

continued **Tab. 4.4:** Table of Characteristics of DUS Test Guidelines in *Phalaenopsis*

Sl. No	Characteristics	States	Notes	Example Varieties /Hybrids	Type of Assessment
40. QN (c)	Lip: length of apical lobe (cm)	Short (<1)	3	--	MG
		Medium (1-2)	5	Dtps. Plum Rose x Ox Black Jack, Dtps. Younghome Orange Lip, Dtps. Hsin Yang Fortune, Dtps. Mount Lip Taisuco, P. Champion Porter, P. Amabilis, P. Strawberry, Dtps.Shih Hua Long First Love, P. Taida Salu, Dtps. Tying Shin Zebra, Dtps. Taida Salu Red, P. Kaleidoscope, P. Jin Cheng Sun, P. Sin Yuan Golden Beauty, Dtps. Chian Xen Magpie, P. Little Gem Stripe, P. Surf Song, P. Timothy Christopher Dtps. Happy UFO, P. Big White Pink Stripe, P. Medium Pink, P. Leodora, P. Magic Kiss, Goldie, P. Lucyna, P. Memoria Francis Hunter, Dtps. Leopard Princess M611, P. Ho's Little Caroline, Dtps. Sogo Soft, Dtps. Chin Ann Diamond 'Alisun', Dtps. Shulong Sun Beauty, Dtps. Shu Long Pearl, Taida Saln. Red	
		Long (>2)	7	Dtps. Acker's Sweetie, Dtps. Big Red Robe, P. Miki Wata Nabe, P. Luchia Pink, Dtps. Lian Her Happy, P. Roxanne, P. Ivory pearl, P. Brother Girl, Dtps. Ox Prince Thunder, P. Carlotta, Dtps. Gan Lin Fairy 'GL', P. Chian Xen Pearl 'Cx#1', P. Shu Long Spotted Deer, Dtps. Gan Lin Diamond, Dtps. Gan Lin Beauty Hsiung	
41. QN (c)	Lip: width of apical lobe (cm)	Narrow (<1)	3	--	MG
		Medium (1-2)	5	Dtps. Younghome Orange Lip, Dtps. Hsin Yang Fortune, P. Miki Wata Nabe, P. Champion Porter, P. Strawberry, Dtps. Shih Hua Long First Love, P. Taida Salu, Dtps. Tying Shin Zebra, Dtps. Taida Salu Red, P. Kaleidoscope, P. Little Gem Stripe, P. Surf Song, P. Timothy Christopher Dtps. Happy UFO, P. Medium Pink, P. Leodora, P. Magic Kiss, P. Goldie, P. Roxanne, P. Ivory Pearl, P. Lucyna, P. Brother Girl, P. Memoria Francis Hunter P. Carlotta, Dtps. Leopard Princess M611, P. Ho's Little Caroline, Dtps. Sogo Soft, Dtps. Chin Ann Diamond 'Alisun', Dtps. Shulong Sun Beauty, Dtps. Shu Long Pearl	
		Broad (>2)	7	Dtps. Plum Rose x Ox Black Jack, Dtps. Acker's Sweetie, Dtps. Big Red Robe, Dtps. Mount Lip Taisuco, P. Amabilis, P. Luchia Pink, P. Sin Yuan Golden Beauty, Dtps. Chian Xen Magpie, Dtps. Lian Her Happy, P. Big White Pink Stripe, Dtps. Ox Prince Thunder, Dtps. Gan Lin Fairy 'GL', P. Chian Xen Pearl 'Cx#1', P. Shu Long Spotted Deer, Dtps. Gan Lin Diamond, Dtps. Gan Lin Beauty Hsiung	
42. (+) PQ (c)	Lip: presence of whiskers	Absent	1	P. Surf Song, P. Leodora, P. Goldie	VG
		Present	9	Dtps. Plum Rose x Ox Black Jack, Dtps. Younghome Orange Lip, Dtps. Acker's Sweetie, Dtps. Hsin Yang Fortune, Dtps. Big Red Robe, Dtps. Mount Lip Taisuco, P. Miki Wata Nabe, P. Champion Porter, P. Amabilis, P. Strawberry, Dtps.Shih Hua Long First Love, P. Taida Salu, Dtps. Tying Shin Zebra, Dtps. Taida Salu Red, P. Kaleidoscope, P. Luchia Pink, P. Jin Cheng Sun, P. Sin Yuan Golden Beauty, Dtps. Chian Xen Magpie, P. Little Gem Stripe, Dtps. Lian Her Happy, P. Timothy Christopher Dtps. Happy UFO, P. Big White Pink Stripe, P. Medium Pink, P. Magic Kiss, P. Roxanne, P. Ivory pearl, P. Lucyna, P. Brother Girl, P. Memoria Francis Hunter, Dtps. Ox Prince Thunder, P. Carlotta, Dtps. Leopard Princess M611, Dtps. Gan Lin Fairy 'GL', P. Shu Long Spotted Deer, P. Ho's Little Caroline, Dtps. Sogo Soft, Dtps. Gan Lin Beauty Hsiung, Dtps. Chin Ann Diamond 'Alisun', Dtps. Gan Lin Diamond, Dtps. Shulong Sun Beauty, P. Chian Xen Pearl 'Cx#1', Dtps. Shu Long Pearl	

continued **Tab. 4.4:** Table of Characteristics of DUS Test Guidelines in *Phalaenopsis*

Sl. No	Characteristics	States	Notes	Example Varieties /Hybrids	Type of Assessment
43. PQ (c)	Lip: length of whiskers relative to length of apical lobe	Smaller	3	Dtps. Plum Rose x Ox Black Jack, Dtps. Younghome Orange Lip, Dtps. Acker's Sweetie, Dtps. Hsin Yang Fortune, Dtps. Big Red Robe, Dtps. Mount Lip Taisuco, P. Miki Wata Nabe, P. Champion Porter, P. Amabilis, P. Strawberry, Dtps.Shih Hua Long First Love, P. Taida Salu, Dtps. Tying Shin Zebra, Dtps. Taida Salu Red, P. Kaleidoscope, P. Luchia Pink, P. Jin Cheng Sun, P. Sin Yuan Golden Beauty, Dtps. Chian Xen Magpie, P. Little Gem Stripe, P. Surf Song, Dtps. Lian Her Happy, P. Timothy Christopher, Dtps. Happy UFO, P. Big White Pink Stripe, P. Medium Pink, P. Leodora, P. Magic Kiss, P. Goldie, P. Roxanne, P. Lucyna, P. Brother Girl, P. Memoria Francis Hunter, Dtps. Ox Prince Thunder, P. Carlotta, Dtps. Leopard Princess M611, Dtps. Gan Lin Fairy 'GL', P. Shu Long Spotted Deer, P. Ho's Little Caroline, Dtps. Sogo Soft, Dtps. Gan Lin Beauty Hsiung, Dtps. Chin Ann Diamond 'Alisun', Dtps. Gan Lin Diamond, Dtps. Shulong Sun Beauty, P. Chian Xen Pearl 'Cx#1', Dtps. Shu Long Pearl	VG
		Equal	5	P. Ivory Pearl	
		Larger	7	--	
44. (*) (+) PQ	Lip: shape of apical lobe	Ovate	1	P. Ho's Little Caroline	VS
		Elliptic	2	Dtps. Hsin Yang Fortune, Dtps. Tying Shin Zebra, P. Surf Song, P. Goldie, P. Carlotta	
		Obovate	3	--	
		Semi-circular	4	--	
		Deltoid	5	P. Leodora	
		Obdeltoid	6	P. Champion Porter, P. Memoria Francis Hunter, Dtps. Ox Prince Thunder	
		Rhombic	7	Dtps. Plum Rose x Ox Black Jack, Dtps. Younghome Orange Lip, Dtps. Acker's Sweetie, Dtps. Big Red Robe, Dtps. Mount Lip Taisuco, P. Miki Wata Nabe, P. Amabilis, Dtps. Shih Hua Long First Love, P. Taida Salu, Dtps. Taida Salu Red, P. Kaleidoscope, P. Luchia Pink, P. Jin Cheng Sun, P. Sin Yuan Golden Beauty, Dtps. Chian Xen Magpie, P. Little Gem Stripe, Dtps. Lian Her Happy, P. Timothy Christopher Dtps. Happy UFO, P. Big White Pink Stripe, P. Medium Pink, P. Magic Kiss, P. Roxanne, P. Ivory pearl, P. Lucyna, P. Brother Girl, Dtps. Leopard Princess M611, Dtps. Gan Lin Fairy 'GL', P. Shu Long Spotted Deer, Dtps. Sogo Soft, Dtps. Gan Lin Beauty Hsiung, Dtps. Chin Ann Diamond 'Alisun', Dtps. Gan Lin Diamond, Dtps. Shulong Sun Beauty, P. Chian Xen Pearl 'Cx#1', Dtps. Shu Long Pearl, P. Strawberry	
		Orbicular	8	--	

continued **Tab. 4.4:** Table of Characteristics of DUS Test Guidelines in *Phalaenopsis*

Sl. No	Character-istics	States	Notes	Example Varieties /Hybrids	Type of Assess-ment
45. (*) (+) PQ	Lip: bump and ridge on apical lobe	Absent	1	Dtps. Plum Rose x Ox Black Jack, Dtps. Acker's Sweetie, Dtps. Big Red Robe, Dtps. Mount Lip Taisuco, P. Miki Wata Nabe, Dtps.Shih Hua Long First Love, P. Amabilis, P. Luchia Pink, Dtps. Lian Her Happy, P. Big White Pink Stripe, P. Medium Pink, P. Memoria Francis Hunter, Dtps. Ox Prince Thunder, P. Shu Long Spotted Deer, Dtps. Chin Ann Diamond 'Alisun', Dtps. Gan Lin Diamond, Dtps. Gan Lin Beauty Hsiung	VG
		Present	9	Dtps. Younghome Orange Lip, Dtps. Hsin Yang Fortune, P. Champion Porter, P. Strawberry, P. Taida Salu, Dtps. Tying Shin Zebra, Dtps. Taida Salu Red, P. Kaleidoscope, P. Jin Cheng Sun, P. Sin Yuan Golden Beauty, Dtps. Chian Xen Magpie, P. Little Gem Stripe, P. Surf Song, P. Timothy Christopher Dtps. Happy UFO, P. P. Leodora, P. P. Magic Kiss, P. Goldie, P. P. Roxanne, P. Ivory pearl, P. Lucyna, P. Brother Girl, P. Carlotta, Dtps. Leopard Princess M611, Dtps. Gan Lin Fairy 'GL', P. Ho's Little Caroline, Dtps. Sogo Soft, Dtps. Shulong Sun Beauty, P. Chian Xen Pearl 'Cx#1', Dtps. Shu Long Pearl	
46. (+) PQ	Lip: shape of lateral lobe	Semi-circular	1	P. Leodora	VG
		Oblong	3		
		Oblanceolate	5	Dtps. Tying Shin Zebra, P. Surf Song	
		Obtriangle	7	Dtps. Hsin Yang Fortune, Dtps.Shih Hua Long First Love, P. Champion Porter, P. Taida Salu, P. Kaleidoscope, P. Little Gem Stripe, P. Timothy Christopher Dtps. Happy UFO, P. Magic Kiss, Goldie, P. Lucyna, P. Carlotta, Dtps. Leopard Princess M611, P. Ho's Little Caroline, Dtps. Sogo Soft	
		Sub-orbicular	9	Dtps. Plum Rose x Ox Black Jack, Dtps. Younghome Orange Lip, Dtps. Acker's Sweetie, Dtps. Big Red Robe, Dtps. Mount Lip Taisuco, P. Miki Wata Nabe, P. Amabilis, P. Strawberry, Dtps. Taida Salu Red, P. Luchia Pink, P. Jin Cheng Sun, P. Sin Yuan Golden Beauty, Dtps. Chian Xen Magpie, Dtps. Lian Her Happy, P. Big White Pink Stripe, P. Medium Pink, P. Roxanne, P. Ivory pearl, P. Brother Girl, P. Memoria Francis Hunter, Dtps. Ox Prince Thunder Dtps. Gan Lin Fairy 'GL', P. Shu Long Spotted Deer, Dtps. Gan Lin Beauty Hsiung, Dtps. Chin Ann Diamond 'Alisun', Dtps. Gan Lin Diamond, Dtps. Shulong Sun Beauty, P. Chian Xen Pearl 'Cx#1', Dtps. Shu Long Pearl.	
47. (+) PQ	Lip: curvature of lateral lobe	Slightly incurved	3	Dtps. Younghome Orange Lip, Dtps. Hsin Yang Fortune, P. Miki Wata Nabe, P. Amabilis, P. Taida Salu, Dtps. Tying Shin Zebra, Dtps. Taida Salu Red, P. Kaleidoscope, P. Jin Cheng Sun, P. Sin Yuan Golden Beauty, Dtps. Chian Xen Magpie, P. Surf Song, Dtps. Lian Her Happy, P. Timothy Christopher Dtps. Happy UFO, P. Big White Pink Stripe, P. Medium Pink, P. Leodora, P. Magic Kiss, P. Goldie, P. Roxanne, P. Ivory pearl, P. Brother Girl, P. Memoria Francis Hunter, Dtps. Ox Prince Thunder, Dtps. Leopard Princess M611, P. Chian Xen Pearl 'Cx#1', P. Shu Long Spotted Deer, Dtps. Sogo Soft, Dtps. Gan Lin Beauty Hsiung.	VG
		Strongly incurved	7	Dtps. Plum Rose x Ox Black Jack, Dtps. Acker's Sweetie, Dtps. Big Red Robe, Dtps. Mount Lip Taisuco, P. Champion Porter, P. Strawberry, Dtps.Shih Hua Long First Love, P. Luchia Pink, P. Little Gem Stripe P. Lucyna, P. Carlotta, Dtps. Gan Lin Fairy 'GL', P. Ho's Little Caroline, Dtps. Chin Ann Diamond 'Alisun', Dtps. Gan Lin Diamond, Dtps. Shulong Sun Beauty, Dtps. Shu Long Pearl	

continued Tab. 4.4: Table of Characteristics of DUS Test Guidelines in *Phalaenopsis*

Sl. No	Character-istics	States	Notes	Example Varieties /Hybrids	Type of Assess-ment
48. PQ (c)	Lip: size of lateral lobe relative to apical lobe	Smaller	3	Dtps. Younghome Orange Lip, Dtps. Acker's Sweetie, Dtps. Hsin Yang Fortune, Dtps. Big Red Robe, P. Champion Porter, P. Amabilis, P. Strawberry, Dtps.Shih Hua Long First Love, P. Taida Salu, Dtps. Tying Shin Zebra, Dtps. Taida Salu Red, P. Kaleidoscope, P. Jin Cheng Sun, P. Sin Yuan Golden Beauty, Dtps. Chian Xen Magpie, P. Little Gem Stripe, P. Surf Song, Dtps. Lian Her Happy, P. Timothy Christopher Dtps. Happy UFO, P. Big White Pink Stripe, P. Leodora, P. Magic Kiss, P. G oldie, P. Roxanne, P. Ivory pearl, P. Lucyna, P. Brother Girl, P. Memoria Francis Hunter, Dtps. Ox Prince Thunder P. Carlotta, Dtps. Leopard Princess M611, Dtps. Gan Lin Fairy 'GL', P. Shu Long Spotted Deer, P. Ho's Little Caroline, Dtps. Sogo Soft, Dtps. Gan Lin Beauty Hsiung, Dtps. Chin Ann Diamond 'Alisun', Dtps. Gan Lin Diamond, Dtps. Shulong Sun Beauty, P. Chian Xen Pearl 'Cx#1', Dtps. Shu Long Pearl.	VS
		Equal	5	Dtps. Mount Lip Taisuco, P. Miki Wata Nabe, P. Luchia Pink.	
		Larger	7	Dtps. Plum Rose x Ox Black Jack, P. Medium Pink	
49. QL (d)	Lip: number of colours	One	1	--	VG
		Two	3	--	
		Three or more	5	Dtps. Acker's Sweetie, Dtps. Big Red Robe, Dtps. Mount Lip Taisuco, P. Miki Wata Nabe, P. Champion Porter, P. Amabilis, P. Strawberry, P. Taida Salu, Dtps. Taida Salu Red, P. Luchia Pink, P. Jin Cheng Sun, P. Surf Song, Dtps. Lian Her Happy, P. Timothy Christopher Dtps. Happy UFO, P. Big White Pink Stripe, P. Medium Pink, P. Magic Kiss, Goldie, P. Lucyna, Brother Girl, P. Memoria Francis Hunter, P. Carlotta, Dtps. Leopard Princess M611, Dtps. Gan Lin Fairy 'GL', P. Shu Long Spotted Deer, P. Ho's Little Caroline, Dtps. Gan Lin Beauty Hsiung, Dtps. Chin Ann Diamond 'Alisun', Dtps. Gan Lin Diamond, Dtps. Shulong Sun Beauty, P. Chian Xen Pearl 'Cx#1', Dtps. Plum Rose x Ox Black Jack, Dtps. Young-home Orange Lip, Dtps. Hsin Yang Fortune, Dtps.Shih Hua Long First Love, Dtps. Tying Shin Zebra, P. Kaleidoscope, P. Sin Yuan Golden Beauty, Dtps. Chian Xen Magpie, P. Little Gem Stripe, P. Leodora, P. Roxanne, P. Ivory pearl, Dtps. Ox Prince Thunder, Dtps. Sogo Soft, Dtps. Shu Long Pearl	

continued **Tab. 4.4:** Table of Characteristics of DUS Test Guidelines in *Phalaenopsis*

Sl. No	Character-istics	States	Notes	Example Varieties /Hybrids	Type of Assess-ment
50. (*) QL (d)	Lip: color pattern of apical lobe	Uniform	1	Dtps. Plum Rose x Ox Black Jack, Dtps. Acker's Sweetie, P. Champion Porter, P. Strawberry, P. Big White Pink Stripe, P. Medium Pink, P. Brother Girl, P. Carlotta, Dtps. Gan Lin Fairy 'GL'	VG
		Shaded	2	Dtps. Younghome Orange Lip, Dtps. Hsin Yang Fortune, Dtps. Big Red Robe, Dtps. Mount Lip Taisuco, P. Miki Wata Nabe, P. Amabilis, Dtps.Shih Hua Long First Love, P. Taida Salu, Dtps. Tying Shin Zebra, Dtps. Taida Salu Red, P. Kaleidoscope, P. Luchia Pink, P. Jin Cheng Sun, P. Sin Yuan Golden Beauty, Dtps. Chian Xen Magpie, P. Little Gem Stripe, Dtps. Lian Her Happy, P. Leodora, P. Magic Kiss, P. Roxanne, P.Ivory pearl, P. Memoria Francis Hunter, Dtps. Ox Prince Thunder, Dtps. Leopard Princess M611, P. Shu Long Spotted Deer, P. Ho's Little Caroline, Dtps. Sogo Soft, Dtps. Gan Lin Beauty Hsiung, Dtps. Shulong Sun Beauty, P. Chian Xen Pearl 'Cx#1', Dtps. Shu Long Pearl	
		Edged	3	P. Timothy Christopher Dtps. Happy UFO	
		Striped	4	--	
		Netted	5	--	
		Spotted	6	P. Surf Song, Goldie, P. Lucyna, Dtps. Chin Ann Diamond 'Alisun', Dtps. Gan Lin Diamond.	
		Mixed (specify)	7	--	
51. (*) QL (d)	Lip: pre-dominant colour of apical lobe as per RHS colour chart	Green	1	--	VG
		White	2	P. Miki Wata Nabe, P. Miki Wata Nabe	
		Pink	3	--	
		Yellow	4	P. Memoria Francis Hunter	
		Red	5	Dtps. Hsin Yang Fortune	
		Purple	6	Dtps. Taida Salu Red, P. Kaleidoscope, P. Strawberry, Dtps. Shih Hua Long First Love, Dtps. Ox Prince Thunder, Dtps. Chian Xen Magpie, Dtps. Plum Rose x Ox Black Jack	
		Blue	7	--	
		Violet	8	--	

continued **Tab. 4.4:** Table of Characteristics of DUS Test Guidelines in *Phalaenopsis*

Sl. No	Character-istics	States	Notes	Example Varieties /Hybrids	Type of Assess-ment
52. (*) QL (d)	Lip: colour pattern of lateral lobe	Uniform	1	--	VG
		Shaded	2	Dtps. Lian Her Happy, P. Roxanne, P. Ivory pear, P. Memoria Francis Hunter, Dtps. Leopard Princess M611, Dtps. Shulong Sun Beauty, P. Shu Long Spotted Deer, Dtps. Sogo Soft.	
		Edged	3	--	
		Striped	4	--	
		Netted	5	--	
		Spotted	6	--	
		Mixed (specify)	7	Dtps. Plum Rose x Ox Black Jack, Dtps. Younghome Orange Lip, Dtps. Acker's Sweetie, Dtps. Hsin Yang Fortune, Dtps. Big Red Robe, Dtps. Mount Lip Taisuco, P. Miki Wata Nabe, P. Champion Porter, P. Amabilis, P. Strawberry, Dtps.Shih Hua Long First Love, P. Taida Salu, Dtps. Tying Shin Zebra, Dtps. Taida Salu Red, P. Kaleidoscope, P. Luchia Pink, P. Jin Cheng Sun, P. Sin Yuan Golden Beauty, Dtps. Chian Xen Magpie, P. Little Gem Stripe, P. Surf Song, P. Timothy Christopher Dtps. Happy UFO, P. Big White Pink Stripe, P. Medium Pink, P. Leodora, P. Magic Kiss, Goldie, P. Lucyna, P. Brother Girl, Dtps. Ox Prince Thunder, P. Carlotta, Dtps. Gan Lin Fairy 'GL', P. Ho's Little Caroline, Dtps. Gan Lin Beauty Hsiung, Dtps. Chin Ann Diamond 'Alisun', Dtps. Gan Lin Diamond, P. Chian Xen Pearl 'Cx#1', Dtps. Shu Long Pearl (Shaded and spotted)	
53. (*) QL	Lip: domin-ant colour of lateral lobe as per RHS colour chart	Green	1	--	VS
		White	2	P. Miki Wata Nabe	
		Pink	3	--	
		Yellow	4	P. Memoria Francis Hunter	
		Red	5	--	
		Purple	6	Dtps. Taida Salu Red, P. Kaleidoscope, P. Strawberry, Dtps. Hsin Yang Fortune, Dtps.Shih Hua Long First Love, Dtps. Ox Prince Thunder, Dtps. Chian Xen Magpie, Dtps. Plum Rose x Ox Black Jack	
		Blue	7	--	
		Violet	8	--	

continued **Tab. 4.4:** Table of Characteristics of DUS Test Guidelines in *Phalaenopsis*

Sl. No	Character-istics	States	Notes	Example Varieties /Hybrids	Type of Assess-ment
54. (+) PQ	Lip: callus	Prominent	1	Dtps. Plum Rose x Ox Black Jack, Dtps. Younghome Orange Lip, Dtps. Acker's Sweetie, Dtps. Hsin Yang Fortune, Dtps. Big Red Robe, Dtps. Mount Lip Taisuco, P. Miki Wata Nabe, P. Champion Porter, P. Amabilis, P. Strawberry, Dtps.Shih Hua Long First Love, P. Taida Salu, Dtps. Tying Shin Zebra, Dtps. Taida Salu Red, P. Kaleidoscope, P. Luchia Pink, P. Jin Cheng Sun, P. Sin Yuan Golden Beauty, Dtps. Chian Xen Magpie, P. Little Gem Stripe, P. Surf Song, Dtps. Lian Her Happy, P. Timothy Christopher Dtps. Happy UFO, P. Big White Pink Stripe, P. Medium Pink, P. Leodora, P. Magic Kiss, P. Goldie, P. Roxanne, P. Ivory pearl, P. Lucyna, P. Brother Girl, P. Memoria Francis Hunter, Dtps. Ox Prince Thunder, P. Carlotta, Dtps. Leopard Princess M611, Dtps. Gan Lin Fairy 'GL', P. Shu Long Spotted Deer, P. Ho's Little Caroline, Dtps. Sogo Soft, Dtps. Gan Lin Beauty Hsiung, Dtps. Chin Ann Diamond 'Alisun', Dtps. Gan Lin Diamond, Dtps. Shulong Sun Beauty, P. Chian Xen Pearl 'Cx#1', Dtps. Shu Long Pearl	VS
		Flat	9	--	
55. PQ	Lip: pubes-cence	Absent	1	Dtps. Plum Rose x Ox Black Jack, Dtps. Younghome Orange Lip, Dtps. Acker's Sweetie, Dtps. Hsin Yang Fortune, Dtps. Big Red Robe, Dtps. Mount Lip Taisuco, P. Miki Wata Nabe, P. Champion Porter, P. Amabilis, P. Strawberry, Dtps.Shih Hua Long First Love, P. Taida Salu, Dtps. Tying Shin Zebra, Dtps. Taida Salu Red, P. Kaleidoscope, P. Luchia Pink, P. Jin Cheng Sun, P. Sin Yuan Golden Beauty, Dtps. Chian Xen Magpie, P. Little Gem Stripe, P. Surf Song, Dtps. Lian Her Happy, P. Timothy Christopher Dtps. Happy UFO, P. Big White Pink Stripe, P. Medium Pink, P. Leodora, P. Magic Kiss, P. Goldie, P. Roxanne, P. Ivory Pearl, P. Lucyna, P. Brother Girl, P. Memoria Francis Hunter, Dtps. Ox Prince Thunder, P. Carlotta, Dtps. Leopard Princess M611, Dtps. Gan Lin Fairy 'GL', P. Shu Long Spotted Deer, P. Ho's Little Caroline, Dtps. Sogo Soft, Dtps. Gan Lin Beauty Hsiung, Dtps. Chin Ann Diamond 'Alisun', Dtps. Gan Lin Diamond, Dtps. Shulong Sun Beauty, P. Chian Xen Pearl 'Cx#1', Dtps. Shu Long Pearl	VS
		Present	9	--	

continued Tab. 4.4: Table of Characteristics of DUS Test Guidelines in *Phalaenopsis*

Sl. No	Character-istics	States	Notes	Example Varieties /Hybrids	Type of Assess-ment
56. QN (c)	Column: length (cm)	Short (<0.5)	3	--	MS
		Medium (0.5-1.0)	5	Dtps. Plum Rose x Ox Black Jack, Dtps. Younghome Orange Lip, Dtps. Acker's Sweetie, Dtps. Hsin Yang Fortune, Dtps. Mount Lip Taisuco, Dtps. Big Red Robe, P. Miki Wata Nabe, Dtps.Shih Hua Long First Love, P. Taida Salu, P. Champion Porter, P. Amabilis, P. Strawberry, Dtps. Tying Shin Zebra, Dtps. Taida Salu Red, P. Luchia Pink, P. Sin Yuan Golden Beauty, Dtps. Chian Xen Magpie, P. Kaleidoscope, P. Little Gem Stripe, P. Surf Song, Dtps. Lian Her Happy, P. Timothy Christopher Dtps. Happy UFO, P. Big White Pink Stripe, P. Medium Pink, P. Roxanne, P. Lucyna, Brother Girl, P. Memoria Francis Hunter, Dtps. Ox Prince Thunder, Dtps. Leopard Princess M611, Dtps. Gan Lin Fairy 'GL', P. Shu Long Spotted Deer, P. Ho's Little Caroline, Dtps. Sogo Soft, Dtps. Gan Lin Beauty Hsiung, Dtps. Gan Lin Diamond, Dtps. Shulong Sun Beauty, P. Chian Xen Pearl 'Cx#1', Dtps. Shu Long Pearl	
		Long (>1.0-1.5)	7	P. Jin Cheng Sun, P. Magic Kiss, P. Leodora, Goldie, P. Ivory Pearl, P. Carlotta, Dtps. Chin Ann Diamond 'Alisun'	
		Extra long (>1.5)	9	--	
57. (*) QL (e)	Column: colour as per RHS colour chart	Green	1	--	VS
		White	2	Dtps. Taida Salu Red, P. Kaleidoscope, P. Miki Wata Nabe, Dtps. Hsin Yang Fortune, Dtps.Shih Hua Long First Love, P. Memoria Francis Hunter, Dtps. Ox Prince Thunder, Dtps. Chian Xen Magpie	
		Pink	3	--	
		Yellow	4	--	
		Red	5	--	
		Purple	6	P. Strawberry, Dtps. Plum Rose x Ox Black Jack	
		Blue	7	--	
		Violet	8	--	
58. QN	Flower longev-ity on the plant (days)	Short (<15)	3	P. Strawberry, P. Memoria Francis Hunter.	VS
		Medium (15-30)	5	P. Lucyna, Dtps. Younghome Orange Lip, P. Ivory Pearl, Dtps. Plum Rose x Ox Black Jack	
		Long (>30)	7	Dtps.Shih Hua Long First Love, Dtps. Tying Shin Zebra, Dtps. Taida Salu Red, P. Timothy Christopher Dtps. Happy UFO, P. Surf Song, P. Medium Pink, P. Leodora, Dtps. Ox Prince Thunder, P. Jin Cheng Sun, P. Magic Kiss, P. Little Gem Stripe, Dtps. Hsin Yang Fortune, P. Kaleidoscope, P. Miki Wata Nabe, Dtps. Chian Xen Magpie	

4.2.7 Explanation on the Table of Characteristics

4.2.7.1 Guidelines for Recording the Observations of Vegetative and Flowering Characteristics

Characteristics indicated with (a), (b), (c), (d) and (e) in the first column of the Table of Characteristics should be examined as indicated below:

a) Observations on the leaf should made on the longest leaf of the flowering plant.
b) Observations on the inflorescence and the flower shall be made at the time when 50% of the flowers on the inflorescence have opened and the most recently fully opened flower on the inflorescence before the color starts to fade.
c) Observations on the length and width of the flower and parts of the flower shall be made on the spread out positions.
d) Observations on the color of the sepal, the petal and the lip shall be made on inner side at apex, mid and base portion.
e) Observations on the colour of column shall be made on inner side at apex, mid and basal region.

4.2.7.2 Explanation for Individual Characteristics

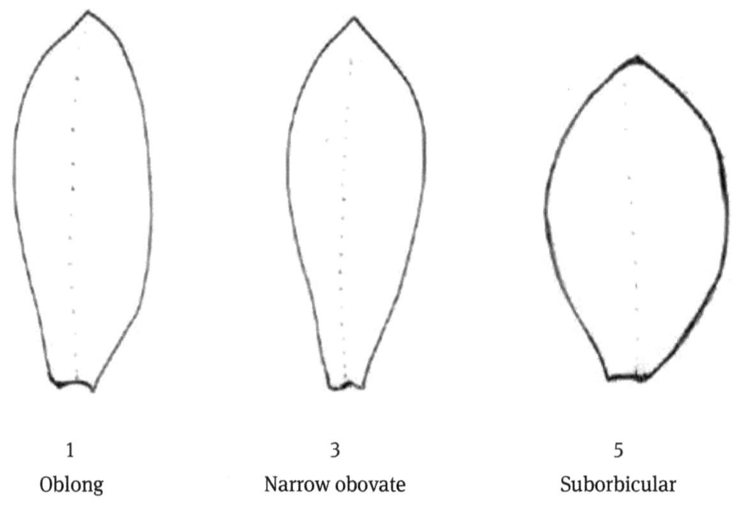

Fig. 4.6: Leaf shape

1 Acute | 3 Notched | 5 Obtuse

Fig. 4.7: Leaf apex

1 Solitary | 3 Raceme | 5 Panicle

Fig. 4.8: Inflorescence type

1 Open | 3 Touching | 5 Overlapping

Fig. 4.9: Flower arrangement of petals

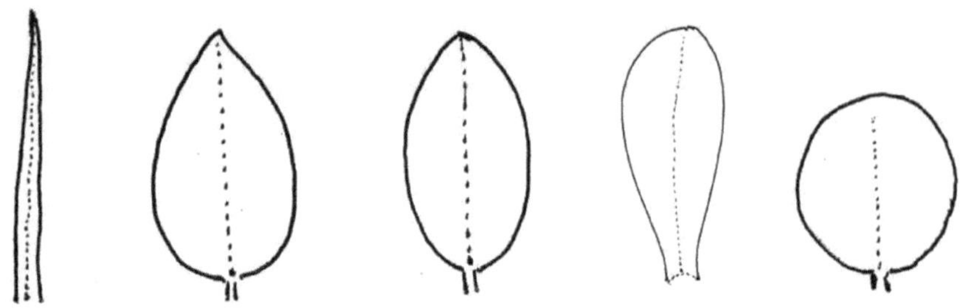

1 Linear | 3 Ovate | 5 Elliptic | 7 Obovate | 9 Round

Fig. 4.10: Shape of Dorsal sepal (20) and Lateral sepal (28)

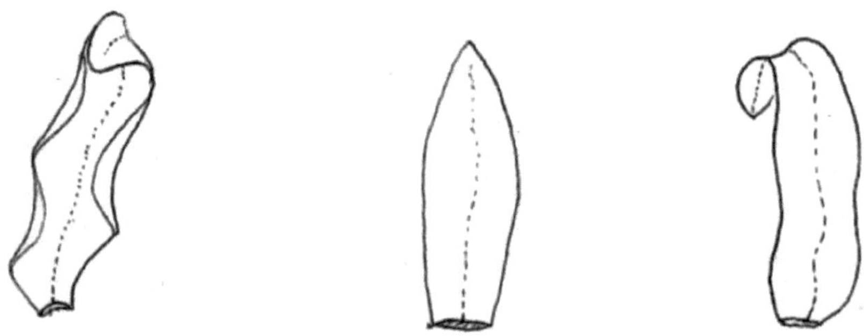

1 Incurved | 3 Straight | 5 Reflexed

Fig. 4.11: Curvature of Dorsal sepal (21) and petal (35)

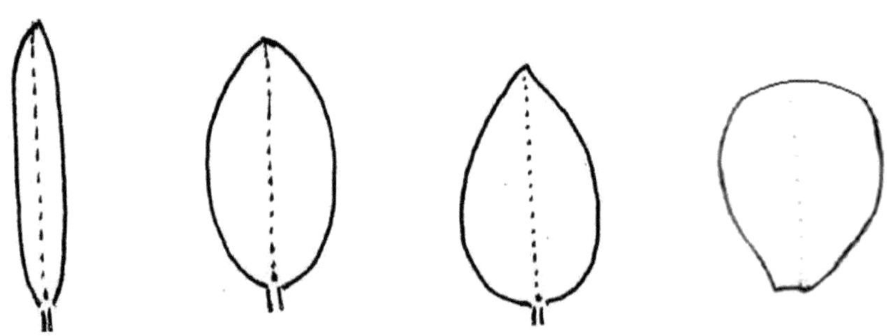

1 Oblong | 3 Elliptic | 5 Ovate | 7 Suborbicular

Fig. 4.12: Petal shape

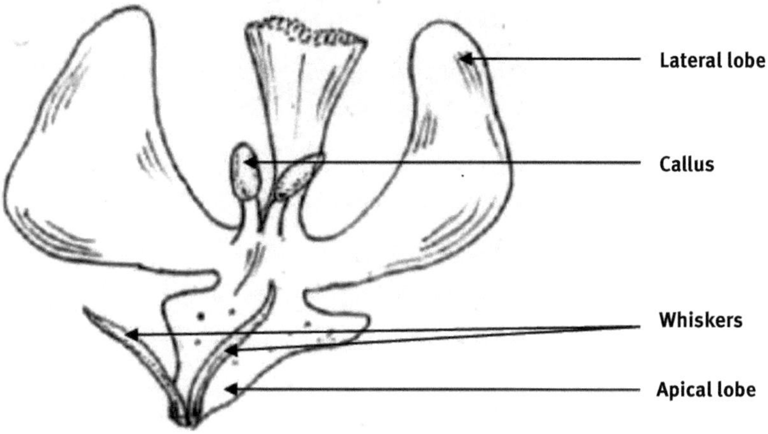

Fig. 4.13: Lip: presence of whiskers (42) and callus (54)

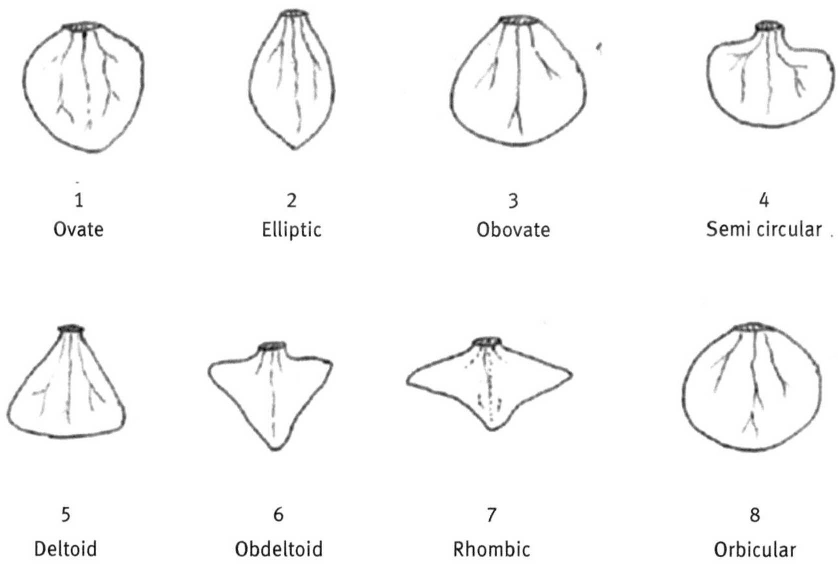

Fig. 4.14: Lip: shape of apical lobe

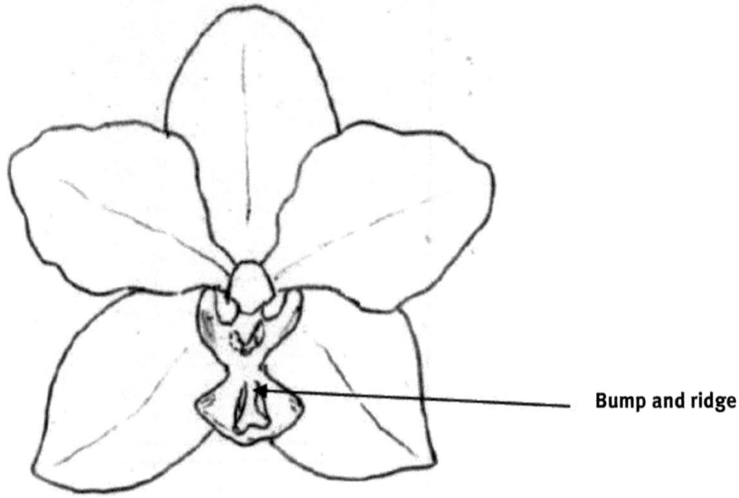

Fig. 4.15: Lip: bump and ridge on apical lobe

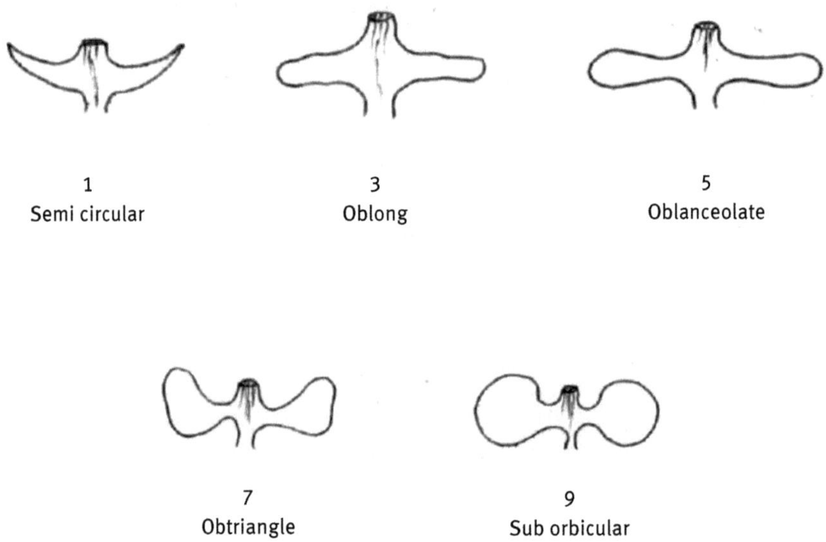

Fig. 4.16: Lip: shape of lateral lobe

3
Slightly incurved

7
Strongly incurved

Fig. 4.17: Curvature of lateral lobe

4.3 Cattleya Lindl. (PVJ, Nov, 2012)

4.3.1 Subject

These test guidelines apply to all vegetatively propagated varieties of *Cattleya* Lindl. and alliance of the family Orchidaceae.

4.3.2 Plant Material Required

- The Protection of Plant Varieties and Farmers' Rights Authority (PPV & FRA) shall decide when, where and in what quantity and quality of plant material are required for testing of a variety denomination for registration under the Protection of Plant Varieties and Farmers' Rights (PPV & FR) Act, 2001. Applicants submitting such plant material from a country other than India shall make sure that all customs and quarantine requirements stipulated under relevant national legislations and regulations are complied with.
- For all varieties, 20 plants (10 for each Centre) that are two to three years old with at least two shoots shall be required for DUS testing.
- The plant material supplied should be visibly healthy, not lacking in vigour nor affected by any pests or diseases or mechanical damage.
- Plant material shall not have undergone any chemical or bio-physical treatment unless the competent authority allow or request such treatment. If it has been treated, details of the treatment must be given.

4.3.3 Conduct of Test

- The minimum duration of test should normally be two similar flowering seasons.
- Test shall normally be conducted at two places. If any essential characteristic of the variety is not expressed for visual observations at these places, the variety

shall be considered for further examination at another appropriate test site or under special test protocol on request of the applicant.

- The test should be carried out under greenhouse conditions, ensuring satisfactory growth for the expression of the relevant characteristics of the variety and for the conduct of the examination.
- The design of the the test should be such that the plants or parts of the plants may be removed for measurement and counting without prejudice to the observations, which must be made up to the end of the flowering period. Each test should be designed to include results from a total of at least 10 plants.
- Unless otherwise indicated, all observations determined by measuring or counting should be made on 10 plants or parts taken from each of 10 plants.
- Additional tests for special purposes may be established.
- Normally, growth regulators shall not be used.

4.3.4 Methods and Observations

- The characteristics described in the Table of Characteristics (Tab.4.5) shall be used for the testing of varieties for their DUS.
- For the assessment of Distinctiveness and Stability, all observations shall be made on 10 plants or parts taken from each of 10 plants.
- For the assessment of Uniformity, a population standard of 1% and an acceptance probability of at least 95% shall be applied. In the case of a sample size of 10 plants, the maximum number of off-types allowed would be 1.
- All observations on the shoot shall be made on the flowering shoot.
- All observations on the leaf shall be made on the longest leaf of a flowering shoot.
- All observations on the inflorescence and the flower shall be made at the time when 50% of the flowers on the inflorescence have opened and on the most recently fully opened flower on the inflorescence before fading of colour.
- All observations on the length and width of the flower and parts of the flower shall be made in the spread out position.
- All observations on the colour of sepal, petal, lip and column shall be made on the inner side.
- For the assessment of colour characteristics, the Royal Horticultural Society (RHS) colour chart shall be used.

4.3.5 Grouping of Varieties

- The selection of varieties of common knowledge to be grown in the trial with the candidate varieties and the way in which these varieties are divided into groups to facilitate the assessment of distinctiveness is aided by the use of grouping characteristics.

- Grouping characteristics are those in which the documented states of expression, even where produced at different locations, can be used either individually or in combination with other such characteristics: (a) to select varieties of common knowledge that can be excluded from the growing trial used for examination of distinctiveness; and (b) to organize the growing trial so that similar varieties are grouped together.
- The following have been agreed as useful grouping characteristics:
  - Plant: height (Characteristic 1)
  - Leaf: number/pseudobulb (Characteristic 6)
  - Flower size: width in front view (Characteristic 17)
  - Petal: predominant colour (Characteristic 36)
  - Lip: predominant colour (Characteristic 41)
  - Lip: colour pattern (Characteristic 43)

4.3.6 Characteristics and Symbols

- To assess Distinctiveness, Uniformity and Stability, the characteristics and their states as given in the Table of Characteristics shall be used.
- Notes 1-9 (numbers) shall be used to describe the state of each character for the purpose of electronic data processing.
- Legend

(*) Characteristics that shall be observed during every growing season for varieties and hybrids and shall always be included in the description of the varieties and hybrids, except when the state of expression of any of these characters is rendered impossible by a preceding phenological characteristic or by the environment conditions of the testing region. Under such exceptional situations, adequate explanation shall be provided.

(+) See explanations on the Table of Characteristics

Characteristics denoted with symbols QL, QN and PQ in the first column of the Table of Characteristics shall be indicated as:

QL: Qualitative characteristic

QN: Quantitative characteristic

PQ: Pseudo-qualitative characteristic

- **(a)-(e)** see 4.3.7.1 for explanation
- Type of assessment of characteristics indicated in column six of the Table of Characteristics are as follows:

MG: Measurement by a single observation of a group of plants or parts of plants

MS: Measurement of a number of individual plants or parts of plants

VG: Visual assessment by a single observation of a group of plants or parts of plants

VS: Visual assessment by observations of individual plants or parts of plant

Tab. 4.5: Table of Characteristics of DUS Test Guidelines in *Cattleya*

Sl. No.	Characteristic	State	Note	Example Varieties /hybrids	Type of Assessment
1. (*) QN	Plant height (From base to the tip of the flowering shoot)	Small (<15 cm)	3	Lc 'Manriner Far Horizon x L. anceps coerulea, Lc 'Ahmad Sheikhi'	MS
		Medium (15-30 cm)	5	Lc 'Purple Cascade Fragrant B', Blc 'Mem Ann Balmores Convess', C. 'Queen Sirikhit', Blc 'Hsinging Catherine', Blc 'Chinese Beauty Orchid Queen', Blc 'Guanmiau City'	
		Large (>30 cm)	7	--	
2. (*) (+) PQ	Nature of pseudobulb (at flowering)	Cylindric	1	Blc 'Guanmiau City', C. 'Queen Sirikhit'	VG
		Clavate	3	Lc 'Purple Cascade Fragrant B', Blc 'Mem Ann Balmores Convess', Blc 'Hsinging Catherine', Blc 'Chinese Beauty Orchid Queen'	
		Globular / Ovoid	5	Lc 'Ahmad Sheikhi'	
3. QN	Pseudobulb length (at flowering)	Short (<10 cm)	3	Blc 'Guanmiau City', Lc 'Manriner Far Horizon x L. anceps coerulea, Lc 'Ahmad Sheikhi'	MS
		Medium (10-20 cm)	5	Lc 'Purple Cascade Fragrant B', Blc 'Mem Ann Balmores Convess', C. 'Queen Sirikhit', Blc 'Hsinging Catherine', Blc 'Chinese Beauty Orchid Queen'	
		Long (>20 cm)	7	--	
4. QN	Pseudobulb breadth at broadest part (at flowering)	Narrow (<1.0 cm)	3	Lc 'Purple Cascade Fragrant B', Blc 'Guanmiau City',	MS
		Medium (1.0-2.0 cm)	5	Blc 'Mem Ann Balmores Convess', C. 'Queen Sirikhit', Blc 'Hsinging Catherine', Blc 'Chinese Beauty Orchid Queen', Lc 'Manriner Far Horizon x L. anceps coerulea, Lc 'Ahmad Sheikhi'	
		Broad (>2.0 cm)	7	--	
5. QN	Number of shoots / plant	Few (<5)	3	Lc 'Purple Cascade Fragrant B', Blc 'Mem Ann Balmores Convess',	VG
		Many (5 or more)	7	C. 'Queen Sirikhit', Blc 'Hsinging Catherine', Blc 'Chinese Beauty Orchid Queen', Blc 'Guanmiau City', Lc 'Manriner Far Horizon x L. anceps coerulea, Lc 'Ahmad Sheikhi'	

continued Tab. 4.5: Table of Characteristics of DUS Test Guidelines in *Cattleya*

Sl. No.	Character-istic	State	Note	Example Varieties /hybrids	Type of Assess-ment
6. (*) QN (a)	Number of leaves/ pseudobulb	One (Unifoliate)	1	Lc 'Purple Cascade Fragrant B', Blc 'Mem Ann Balmores Convess', Blc 'Hsinging Catherine', Blc 'Chinese Beauty Orchid Queen', Blc 'Guanmiau City', Lc 'Manriner Far Horizon x L. anceps coerulea, Lc 'Ahmad Sheikhi'	VG
		Two (Bifoliate)	9	C. 'Queen Sirikhit'	
7. QN (a)	Leaf length (largest leaf)	Short (<10 cm)	3	--	MS
		Medium (10-15 cm)	5	C. 'Queen Sirikhit', Blc 'Hsinging Catherine', Blc 'Guanmiau City', Lc 'Manriner Far Horizon x L. anceps coerulea, Lc 'Ahmad Sheikhi'	
		Long (>15 cm)	7	Lc 'Purple Cascade Fragrant B', Blc 'Mem Ann Balmores Convess', Blc 'Chinese Beauty Orchid Queen',	
8. QN (a)	Leaf width (largest leaf)	Narrow (<2.5 cm)	3	--	MS
		Medium (2.5-5.0 cm)	5	C. 'Queen Sirikhit', Blc 'Hsinging Catherine', Blc 'Guanmiau City', Lc 'Manriner Far Horizon x L. anceps coerulea	
		Broad (>5.0 cm)	7	Lc 'Purple Cascade Fragrant B', Blc 'Mem Ann Balmores Convess', Blc 'Chinese Beauty Orchid Queen', Lc 'Ahmad Sheikhi'	
9. (+) PQ (a)	Leaf shape (largest leaf)	Narrow oblong	1	Lc 'Manriner Far Horizon x L. anceps coerulea	VG
		Ligulate	3	--	
		Elliptic	5	Lc 'Purple Cascade Fragrant B', Blc 'Mem Ann Balmores Convess', Blc 'Hsinging Catherine', Blc 'Chinese Beauty Orchid Queen', Blc 'Guanmiau City', Lc 'Ahmad Sheikhi'	
		Ovate	7	C. 'Queen Sirikhit'	
10. PQ (a)	Leaf colour pattern	Uniformly green on both sides	1	C. 'Queen Sirikhit', Lc 'Purple Cascade Fragrant B', Blc 'Hsinging Catherine'	VG
		Green upside/ purple beneath	3	Lc 'Ahmad Sheikhi'	
		Spotted	5	Blc 'Chinese Beauty Orchid Queen',	
		Streaked	7	--	

continued **Tab. 4.5:** Table of Characteristics of DUS Test Guidelines in *Cattleya*

Sl. No.	Character-istic	State	Note	Example Varieties /hybrids	Type of Assess-ment
11. QN (b)	Inflores-cence: number / pseudobulb at a time	One	1	Lc 'Purple Cascade Fragrant B', Blc 'Mem Ann Balmores Convess', C. 'Queen Sirikhit', Blc 'Hsinging Catherine', Blc 'Chinese Beauty Orchid Queen', Blc 'Guanmiau City', Lc 'Manriner Far Horizon x L. anceps coerulea, Lc 'Ahmad Sheikhi'	VG
		Two or more	5	--	
12. QN (b)	Inflores-cence length (base of peduncle to the tip of the flowers)	Short (<15 cm)	3	Lc 'Manriner Far Horizon x L. anceps coerulea, Lc 'Ahmad Sheikhi'	MS
		Medium (15-30 cm)	5	Lc 'Purple Cascade Fragrant B', Blc 'Mem Ann Balmores Convess', C. 'Queen Sirikhit', Blc 'Hsinging Catherine', Blc 'Chinese Beauty Orchid Queen', Blc 'Guanmiau City'	
		Long (>30 cm)	7	--	
13. QN (b)	Peduncle sheath length	Short (<3 cm)	3	--	MG
		Medium (3-5 cm)	5	C. 'Queen Sirikhit', Blc 'Guanmiau City', Lc 'Ahmad Sheikhi'	
		Long (>5 cm)	7	Lc 'Purple Cascade Fragrant B', Blc 'Mem Ann Balmores Convess', Blc 'Hsinging Catherine', Blc 'Chinese Beauty Orchid Queen', Lc 'Manriner Far Horizon' x L. anceps coerulea	
14. QN (b)	Peduncle length (base of inflores-cence to the 1st Flower)	Short (<5 cm)	3	Lc 'Manriner Far Horizon' x L. anceps coerulea, Lc 'Ahmad Sheikhi'	MS
		Medium (5-10 cm)	5	Lc 'Purple Cascade Fragrant B', Blc 'Mem Ann Balmores Convess', C. 'Queen Sirikhit', Blc 'Hsinging Catherine', Blc 'Chinese Beauty Orchid Queen'	
		Long (>10 cm)	7	Blc 'Guanmiau City',	
15. QN (b)	Flower number per peduncle	Few (up to three)	3	Blc 'Mem Ann Balmores Convess', Blc 'Hsinging Catherine', Blc 'Chinese Beauty Orchid Queen', Lc 'Manriner Far Horizon x L. anceps coerulea, Lc 'Ahmad Sheikhi'	MG
		Many (more than three)	5	C. 'Queen Sirikhit', Blc 'Guanmiau City', Lc 'Purple Cascade Fragrant B'	

continued Tab. 4.5: Table of Characteristics of DUS Test Guidelines in *Cattleya*

Sl. No.	Characteristic	State	Note	Example Varieties /hybrids	Type of Assessment
16. QN (c)	Flower length (Tip of dorsal sepal to tip of lip)	Short (<8 cm)	3	C. 'Queen Sirikhit'	MS
		Medium (8-16 cm)	5	Lc 'Purple Cascade Fragrant B', Blc 'Mem Ann Balmores Convess', Blc 'Hsinging Catherine', Blc 'Chinese Beauty Orchid Queen', Blc 'Guanmiau City', Lc 'Manriner Far Horizon x L. anceps coerulea, Lc 'Ahmad Sheikhi'	
		Long (>16 cm)	7	--	
17. (*) QN (c)	Flower width in front view (distance between tips of two lateral petals)	Narrow (<8 cm)	3	C. 'Queen Sirikhit'	MS
		Medium (8-16 cm)	5	Lc 'Purple Cascade Fragrant B', Blc 'Mem Ann Balmores Convess', Blc 'Hsinging Catherine', Blc 'Chinese Beauty Orchid Queen', Blc 'Guanmiau City', Lc 'Manriner Far Horizon x L. anceps coerulea, Lc 'Ahmad Sheikhi'	
		Broad (>16 cm)	7	--	
18. QN (b)	Flower longevity on the plant	Short (<15 days)	3	--	VG
		Medium (15-30 days)	5	Blc 'Chinese Beauty Orchid Queen', Blc 'Guanmiau City', Lc 'Purple Cascade Fragrant B', Blc 'Mem Ann Balmores Convess'	
		Long (>30 days)	7	C. 'Queen Sirikhit', Blc 'Hsinging Catherine', Lc 'Manriner Far Horizon x L. anceps coerulea, Lc 'Ahmad Sheikhi'	
19. QN (c)	Dorsal sepal length	Short (<5 cm)	3	C. 'Queen Sirikhit'	MS
		Medium (5-10 cm)	5	Lc 'Purple Cascade Fragrant B', Blc 'Mem Ann Balmores Convess', Blc 'Hsinging Catherine', Blc 'Chinese Beauty Orchid Queen', Blc 'Guanmiau City', Lc 'Manriner Far Horizon x L. anceps coerulea, Lc 'Ahmad Sheikhi'	
		Long (>10 cm)	7	--	

continued **Tab. 4.5:** Table of Characteristics of DUS Test Guidelines in *Cattleya*

Sl. No.	Character-istic	State	Note	Example Varieties /hybrids	Type of Assess-ment
20. QN (c)	Dorsal sepal width (at middle)	Narrow (<2 cm)	3	C. 'Queen Sirikhit', Blc 'Guanmiau City', Lc 'Manriner Far Horizon x L. anceps coerulea	MS
		Medium (2-3 cm)	5	Lc 'Ahmad Sheikhi'	
		Broad (>3 cm)	7	Lc 'Purple Cascade Fragrant B', Blc 'Mem Ann Balmores Convess', Blc 'Hsinging Catherine', Blc 'Chinese Beauty Orchid Queen'	
21. (+) PQ	Dorsal sepal shape	Oblong	1	Lc 'Purple Cascade Fragrant B', Blc 'Chinese Beauty Orchid Queen', Lc 'Ahmad Sheikhi'	VG
		Lanceolate	3	Blc 'Guanmiau City'	
		Elliptic	5	Blc 'Mem Ann Balmores Convess', C. 'Queen Sirikhit', Lc 'Manriner Far Horizon x L. anceps coerulea	
		Ovate	7	Blc 'Hsinging Catherine'	
22. (+) PQ	Dorsal sepal apex	Acute	1	Lc 'Purple Cascade Fragrant B', Blc 'Mem Ann Balmores Convess', C. 'Queen Sirikhit', Blc 'Chinese Beauty Orchid Queen', Blc 'Guanmiau City', Lc 'Manriner Far Horizon x L. anceps coerulea	VG
		Notched	3	Blc 'Hsinging Catherine'	
		Obtuse	5	Lc 'Ahmad Sheikhi'	
23. (+) PQ	Dorsal sepal curvature	Incurved	1	Lc 'Purple Cascade Fragrant B', C. 'Queen Sirikhit', Blc 'Chinese Beauty Orchid Queen', Blc 'Guanmiau City',	VG
		Straight	3	Blc 'Mem Ann Balmores Convess', Blc 'Hsinging Catherine', Lc 'Manriner Far Horizon x L. anceps coerulea	
		Reflexed	5	Lc 'Ahmad Sheikhi'	
24. QN (c)	Lateral sepal length	Short (<5 cm)	3	Lc 'Purple Cascade Fragrant B', C. 'Queen Sirikhit', Blc 'Guanmiau City', Lc 'Manriner Far Horizon x L. anceps coerulea	MS
		Medium (5-10 cm)	5	Blc 'Mem Ann Balmores Convess', Blc 'Hsinging Catherine', Blc 'Chinese Beauty Orchid Queen', Lc 'Ahmad Sheikhi'	
		Long (>10 cm)	7	--	

continued Tab. 4.5: Table of Characteristics of DUS Test Guidelines in *Cattleya*

Sl. No.	Character-istic	State	Note	Example Varieties /hybrids	Type of Assess-ment
25. QN (c)	Lateral sepal width	Narrow (<2 cm)	3	Lc 'Purple Cascade Fragrant B', C. 'Queen Sirikhit', Blc 'Guanmiau City', Lc 'Manriner Far Horizon x L. anceps coerulea	MS
		Medium (2-3 cm)	5	Lc 'Ahmad Sheikhi'	
		Broad (>3 cm)	7	Blc 'Mem Ann Balmores Convess', Blc 'Hsinging Catherine', Blc 'Chinese Beauty Orchid Queen'	
26. (+) PQ	Lateral sepal shape	Oblong	1	Lc 'Purple Cascade Fragrant B'	VG
		Lanceolate	3	Blc 'Guanmiau City'	
		Elliptic	5	C. 'Queen Sirikhit', Blc 'Chinese Beauty Orchid Queen', Lc 'Manriner Far Horizon x L. anceps coerulea, Lc 'Ahmad Sheikhi'	
		Ovate	7	Blc 'Mem Ann Balmores Convess', Blc 'Hsinging Catherine'	
27. (+) PQ	Lateral sepal apex	Acute	1	Lc 'Purple Cascade Fragrant B', Blc 'Mem Ann Balmores Convess', C. 'Queen Sirikhit', Blc 'Chinese Beauty Orchid Queen', Blc 'Guanmiau City'	MG
		Notched	3	Blc 'Hsinging Catherine', Lc 'Ahmad Sheikhi'	
		Obtuse	5	--	
28. (+) PQ	Lateral sepal curvature	Incurved	1	Lc 'Purple Cascade Fragrant B', Blc 'Mem Ann Balmores Convess', C. 'Queen Sirikhit', Blc 'Chinese Beauty Orchid Queen', Lc 'Manriner Far Horizon x L. anceps coerulea	VG
		Straight	3	Blc 'Guanmiau City', Lc 'Ahmad Sheikhi'	
		Reflexed	5	Blc 'Hsinging Catherine'	
29. QL (d)	Sepal dom-inant colour	Green	1	--	VG
		White	2	C. 'Queen Sirikhit', Blc 'Hsinging Catherine'	
		Pink	3	--	
		Yellow	4	Blc 'Mem Ann Balmores Convess', Lc 'Ahmad Sheikhi'	
		Red	5	Blc 'Chinese Beauty Orchid Queen'	
		Purple	6	Lc 'Purple Cascade Fragrant B', Lc 'Man-riner Far Horizon x L. anceps coerulea	
		Blue	7	--	
		Violet	8	Blc 'Guanmiau City'	

continued **Tab. 4.5:** Table of Characteristics of DUS Test Guidelines in *Cattleya*

Sl. No.	Character-istic	State	Note	Example Varieties /hybrids	Type of Assess-ment
30. QL (d)	Sepal colour pattern inside	Uniform	1	Lc 'Purple Cascade Fragrant B', Blc 'Mem Ann Balmores Convess', C. 'Queen Sirikhit', Blc 'Hsinging Catherine', Lc 'Manriner Far Horizon x L. anceps coerulea, Lc 'Ahmad Sheikhi'	MG
		Mixed	2	--	
		Spotted	3	Blc 'Guanmiau City'	
		Blotched	4	--	
		Shaded / Striped	5	Blc 'Chinese Beauty Orchid Queen'	
		Netted	6	--	
31. QN (c)	Petal length	Short (<5 cm)	3	Lc 'Purple Cascade Fragrant B', C. 'Queen Sirikhit', Blc 'Guanmiau City'	MS
		Medium (5-10 cm)	5	Blc 'Mem Ann Balmores Convess', Blc 'Hsinging Catherine', Blc 'Chinese Beauty Orchid Queen', Lc 'Manriner Far Horizon x L. anceps coerulea, Lc 'Ahmad Sheikhi'	
		Long (>10 cm)	7	--	
32. QN (c)	Petal width (at middle)	Narrow (<2 cm)	3	Lc 'Purple Cascade Fragrant B', Blc 'Guanmiau City'	MS
		Medium (2-5 cm)	5	C. 'Queen Sirikhit', Lc 'Manriner Far Horizon x L. anceps coerulea, Lc 'Ahmad Sheikhi'	
		Broad (>5 cm)	7	Blc 'Mem Ann Balmores Convess', Blc 'Hsinging Catherine', Blc 'Chinese Beauty Orchid Queen'	
33. (+) PQ	Petal shape	Oblong	1	--	VG
		Lanceolate	3	--	
		Elliptic	5	Lc 'Purple Cascade Fragrant B', Blc 'Guanmiau City', Lc 'Ahmad Sheikhi'	
		Ovate	7	Blc 'Mem Ann Balmores Convess', C. 'Queen Sirikhit', Blc 'Hsinging Catherine', Blc 'Chinese Beauty Orchid Queen', Lc 'Manriner Far Horizon x L. anceps coerulea	
		Round	9	--	
34. (+) PQ	Petal curvature	Incurved	1	Lc 'Purple Cascade Fragrant B', C. 'Queen Sirikhit', Lc 'Manriner Far Horizon x L. anceps coerulea	VG
		Straight	3	Blc 'Mem Ann Balmores Convess', Lc 'Ahmad Sheikhi'	
		Reflexed	5	Blc 'Hsinging Catherine', Blc 'Chinese Beauty Orchid Queen', Blc 'Guanmiau City'	

continued Tab. 4.5: Table of Characteristics of DUS Test Guidelines in *Cattleya*

Sl. No.	Character-istic	State	Note	Example Varieties /hybrids	Type of Assess-ment
35. (+) PQ	Petal margin	Entire	1	Blc 'Guanmiau City'	VG
		Undulate	3	Lc 'Purple Cascade Fragrant B', Lc 'Ahmad Sheikhi'	
		Undulate crisped	5	Blc 'Mem Ann Balmores Convess', Blc 'Hsinging Catherine', Blc 'Chinese Beauty Orchid Queen', C. 'Queen Sirikhit'	
36. (*) QL (d)	Petal predominant colour	Green	1	--	VG
		White	2	C. 'Queen Sirikhit', Blc 'Hsinging Catherine'	
		Yellow	3	Blc 'Mem Ann Balmores Convess', Blc 'Chinese Beauty Orchid Queen', Lc 'Ahmad Sheikhi'	
		Pink	4	--	
		Red	5	--	
		Purple	6	Lc 'Purple Cascade Fragrant B', Lc 'Man-riner Far Horizon x L. anceps coerulea	
		Blue	7	--	
		Violet	8	Blc 'Guanmiau City'	
37. QL (d)	Petal colour pattern	Uniform	1	Lc 'Purple Cascade Fragrant B', Blc 'Mem Ann Balmores Convess', C. 'Queen Sirikhit', Blc 'Hsinging Catherine', Lc 'Manriner Far Horizon x L. anceps coerulea, Lc 'Ahmad Sheikhi'	VG
		Mixed	3	--	
		Spotted	5	Blc 'Guanmiau City'	
		Striped/ shaded	7	Blc 'Chinese Beauty Orchid Queen'	
		Netted	9	--	
38. QN (c)	Lip length	Short (<4 cm)	3	Lc 'Purple Cascade Fragrant B'	MG
		Medium (4-7 cm)	5	C. 'Queen Sirikhit', Blc 'Chinese Beauty Orchid Queen', Blc 'Guanmiau City', Lc 'Manriner Far Horizon x L. anceps coerulea, Lc 'Ahmad Sheikhi'	
		Long (>7 cm)	7	Blc 'Mem Ann Balmores Convess', Blc 'Hsinging Catherine'	
39. QN (c)	Lip width	Narrow (<4 cm)	3	Lc 'Purple Cascade Fragrant B', C. 'Queen Sirikhit', Blc 'Guanmiau City'	MG
		Medium (4-7 cm)	5	Blc 'Hsinging Catherine', Blc 'Chinese Beauty Orchid Queen', Lc 'Manriner Far Horizon x L. anceps coerulea, Lc 'Ahmad Sheikhi'	
		Broad (>7 cm)	7	Blc 'Mem Ann Balmores Convess'	

continued **Tab. 4.5:** Table of Characteristics of DUS Test Guidelines in *Cattleya*

Sl. No.	Characteristic	State	Note	Example Varieties /hybrids	Type of Assessment
40. (+) PQ	Lip shape	Oblong	1	Blc 'Hsinging Catherine', Lc 'Ahmad Sheikhi'	VG
		Lanceolate	3	--	
		Elliptic	5	--	
		Ovate	7	Lc 'Purple Cascade Fragrant B', Blc 'Mem Ann Balmores Convess', Blc 'Chinese Beauty Orchid Queen', Lc 'Manriner Far Horizon x L. anceps coerulea	
41. (*) QL (d)	Lip predominant colour	Green	1	--	VS
		White	2	--	
		Yellow	3	Blc 'Mem Ann Balmores Convess', C. 'Queen Sirikhit', Lc 'Ahmad Sheikhi'	
		Pink	4	--	
		Red	5	Blc 'Chinese Beauty Orchid Queen',	
		Purple	6	Lc 'Purple Cascade Fragrant B', Blc 'Hsinging Catherine', Blc 'Guanmiau City', Lc 'Manriner Far Horizon x L. anceps coerulea	
		Blue	7	--	
		Violet	8	--	
42. QL (d)	Lip colour (nos.)	One	1	--	VS
		Two	3	Lc 'Purple Cascade Fragrant B', (white & purple), C. 'Queen Sirikhit'(yellow & green), Blc 'Hsinging Catherine'(yellow & purple), Blc 'Chinese Beauty Orchid Queen', (red & purple), Lc 'Ahmad Sheikhi' (yellow & purple)	
		More than two	5	Blc 'Mem Ann Balmores Convess', (green, yellow & orange), Blc 'Guanmiau City', (yellow, purple & violet), Lc 'Manriner Far Horizon x L. anceps coerulea (purple, yellow & green)	
43. (*) QL (d)	Lip colour pattern	Uniform	1	--	VG
		Mixed	3	--	
		Spotted	5	Blc 'Guanmiau City', Lc 'Ahmad Sheikhi'	
		Striped/ Shaded	7	Lc 'Purple Cascade Fragrant B', Blc 'Mem Ann Balmores Convess', C. 'Queen Sirikhit', Blc 'Hsinging Catherine', Blc 'Chinese Beauty Orchid Queen', Lc 'Manriner Far Horizon x L. anceps coerulea	
		Netted	9	--	

continued **Tab. 4.5:** Table of Characteristics of DUS Test Guidelines in *Cattleya*

Sl. No.	Character-istic	State	Note	Example Varieties /hybrids	Type of Assess-ment
44. (+) PQ	Lip lateral lobe shape	Oblong	1	Blc 'Guanmiau City', C. 'Queen Sirikhit', Blc 'Chinese Beauty Orchid Queen'	VG
		Lanceolate	3	--	
		Elliptic	5	--	
		Ovate	7	--	
		Round	9	Blc 'Hsinging Catherine'	
45. (+) PQ	Lip lateral lobe margin	Entire	1	Lc 'Purple Cascade Fragrant B', Lc 'Ahmad Sheikhi'	VG
		Undulate	3	Blc 'Guanmiau City', C. 'Queen Sirikhit', Blc 'Chinese Beauty Orchid Queen', Blc 'Hsinging Catherine'	
		Undu-late-crisped	5	Blc 'Mem Ann Balmores Convess'	
46. (+) PQ	Lip mid lobe shape	Oblong	1	Lc 'Purple Cascade Fragrant B', Lc 'Ahmad Sheikhi', Blc 'Mem Ann Balmores Convess', Blc 'Chinese Beauty Orchid Queen'	VG
		Lanceolate	3	--	
		Elliptic	5	--	
		Ovate	7	--	
		Round	9	Blc 'Guanmiau City', Blc 'Hsinging Catherine'	
47. (+) PQ	Lip mid lobe margin	Entire	1	--	VG
		Undulate	3	Lc 'Purple Cascade Fragrant B'	
		Undu-late-crisped	5	Lc 'Ahmad Sheikhi', Blc 'Mem Ann Balmores Convess', Blc 'Chinese Beauty Orchid Queen', Blc 'Guanmiau City', Blc 'Hsinging Catherine', C. 'Queen Sirikhit'	
48. PQ	Lip surface inside	Glabrous	1	Lc 'Purple Cascade Fragrant B', Blc 'Mem Ann Balmores Convess', Blc 'Guanmiau City', Lc 'Manriner Far Horizon x L. anceps coerulea	VG
		Bump and ridged	3	--	
		Pubescent/ tubercled	5	C. 'Queen Sirikhit', Blc 'Hsinging Cather-ine', Blc 'Chinese Beauty Orchid Queen', Lc 'Ahmad Sheikhi'	

continued **Tab. 4.5:** Table of Characteristics of DUS Test Guidelines in *Cattleya*

Sl. No.	Character-istic	State	Note	Example Varieties /hybrids	Type of Assess-ment
49. QN	Column length	Short (<2 cm)	1	Lc 'Purple Cascade Fragrant B', Blc 'Guanmiau City'	MS
		Long (2 cm or more)	5	Blc 'Mem Ann Balmores Convess', C. 'Queen Sirikhit', Blc 'Hsinging Catherine', Blc 'Chinese Beauty Orchid Queen', Lc 'Manriner Far Horizon x L. anceps coerulea, Lc 'Ahmad Sheikhi'	
50. QL (e)	Column colour (nos.)	One	1	Lc 'Manriner Far Horizon x L. anceps coerulea (white)	VG
		Two or more	9	Lc 'Purple Cascade Fragrant B', (purple & green), Blc 'Mem Ann Balmores Convess' (yellow & green) C. 'Queen Sirikhit'(yellow & green), Blc 'Hsinging Catherine' (yellow, green & white), Blc 'Chinese Beauty Orchid Queen', (red, purple & yellow), Blc 'Guanmiau City'(purple, yellow & white), Lc 'Ahmad Sheikhi' (yellow & green)	
51. QL (e)	Column colour pattern	Uniform	1	Blc 'Mem Ann Balmores Convess', C. 'Queen Sirikhit', Lc 'Manriner Far Horizon x L. anceps coerulea, Lc 'Ahmad Sheikhi'	VG
		Spotted	3	Blc 'Guanmiau City'	
		Blotched	5	--	
		Streaked / striped /shaded	7	Lc 'Purple Cascade Fragrant B', Blc 'Hsinging Catherine', Blc 'Chinese Beauty Orchid Queen'	
52. QN (c)	Length of flower pedicel (with ovary)	Short (<5 cm)	1	Lc 'Purple Cascade Fragrant B', C. 'Queen Sirikhit', Blc 'Guanmiau City'	MS
		Long (5 cm or more)	5	Blc 'Mem Ann Balmores Convess', Blc 'Hsinging Catherine', Blc 'Chinese Beauty Orchid Queen', Lc 'Manriner Far Horizon x L. anceps coerulea, Lc 'Ahmad Sheikhi'	
53. PQ	Flowering season	Winter	1	Blc 'Chinese Beauty Orchid Queen', Blc 'Guanmiau City', Lc 'Manriner Far Horizon x L. anceps coerulea, Lc 'Ahmad Sheikhi'	VG
		Summer	3	--	
		Rainy	5	Lc 'Purple Cascade Fragrant B', Blc 'Mem Ann Balmores Convess', C. 'Queen Sirikhit', Blc 'Hsinging Catherine'	
		Year round	7	--	

4.3.7 Explanation on the Table of Characteristics

4.3.7.1 Guidelines for Recording the Observations of Vegetative and Flowering Characteristics

Characteristics indicated with (a), (b), (c), (d) and (e) in the first column of the Table of Characteristics should be examined as indicated below:

a) Observations on the leaf should be made on the longest leaf of flowering plant.
b) Observations on the inflorescence and the flower shall be made at the time when 50% of the flowers on the inflorescence have opened and the most recently fully opened flower on the inflorescence before the color starts to fade.
c) Observations on the length and width of the flower and parts of the flower shall be made on the spread out positions.
d) Observations on the color of the sepal, the petal and the lip shall be made on inner side at apex, mid and base portion.
e) Observations on the colour of column shall be made on inner side at apex, mid and basal region.

4.3.7.2 Explanation for Individual Characteristics

1 Cylindric | 3 Clavate | 5 Globular/Ovoid

Fig. 4.18: Nature of pseudobulb

1 Narrow oblong | 3 Ligulate | 5 Elliptic | 7 Ovate

Fig. 4.19: Leaf shape

1 Oblong | 3 Lanceolate | 5 Elliptic | 7 Ovate

Fig. 4.20: Dorsal sepal shape (21) and Lateral sepal shape (26)

1 Acute | 3 Notched | 5 Obtuse

Fig. 4.21: Dorsal sepal apex (22) and Lateral sepal apex (27)

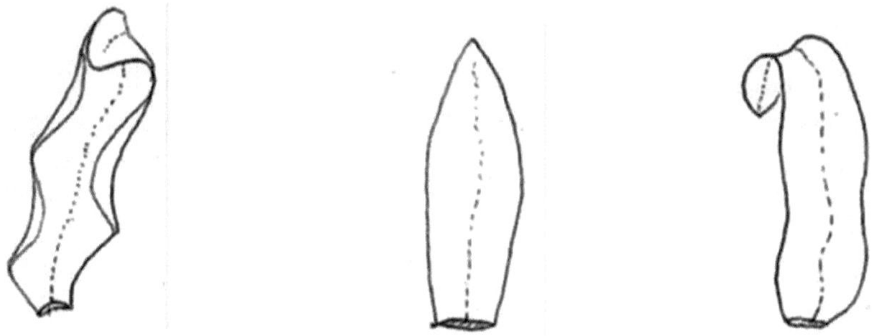

1 Incurved | 3 Straight | 5 Reflexed

Fig. 4.22: Dorsal sepal curvature (23), Lateral sepal curvature (28) and Petal curvature (34)

1 Oblong | 3 Lanceolate | 5 Elliptic | 7 Ovate | 9 Round

Fig. 4.23: Petal shape (33), Lip lateral lobe shape (44) and Lip mid lobe shape (46)

1 Entire | 3 Undulate | 5 Undulate crisped

Fig. 4.24: Petal margin (35), Lip lateral lobe margin (45) and Lip mid lobe margin (47)

4.4 Molecular Characterization of Valuable Species

Molecular markers are of great importance in identifying new desirable genes for crop improvement. Marker assisted selection reduces the time of breeding programme. The utilization of markers also aid in identifying important quantitative trait loci (QTL).

4.4.1 Isozyme Analysis

Proteins are the primary products of structural genes and are very useful for direct genetic study. Changes in coding base sequences also result in corresponding changes in the primary structure of proteins. Even a single amino acid substitution could have significant effects on the migration of proteins under an electric field as in electrophoresis. Isozymes are different forms of an enzyme showing the same catalytic activity but differ in charge and mobility.

4.4.2 DNA Finger Printing

DNA Finger Printing is a technique used to distinguish between individuals of the same species using only samples of their DNA. DNA profiling exploits highly variable repeat sequences called Variable Number Tandem Repeats. These loci are variable enough that two unrelated humans are unlikely to have the same alleles.
Basically there are two types of DNA Finger Printing (Bhattacharjee and De, 2010):

4.4.2.1 Classical Hybridization Based Finger Printing

This technique is practiced by cutting of genomic DNA with a restriction enzyme. In this method, DNA is digested with the restriction enzymes and DNA fragments are separated according to their size by electrophoresis on a gel. The gel is southern blotted in to a membrane and specific fragments are made visible by hybridization with labelled probes.

RFLP (Restriction Fragment Length Polymorphism) analysis is found to be useful for estimating genetic diversity, and to assist in the conservation of endangered species and plant genetic resources. It is also used for plant genome mapping.

4.4.2.2 Polymerase Chain Reaction (PCR) Based Finger Printing

It amplifies the amounts of a specific region of DNA using oligo nucleotide primers and a thermostable DNA polymerase. The amplified products are separated by electrophoresis on agarose gels and detected with ethidium bromide. This method is useful for estimating genetic diversity, identification of species or cultivars, genome mapping, and population genetics etc.

4.4.2.2.1 RAPD (Randomly Amplified Polymorhic DNA)
This is the efficient method for genome mapping and characterization of genetic resources. It is based on repeated amplification of DNA sequences using arbitrary primers to provide DNA fingerprints.

4.4.2.2.2 SPAR (Single SSR Primer Amplification Reaction)
In this case, exponential amplification occurs from the single primer reaction with a particular SSR and polymorphism is SSR based. So, multiple loci are detected from a genome using single PCR Reaction.

4.4.2.2.3 DAF (DNA Amplification Fingerprinting)
This technique exploits single arbitrary primers for amplification of DNA based on PCR. This method is effective for genetic typing and mapping.

4.4.2.2.4 AP-PCR (Arbitrary Primed Polymerase Chain Reaction)
This is one type of RAPD in which discrete amplification patterns are generated through the employment of single primers of 10-50 bases in length.

4.4.2.2.5 RAMPO (Randomly Amplified Microsatellite Polymorphisms)
The method consists of amplification of genomic DNA using arbitrary (RAPD) primers followed by separation with electrophoresis and hybridization of dried gel with micro-satellite oligo-nucleotide probes. It is used in genetic finger printing of closely related species.

4.4.2.2.6 AFLP (Amplified Fragment Length Polymorphism)
This marker is based on polymerase chain reaction and used for rapid screening of genetic diversity. AFLP techniques generate hundreds of highly replicable markers from DNA of any organism and so, allows high resolution genotyping of finger printing quality. It has broad application in systemics, population genetics and mapping of quantitative trait loci.

4.4.2.2.7 ISSR (Inter Simple Sequence Repeat Markers)
In this case, micro-satellite based primers are used to amplify inter-SSR DNA sequences. Here, a number of microsatellite anchored at the $3'$ end are used for amplifying genomic DNA, which increases their specificity. An unlimited number of primers are synthesized for various combinations of di-, tri-, tetra- and penta-nucleotides with an anchor made up of a few bases and has broad range applications in plant species.

4.4.2.2.8 Mitochondrial Analysis

mtDNA is useful in determining the unclear identities. It is one of the important molecular techniques used in studying the extent and distribution of variation in gene pools maintained in various ornamental crops. The marker developed from different molecular techniques can be combined to obtain DNA finger prints in important ornamental corps. DNA finger printing is useful in protecting our indigenous wealth of ornamental plants or varieties developed in our country. It protects intellectual property protection rights of the breeders.

4.4.2.3 Uses of DNA Finger Printing

- To identify misabled plants in the trade.
- Genetic mapping of ornamental plants.
- To assist in cultivar identification, breeding programme and evolutionary research in commercial ornamentals.
- RAPD (Randomly Amplified Polymorphic DNA) markers are utilized to determine the genetic relationship of different cultivars of ornamental crops.
- To analyse rank correlation of results of contributing characters as reported in carnation.
- To assist in intra- and inter-specific breeding of new cultivars of ornamental plants through RAPD markers.

5 Breeding Approaches for Improved Genotypes

Orchids are highly priced in the international market due to their designed spectacular flowers, brilliant colours, delightful appearance, myriad sizes, shapes, forms, and long lasting qualities. The majesty of nature is represented in the most highly evolved family among monocotyledons: orchidaceae. The orchids with their 25,000 to 30,000 species in some 700 to 800 genera constitute the largest family among flowering plants, and exhibit almost innumerable hybrids and inexhaustible varieties.

5.1 Cytogenetics

From the beginning of orchid cytology, conceptions of the chromosome numbers in the family of orchidaceae have changed remarkably. Investigations conducted by Guignard (1982) and Strasburger (1888) observed that an orchid plant has a constant number of chromosomes in reproductive cells. They suggested 16 as a common number of chromosomes for orchids found in reproductive cells of *Gymnadenia conopsea*, *Himantoglossum hircinum*, *Listera ovata*, *Orchis maculata* and *Paphiopedilum barbatum*. Singh and Prakash (1995) reported that the basic numbers and somatic chromosome number show wide variation in orchidaceae; the lowest chromosome number is 2n=10 (*Oncidium pusillum*) and the highest is 2n=200 (*Aeringes* spp.). According to Duncan (1959), in different species and hybrids of orchids chromosome number in meristem (2n) varies: *Cypripedium* (20-22), *Phragmipedium* (20-32), *Paphiopedilum* (26-70), *Ophrys* (36), *Orchis* (20-80), *Daclyorchis* (20-120), *Platanthera* (16-126), *Gymnadenia* (16-80), *Habenaria* (28-42), *Listera* (32-42), *Epipactis* (24-40), *Cephalanthera* (32-36), *Goodyera* (28-42), *Liparis* (30-42), *Epidendrum* (40-160), *Cattleya* (40-102), *Laelia* (40-60), *Laeliocattleya* (40-100), *Brassocattleya* (40-80), *Dendrobium* (38-80), *Cymbidium* (32-80), *Oncidium* (28-56), *Aerides* (38-40), *Vanda* (28-95), and *Phalaenopsis* (38-114).

Triploids are apparently most desirable in commercial *Paphiopedilum* (Duncan, 1947), *Cattleya* (Kamemoto, 1950) and *Vanda* (Storey, 1952). In *Cymbidium*, triploid and tetraploid hybrids allow for improved flower quality (Mehlquist, 1952; Meninger, 1954; Wells, 1956). A pentaploid (Wimber, 1954) and several polysomic individuals (Wimber and Hernlud, 1955) have also been reported in *Cymbidium*. In *Vanda*, diploid and tetraploid species and their hybrids, as well as polyploidy hybrids which have an origin independent of these species, form a hybridization array which extends through pentaploidy (Duncan, 1959).

5.2 Pollination and Fertilization

The morphological structure of orchid flower prevents easy fertilization. Moreover, the lumps of pollen they produce cannot be carried by wind. In nature, insects are usually the pollinators of flowers, but birds are involved in some species. Insects such as bees, moths, butterflies, beetles, ants, flies etc. enter into the lip, search out the nectar, release the cap and pollen masses and lodge on the head. While visiting another flower, the pollinia are pressed against the stigmatic surface. The grains germinate: the pollen tube penetrates the ovary and fertilize the ovules. Male bees are responsible for pollination of many orchids, e.g., *Catasetum, Cycnoches, Gongora, Orchis*. Butterflies and moths also help in the pollination process. *Epidendrum secundatum* is pollinated by butterflies attracted by the nectar. Moths usually pollinate white or light coloured flowers which emit strong odours at night (Bose and Bhattacharjee, 1980). *Epipactis consmilis* is reported to mimic the shape and colour of aphids, which attracts aphidophagous hoverflies for pollination. Flowers of *Oncidium* and *Ophrys* are attracted and pollinated by territory defending male Centries bees. Flowers of *Cypripedium* and *Paphiopedilum* have pouches, and if by chance insects enter into the structure, they must disturb the pollens and fertilization occurs before coming out of the flower. Self pollination occurs in *Cypripedium schlmii, Neottia* spp., *Phaius grandiflorus* etc. because of their inherent structural arrangements which facilitates pollination and fertilization (Bose, *et al*, 1999).

5.3 Natural Hybridization

The presence of natural hybrids evolved from crosses between the species in the wild has been recognized since the early days of orchid collection and cultivation. *Phalaenopsis intermedia,* collected from Philippines, is one of the oldest natural hybrids between *P. aphrodite* and *P. rosea* described correctly by Lindley in 1853. Natural hybridization within the genus *Platanthera* in North America occurred between the closely related species *P. dilatata, P. hyperborea, P. limosa, P. sparsiflora* and *P. stricta*. Two *Masdevallia* species, *M. splendida* and *M. parlaloreana* are recognized as natural hybrids developed from crosses between *M. veitchiana* and *M. barlaeana* which existed in the Andes of Peru. There are also records of inter-generic hybrids evolved under natural conditions and *Laleliocattleya elegans* is one of such examples (Bose *et al*, 1999).

A natural hybrid population of *Oncidium* arising out of a cross between *Onc. staevi* and *Onc. jonesianum* has also been reported. Veitch and Sons (1887-1894) described a complex of natural hybrids among Mexican and Columbian Odontoglossums. At least four different groups of hybrids viz. the *crispo-odoratum* group, *crispo-lindleyanum* group, *crispo-luteopurpureanum* group and *triumphante-prescatoria* group are recognized (Abraham and Vatsala, 1981). Natural interspecific hybrids have been reported in *Cattleya, Laelia, Miltonia* etc. *Cattleya guatemalensis*, collected by Skinner from Guatemala in 1861, was later found to be a natural hybrid between two species from different genera. It was developed from a cross between *Epidendrum auranticum* and *Cattleya skinneri*. At present, this natural hybrid is known under the generic name *Epicattleya*. In Brazil, *Laelio-cattleya elegans* came into existence from a natural cross between *Laelia purpurata* and *Cattleya guttata*, while, *Laeliocattleya scilleriana* is also a natural cross between *Laelia purpurata* and *Laelia intermedia*. Another interspecific natural hybrid *Cattleya intricata* developed from a cross between *C. intermedia* and *C. leopoldii*. One of the most interesting inter-generic natural hybrids spotted in the wild evolved from a cross between *Cattleya warneri* and *Brassovola tuberculata*. Two natural hybrids are also reported in the genus *Ophrys* from central Italy. The earliest known orchid hybrid appears to be *Gymnadenia conopsea* x *Nigritella nigra* found in Alps near Grenoble in 1987 and was described by Villars under the name *Orchis suaveolens*.

5.4 Artificial Hybridization

Orchid growers all over the world have developed crosses between different species and hybrids with varying degrees of success. Mr. John Dominy was the first successful orchid hybridist, who made a cross between *Calanthe masuca* and *Calanthe furcata* which flowered in the year 1856 for the first time. The hybrid was named as *Calanthe* x Domini in honour of its breeder, the first artificially raised hybrid to flower. William Herbert, Dean of Manchester, attempted orchid breeding and was the first person to take a crossed capsule to maturity from a cross between *Orchis* and *Ophrys*. He likely raised few seedlings but they perished after few days (Lenz and Wimber, 1959). John Dominy, in twenty years after his first success, actively engaged in orchid hybridization and was reported to develop at least 25 hybrids. It was only after the discovery of Professor Knudson that orchid seeds could be germinated *in vitro*, that production of orchid hybrids gained its momentum.

The orchid hybrids are offspring derived from the cross between two genetically non-identical individuals. Intraspecific, intragenic and intergeneric hybrids have been obtained in this group of plants. These hybrids have been recorded and registered by Rolfe and Hurst (1909) and by Messers. Sanders (1946 and addenda). Inter-generic crosses are very common in orchids and a large number of crosses involving two genera (bi-generic), three genera (trigeneric), four genera (tetra-generic), five genera (penta generic) hybrids are registered and listed (De and Bhattacharjee, 2011).

5.4.1 Bigeneric Hybrids

Aerdachnis = Aerides x Arachnis
Aeridocentrum = Aerides x Ascocentrum
Aredopsis = Aerides x Phalaenopsis
Ansidium = Anselia x Cymbidium
Aredovanda = Aerides x Vanda
Aranda = Arachnis x Vanda

Aranthera = Arachnis x Renanthera
Ascocenda = Ascocentrum x Vanda
Brassolaelia = Brassovola x Laelia
Doritaenopsis = Doritis x Phalaenopsis
Epicattleya = Epidendrum x Cattleya
Epilaelia = Epidendrum x Laelia
Laeliocattleya = Laelia x Cattleya
Miltonidium = Miltonia x Oncidium
Odontocidium = Odontoglossum x Oncidium
Odontonia = Odontoglossum x Miltonia
Renades = Renanthera x Aerides
Renanopsis = Renanthera x Vandopsis
Renancentrum = Renanthera x Ascocentrum
Renantanda = Renanthera x Vanda
Renanthopsis = Renanthera x Phalaenopsis
Rhynchovanda = Rhyncostylis x Vanda
Sophrocattleya = Sophronitis x Cattleya
Sophrolaelia = Sophronitis x Laelia
Vandoritis = Vanda x Doritis

5.4.2 Trigeneric Hybrids

Brassolaeliocattleya = Brassavola x Laelia x Cattleya
Colmanara = Miltonia x Odontoglossum x Oncidium
Dekensara = Brassovola x Cattleya x Schomburgkia
Dialaeliocattleya = Diacrinum x Laelia x Cattleya
Epilaeliocattleya = Epidendrum x Laelia x Cattleya
Hartara = Broughtonia x Laelia x Sophronitis
Laeliocattkeria = Laelia x Cattleya x Barkeria
Laycockara= Arachnis x Phalaenopsis x Vandopsis
Lowara = Brassavola x Laelia x Sophronitis
Lyonara = Cattleya x Laelia x Schomburgkia
Mizulara = Cattleya x Diacrinum x Schomburgkia
Moirara = Phalaenopsis x Renanthera x Vanda
Mokara = Vanda x Arachnis x Ascocentrum
Nakamotora = Ascocentrum x Neofinetina x Vanda
Osmentara = Broughtonia x Cattleya x Laeliopsis
Rhyndoropsis = Rhyncostylis x Doritis x Phalaenopsis
Sappanara = Arachnis x Phalaenopsis x Renanthera
Sophrolaeliocattleya = Sophronitis x Laelia x Cattleya
Trevorara = Arachnis x Phalaenopsis x Vanda

Vascostylis = Vanda x Ascocentrum x Rhyncostylis
Wilsonara = Cochlioda x Odontoglossum x Oncidium
Yapara = Phalaenopsis x Rhyncostylis x Vanda

5.4.3 Tetrageneric Hybrids

Iwanagara = Brassavola x Cattleya x Diacrinum x Laelia
Kirchara = Cattleya x Epidendrum x Laelia x Sophronitis
Potinara = Brassavola x Cattleya x Laelia x Sophronitis
Recchara = Brassavola x Cattleya x Laelia x Schomburgkia
Withnearara= Aspasia x Miltonia x Odontoglossum x Oncidium
Yamadara = Brassavola x Cattleya x Epidendrum x Laelia

5.4.4 Pentageneric Hybrids

Goodlera = Brassia x Cochlioda x Miltonia x Odontoglossum x Oncidium
Hasegawara = Cattleya x Brassavola x Broughtonia x Laelia x Sophronitis

5.4.5 Hexageneric Hybrids

Brilliabdera = Aspasia x Brassia x Cochlioda x Odontoglossum x Oncidium x Miltonia

It is not possible to make crosses between any two genera, though free breeding is common in orchids. The majority of the success in orchid breeding was brought out by the art of patient breeding, intuition and perseverance of the orchid breeders and on several occasions merely by luck (Arditti, 1992). It requires several years to raise progeny from seeds to flowering stage. Unlike other crops, orchid seeds can not be germinated without special facilities. It takes a long time for the seeds to mature. Moreover, the number of seeds produced in a capsule is so high that to get a representative sample of the progeny that will be required to draw any valid genetic inferences will be very large and may be impossible on most occasions. Hence, information on combining ability and inheritance of characteristics in orchids is scanty.

5.5 Mutagenesis

Hybridization technology has proven very reliable and easy to use and has produced a wide range of successful cultivars with attractive combinations of spray length, bud number, flower colour and form, vase life, fragrance, seasonality and compactness.

By introducing mutagenesis, however, wide variations of flower colours, form, and size can still be obtained in addition to overcoming incompatibility and sterility. In addition, complementary use of molecular techniques will allow breeders to target more specific characteristic changes and cut short breeding time. PCR-based techniques used to analyse the DNA of mutagenic clones found polymorphic fragments that can be developed as molecular markers (Basiran *et al*, 2002).

5.6 Breeding Objectives

The breeder's goal is to produce commercially important hybrids that have market demand and are liked by the consumers. The concept behind development of hybrids in orchids may vary according to the genus and species. The generalized objectives as stated by Bhattacharjee and Das (2008) are given below:

- To breed for better colour, size, and substance of the flower.
- To introduce perfect blending of colours in sepals, petals and lip.
- To create round and full form of sepals and petals with minimum fenestration and twists.
- To increase the length of inflorescence.
- To increase the number of flowers / inflorescence.
- To achieve compactness in flower facing on the spike.
- To develop hybrids showing correct mode of display.
- To extend blooming period.
- To produce miniature forms.
- To produce fragrant varieties.
- To produce flowers with longer vase life.
- To develop types suitable as pot plants.
- To develop hybrids insensitive to strict climatic regime.
- To develop hybrids resistance to biotic stress like diseases particularly to viruses.

5.7 Selection of Parents

Healthy parent plants are to be selected so that they can produce healthy seed pods with innumerable seeds. Selection of parents depend upon the aim of breeding and genetic affinity of the parental lines. Information on whether the parent performs well as pollen parent or seed parent, as well as the vigour and floriferousness of the parent are required as pre-breeding criteria. Knowledge of compatibility, chromosome number and behaviour of the parents are essential. Known ability of the parents to influence the character in its progeny will also be helpful. Very young plant or seedlings blooming first time should not be selected as the mother plant to bear the seed pod. A very healthy and vigorous plant can bear two to three seed pods without affecting the normal health of plant. A cross becomes successful if two different types of orchids are genetically related.

5.8 Compatibility Analysis

A thorough understanding of the compatibility relationships of the genera as well as species is essential for successful hybrid development. According to Lenz and Wimbler, 1959, many of the cases of the apparent self-incompatibility and cross sterility commonly encountered among orchid hybrids could be due to either of two causes hybrid sterility or polyploidy. Duncan and Curtis (1943) observed that the self incompatible orchids always have homomorphic, gametophytic and polyallelic incompatibility with stigmatic inhibition of pollen germination. The failure of fruit development in many reciprocal crosses hints at the operation of an unidirectional incompatibility in orchids (Devi and Deka, 1992).

5.9 Breeding of Hybrids

Many Indian species have obtained worldwide recognition in breeding programmes due to their inherent attractiveness coupled with their ability to transmit desirable characters to hybrids. Some of the leading species are *Aerides multiflorum*, *Cymbidium devonianum*, *C. lowianum*, *C. tracyanum*, *C. elegans*, *Dendrobium aggregatum*, *Den. chrysotoxum*, *Den. formosum*, *Den. nobile*, *Paphiopedilum venustum*, *Vanda coerulea* (Bose and Bhattacharjee, 1980).

5.9.1 Cattleya

The cattleya species like *aurantiaca*, *bicolor*, *bowringiana*, *dowiana*, *forbesii*, *granulosa*, *guttata*, *intermedia*, *labiata*, *loddigesii*, *luteola*, *mossaie*, *triannaei*, *warneri* etc. are extensively used for hybridization work. Moreover, the *Cattleya* has also been crossed with several other genera such as *Berkeria*, *Brassavola*, *Broughtonia*, *Diacrinum*, *Domingoa*, *Epidendrum*, *Laelia*, *Laeliopsis*, *Schomburgkia*, *Sophronitis* and produce multi-generic hybrids. The cattleya cultivars like 'Bob Betts', 'Bow Bells', 'Claesiana', 'Empress Bells', 'Enid', 'Estette', 'General Patton', 'Henrietta Japhet', 'Karae Lyn Sugiyama', 'Margaret Stewart', 'Nellie Roberts', 'Nigritian', 'Pearl Harbour', ' Portica', 'Primma Donna'. 'Vesper Bells' are considered famous and are proven mother plants for breeding work. *Cattleya dowiana* and its variety 'aurea' have been of great importance in the development of the beautiful yellow flowered hybrids. In crosses with the coloured or alba forms of the *Cattleya* species, the yellow colour of *C. dowiana* acts as a recessive trait and does not appear in the first generation, the flower being coloured with anthocyanin pigments. In certain F_2 populations involving *C. dowiana*, the plants segregate to give yellow, cream and white coloured flowers (Fenton, 1951). Mehlquist (1958) studied the inheritance of white flowers with coloured lip. He suggested a gene P to be responsible for this type of colouration due to interact with C and R genes.

The *Cattleya* alliance *Laelia* and *Sophronitis* are known for their bright colour contribution. The redness of many of the orchid hybrids were introduced from *Sophronitis grandiflora*. The flowers of this *Sophronitis* species are small and many of the resulting *Sophrocattleya*, *Sophrolaelia* and *Sophrolaeliocattleya* have comparatively small flowers. Northen (1949) observed that the gene for red is dominant, and when present in a homozygous condition it gives a beautiful clear red colour.

Intergeneric hybrids involving *Brassavola* give a very attractive fimbriated lip. The first *Brassocattleya* was bred as early as 1889 by Maron, in which *Brassavola digbyana* was combined with *Cattleya*, while the first *Laeliocattleya* was bred by Sanders in 1903 from a cross between *Cattleya hardyana* and *Laelia pumilapratens*. The first trigeneric hybrid *Sophrolaeliocattleya* was bred by Holfred in 1907 from a cross between *Laeliocattleya* Aureole and *Sophronitis coccinea*. A hybrid genera *Potinara* (which is a tetrageneric hybrid combining *Cattleya, Laelia, Brassavola* and *Sophronitis*) was registered for the first time in 1922 (Bhattacharjee and Das, 2008). The introduction of blue colour in *Cattleya* breeding was discussed by Granier, 2002. One of the prominent hybrids used extensively for blue *Cattleya* breeding is *Laeliocattleya* 'Canhamiana', a cross between *C. mossiae* and *Laelia purpurea*. Among the genus *Brassavola, B. cucullata, B. cordata, B. nodosa* and *B. perrinii* are extensively used in hybridization (Mathews, 1996). The large flowered *Cattleya* hybrids are the results of breeding involving fifteen *Cattleya* species, two *Laelia* and *Brassavola digbyana* (Herman, 1997). In *Brassocattleya* the varieties 'Deesse' and 'Hortland' are reported to produce many hybrids in *Cattleya* alliance. The trigeneric hybrid *Brassolaeliocattleya* are also well known for their varieties like 'Edwin Chong', 'Golden Myth', 'Herons Ghyll', 'Jane Helton', 'Malvern', 'Marjorie Frey', 'Mellow Glow', Molflora', 'Norman's Bay', 'Nugget', etc which have produced many interesting hybrids. In *Laeliocattleya* are the varieties 'Charlesworthii', 'Easier Bonnet', 'Edgard Van Belle', 'Elissa', 'Grandee', 'Harold J Peterson', 'Ishtar', 'Morro Rock', 'Nugget', 'Pacific' 'Paradasio', 'Princess Margaret', 'S.J. Bracy', 'South Esk', 'Supervia', 'Twinkle Star', which are famous proven parent plants for hybridization work.

5.9.2 Cymbidium

Cymbidiums are highly valued as cut flowers as well as pot plants in the international florist trade. Innumerable numbers of hybrids have been evolved. The species like *Cymbidium devonianum, C. ensifolium, C. erythrostylum, C. grandiflorum, C. hoosai, C. insigne, C. madidum, C. pumilum* and *C. tracyanum* are successfully utilized for the development of many hybrids. Some of the hybrids like 'Balkis', 'Cleo

Sheraton', 'Desiree A'logann', 'Early Bird', 'Joan of Arc', 'Kurun', 'Lucy', 'Lustrous', 'Mayfair', 'Miretta', 'Nam Khan', 'October','Oiso', 'Ortin', 'Redwood', 'Remus', 'Rio Rita', 'Rosanna', 'Shiraj', 'Stanley Fouraker', 'Swallow' and 'Vieux Rose' are outstanding and largely utilized as parent plants for production of many spectacular hybrids.

The species like *C. pumilum*, *C. devonianum* and *C. ensifolium* are small flowered types and cross easily with the large flowered Himalayan species. Most of the hybrids in *Cymbidium* evolved through the utilization eight of the large flowered species viz. *Cymbidium iridioides* (syn. *C. giganteum*), *C. eburneum*, *C. hookerianum* (syn. *C. grandiflorum*), *C. sanderae*, *C. lowianum*, *C. tracyanum*, *C. insigne* and *C. erythrostylum*).

In the case of *Cymbidium* 'Alexanderi', a cross between *C.* 'Eburneo-Lowianum' and *C. insigne*, the progenies of the cross were diploid, but the clone of the cross 'Westtonbrit' proved to be tetraploid. In fact this hybrid changed the world of cut flower cymbidiums. *Cymbidium* 'Alexanderi' was producing everything that was required where quality was concerned (Keith, 2000). This hybrid became popular because of its large white flowers and as a parent for pestal coloured popular hybrids.

The Japanese and Chinese species of *Cymbidium* are utilized for the breeding of miniature types. Miniature hybrids are good as pot plants and slightly tolerant to warmer conditions. The first miniature hybrid in *Cymbidium* evolved in England in the year 1944, and was a cross between C. 'Lousie Sander' x C. pumilum. The variety 'Lousie Sander' is a cross between 'Alexanderi' x 'Ceres'. *C. munronianum* has been used as parent in several breeding programmes for contributing scent characteristics to the offspring.

5.9.3 Dendrobium

Dendrobiums are very popular among the orchids throughout the world for higher productivity and floriferousness. These are well known as cut flowers and potted plants.

The *Dendrobium* species like *aemulum, affine, aggregatum, aries, canaliculatum, chrysotoxum, compactum, d'albertisii, dalhousieanum, delicatum, dicuphum, forbesii, formosum, goldiei, gracilicaule, gracillimum, grantii, johnsoniae, kingianum, laxianthera, laxiflorum, leporium, macrophyllum, mirbellianum, nobile, odoardi, ostrinoglossum, phalaenopsis, primulinum, superbum, taurianum, tetragonum, toftii, tokai, trilamellatum, undulatum, veratifolum, violaceo-flavens* and *williamsonii* are successfully utilized for hybridization programmes and produce many outstanding hybrids of special merit. Vacherot and Lecouffle of France were the pioneers of Dendrobium breeding. The nobile type (narrow petals) Dendrobiums of Eastern Himalayas and *D. phalaenopsis* (rounded petals) of eastern Asia were the most frequently used parents. Kuehnle *et al* (1997) observed that pigments comprised of cyanidin glycosides are the predominant pigments extracted from the lavender and purple Dendrobium. Kamemoto and Amore (1990) reported that a dominant gene P is responsible for lip colour or the semi-alba trait of white petals and sepals with coloured labellum, based on crosses between semi-alba *D. dicuphum* and white *D. affine* or white *D. phalaenopsis* var. *compactum* 'Mauna Kea'. According to Thammasiri *et al* (1986) yellows and greens in Dendrobium are due to carotenoids and chlorophylls. Improved yellow cultivars are contributed by triploidy. Kamemoto *et al* (1999) reported inbreeding depression, which is expressed as a loss in vigour and pollen degeneration in *D. phalaenopsis* and *D. biggibum*, but not found in *D. antennatum*. Several cycles of inbreeding and selection using an amphidiploid Dendrobium 'Jaquelyn Thomas' proved useful to increase flower size and lighten colour (Bobisud and Kamemoto, 1982). It also produced valuable stud plants when used subsequently in outcross to restore vigour. Colour has always been of prime importance in *Dendrobium* breeding ranging from chalky white to yellow, brown and intense crimson (Abraham and Vatsala, 1981).

For producing compact hybrids with short pseudobulbs to breed for potted plants *D. carronii, D. canaliculatum, D. phalaenopsis, D.* var. *compactum, D. bigibbum* var. *compactum, Dendrobium* 'Mini Gem' are suitable (Kuehnle, 2006). Kaiser (1993) suggested one parent of possible interest for breeding fragrant potted plants is *D. d'albertisii*; and out of the 140 species evaluated, 40 percent produced scents ranging from floral to fruity to herbaceous. Some of the outstanding varieties of *Dendrobium* used in further hybridization program and are recognized as proven parents are 'Agnes', 'Ann', 'American Beauty', Amethyst', 'Anouk', 'Artur Elle', 'Black Bountain', 'Caesar',

'Carol Ann', 'Claire Ayau', 'Concert', 'Constance', 'Dixon', 'Endeavour', 'Ethreal Kawamoto', 'Ewa', 'Gold Flush', 'Gold Twist', 'Helen Fukumara', 'Hula Girl', 'Jaquelyn Thomas', 'karen Ono', 'Khaw Young Hong', 'Lady Fay', 'Lady Hamilton', 'Lady Nui', 'Lim Chong Min', 'Lynn Takiguchi', 'Maloe Kanya', 'Manoa Gold', 'Margi Thomas', 'Main Beauty', May Neal', 'Mustard', 'New Hawaii', 'Pale Face', 'Pompadour', 'Rose Chong', 'Shangrilla', Ursula', 'Valley King', 'Vera Patterson', 'Yellow Curls', and 'Yellow Jacket'.

5.9.4 Odontoglossum

This genus produces the finest flowers of all cool growing orchids. The *Odontoglossum* species like *cirrhosum, crispum, hallii, pendulum, pulchellum* are used for hybridization work. There are many hybrids of Odontoglossum on record and a large number are made with related genera like *Aspasia, Cochlioda, Miltonia* and *Oncidium*. The outstanding hybrids of *Odontoglossum* widely used as parents for production of further hybrids are 'Carroll', 'Ismene', 'Patnina' etc. (Bhattacharjee and De, 2003).

5.9.5 Oncidium

The orchid plants under this genus are known as 'Dancing Ladies' and 'Golden Showers' because of their showy alternatively shaped blossoms. Large numbers of

Oncidium species are utilized for production of spectacular hybrids and most important species in this respect are *henekenii, intermedium, lanceanum, leucochilum, marshallianum, pulchellum, retemeyerianum, splendidum, sylvestre, triquetrum, vericosum* and *variegatum*. The genus is cross compatible with other genera like *Aspasia, Brassia, Cochlioda, Comparettia, Gomesa, Macradenia, Miltonia, Odontoglossum, Rodriguezia, Trichocentrum, Trichopilia* etc. and hence, many multi-generic hybrids have been evolved. Some of the outstanding hybrids which have proven worth as parent plants for production of more and more attractive hybrids are 'Angnes Ann', 'Ann Rosa', 'Catherine Wilson', 'Delight', 'Golden Glow', 'Helen Brown', 'Lovely', 'Organ Mountains', 'St. Anne', and 'Waikiki Sunset' (Bhattacharjee and De, 2003).

5.9.6 Paphiopedilum

This is a remarkable genus of a magnificent group of orchids, commonly called 'Lady's Slipper' orchids. The cut flowers remain fresh in plain water for up to six weeks. *Paphiopedilum rothschildianum* is one of the most important species and has produced nearly 200 hybrids. Important species used in hybridization programmes are *bellatulum, godfroyae* and *niveum*. One of the most outstanding hybrids in this genus is 'Rolfei' which is a cross between *Paph. rothschildianum* and *Paph bellatulum*. Another important hybrid is *Paph.* 'Daisy Barclay', which was evolved from a cross between *Paph. rothschildianum* and *Paph. godefroyae*. The cross between *Paph. rothschildianum* and *Paph. niveum* created *Paph. woluwense*. An allelomorph in *Paph. insigne, Paph. spicerianum* and *Paph. boxalli* was reported by Hurst (1925). Rogersen (1991) has suggested that for breeding white coloured *Paphiopedilum*, species like insigne var. *Sanderae, bellatulum* and *godefroyae* are important and described the general principles governing the inheritance of the white flower colour in orchids, and specifically in the genus *Paphiopedilum*. Tu-Fu Philip Liu (2000) stated that the backbone of green and white albino forms was *Paphiopedilum callosum* var *sanderae*; for green breeding, *Paph.* 'Desert Spring, for yellow breeding promising clones are some hybrids of *Paph. sukhakulii* var. *album*, such as *Paph. fairrieanum* var. *album*; for white breeding *Paph. charlesworthii* var. *album*. *Paphiopedilum* is a cross compatible with *Selenipedium* and *Phragmipedium*.

5.9.7 Phalaenopsis

Phalaenopsis is highly valued for its long lasting cut flowers, pot plants and hanging baskets. Several species of *Phalaenopsis* utilized for the production of outstanding hybrids are *amabilis, amboinensis, boxalli, buyssoniane, cornucervi, denevei, equestris, Esmeralda, fasciata, fimbriata, fuscata, grandiflora, lueddemanniana, lindenii, mannii, mariae, parishii, schilleriana, serpentilingua, stuartiana* and *violacea*. Hybridization of *Phalaenopsis* mainly evolved around the development of pure whites, pure pinks and white with pink lips. *P. amabilis*, P. *formosana* and *P. aphrodite* were extensively used for production of outstanding white hybrids. For development of pink coloured hybrids like the *lueddemanniana, sanderiana* and *schileriana* figured prominently. White colours with pink lip was achieved in *P. intermedia* due to natural hybridization between *P. aphrodite* and *P. equestris*. According to Thomas (2001) all modern white Phalaenopsis are descendents of two species, *P. amabilis* and *P. aphrodite*. Phalaenopsis 'Doris' was bred in 1940 and is one of the most influential white hybrids. Yellow and red *Phalaenopsis* breeding started with a cross between *P*. 'Doris', and *P. manni* to produce *P.* 'Golden Louis'. Red *Phalaenopsis* breeding is generally accompanied by problems with fertility, flower size and low flower count. The first fertile and readily available red *Phalaenopsis* was 'Golden Buddha' (Ginsberg, 2000). Another recent trend in *Phalaenopsis* is breeding for blue flowers. The two species that formed the basis of blue *Phalaenopsis* breeding are *Doritis pulcherrima* var. *coerulea* and *Phalaenopsis violacea* var. coerulea (Mountford, 2001). Harper (1993) discussed the contribution made by *Phalaenopsis stuartiana* in the development of multiflora *Phalaenopsis* hybrids. Thomas (2001) reviewed the requirements for flower forms of commercial growers like strong self supporting erect inflorescences, long duration of blooms, compact plant size, wide temperature tolerance, disease resistance, firm substances and consistency of colours.

Selfing and backcrossing techniques have been found immensely beneficial in breeding for superior hybrids. The long continued inbreeding of *Phalaenopsis sanderiana* is a famous example leading eventually to superior clones. Multi-generic hybrids have been evolved by crossing *Phalaenopsis* with other genera like *Aerides, Arachnis, Doritis, Neofinetia, Rhyncostylis, Renanthera* and *Vanda*. Some of the hybrids of *Phalaenopsis* which have served as parent plants for production of many more outstanding hybrids are 'Aalsmeer Rose', 'Ann lovelace', 'Aristocrat', 'Barbara Bred', 'Big Chief', 'Cassango', 'Cast Iron Monarch', 'Chieftain', 'Cindy Brand', 'Doreen', 'Doris', 'Dos Pueblos', 'Elinor Shaffer', 'Elisa', 'Elwy Middleton', 'Fairway Park', 'Fenton Davis Slover', 'Gladys Red', 'Goleta', 'Grace Palm, 'Hollywood', 'Hymen', 'Intermedia', 'Joanna Magale', ' Junita', 'Judy Karleen', 'Kareen', 'Lachesis', 'Lakme', 'Loius Georgianna', 'Louis Market', 'Luzon', 'Margaret Bean', 'Marmouset', 'Martha', 'Mild Red Karleen', 'Mrs' J.W. Veitch', 'New Horizon', 'Palm Beach', 'Pasadena', 'Pink Vision', 'Pink Wave', 'Princes Grabce', 'Queen Emond', 'Radiant Glow', 'Reve Rose', 'Rosewell', Ruby Lips', Seretiny', 'Shivley Temple', 'Susan Market', 'Texas Star' and 'Zada'.

5.9.8 Vanda

This genus bears attractive flowers which are often large, the colour of flower ranges from pure white to variegated pattern of brown, green, and bright pink to blue and purple. These are commercially important cut flowers used in international flower trade. The *Vanda* species which contributed for production of useful and showy hybrids are *amnesiana, batemanni, coerulea, coerulescens, dearci, denisoniana, hookeriana, insignis, lamellata, loatica, limbata, liouvillei, luzonica, marvillii, parviflora, roeblingiana, roxburghii, sanderiana, spathulata, stuavis, sumatrana, teres* and *tricolor.*

The first Vanda hybrid, 'Miss Joaquim', which developed from a cross between *V. teres* and *V. hookeriana* (Teoh Eng Soon, 1998). An albino form of *Vanda* 'Miss Joaquim' was produced by crossing the alba varieties of both the parents. *Vanda* 'Miss Joaquim' is said to have taken the form of *V. hookeriana* and the colour of *V. teres* (Tim Wing Yam, 2001). Fuchs (1997) reported that *Vanda sanderiana* and *V. coerulea* are the two important vandal species found in the background of most of the vandaceous hybrids. *V. sanderiana* gives full form, whereas *V. coerulea* imparts the rich blue violet colouration, lobely tessellation as well as the long inflorescence.

The genus *Vanda* is cross compatible with other allied groups like *Aerides, Arachnis, Ascocentrum, Doritis, Neofinetia, Phalaenopsis, Renanthera, Trichoglottis and Vandopsis.* Some of the important hybrids of Vanda which contributed as parent plants for production of many more hybrids are "Amoene', 'Betsy Summer', 'Bull Sutton', 'Eisenhower', 'Ellen Noa', 'Emily Notley', 'Ernest', Fujinaga', 'Frank Crook', 'Haledena', 'Helen Reynolds', 'Hilo Blue', 'Honolulu', 'Jennie Hashimoto', 'Josephine Van Bero', 'Kapolio', 'Manila', 'Manisaki', 'Miss Joaquim', 'Noel', 'Nora Potter', 'Norbert Alphanso', 'Onomea', 'Poepoe', 'Rubella', 'Ruby Prince', 'Tan Chay Yan', 'Tatzeri', 'Trimerrill', 'Trisher', 'Venus' and 'Waipuna'.

5.10 Polyploidy Breeding

Polyploidy and introgressive hybridization played a major role in the development of orchid hybrids. In some genera like *Cymbidium, Paphiopedilum, Phalaenopsis, Cattleya, Laelia, Sophronitis,* polyploidy coupled with intergeneric compatibility has culminated in the formation of hybrid groups which show both greater size and hybrid vigour as compared to parental species. Most orchids have two basic sets of (diploid 2x) chromosomes. The most common form of polyploidy is the doubling of the chromosome number from diploid (2x) to tetraploid (4x). Tetraploid plants are more fertile and produce flowers of better texture, bigger size and more intense colouration. According to Abraham and Vatsala (1981), infertility in polyploids often results from pairing abnormalities during meiosis. Sterility is most frequently caused by triploidy, commonly encountered in many of the cultivated orchids. Kamemoto *et al* (1999) has described a strategy in which a triploid yellow, *D.* 'Mary Mak' was crossed to a diploid *D. helix.* Among a population of mostly aneuploids, several tetraploids were obtained for use in subsequent breeding.

Hawaiian breeders produced exciting tetraploid clones by repeated selfing of *V.* 'Miss Joaquim', which is a diploid (2n = 38). The most popular clones produced are 'Atherton', 'Juliet', 'Hula Girl', 'Wood Lawn' and 'Douglas'. Most of the subsequent *V.* 'Miss Joaquim', hybrids are based on these tetraploids. About 450 hybrids have been raised from *Vanda*, 'Miss Joaquim' (Tim Wing Yam, 2001). In *Phaius tankervilliae*, treatment of protocorms by colchicine treatment (50ppm) induced polyploidy. Doubling of chromosome number was associated with increased guard cell and stomatal size and height and weight of plants (Devi and Deka, 2000). According to Hedren *et al* (2000) diploid members of the genus *Nigritella* have sexual reproduction, whereas polyploidy members are characterized by agamospermy. The AFLP data support the general picture of polyploidy evolution in *Dactylorhiza* .i.e. that allotetraploid derivatives have arisen repeatedly as a result of hybridization between the two parental groups *D. incarnata* and *D. maculata* group (Hedren *et al*, 2001).

5.11 Genetic Engineering

Genetic engineering coupled with tissue culture technique provides a useful way to introduce specific genes into plants as evidenced by a successful introduction of improved protein quality, novel flower colour and disease, insect and pest resistance into various crops in the relatively short period of time. In orchids, the first report on genetic transformation was made by Kuehnle and Sugii (1992). Several gene transferred methods have been tested with *Dendrobium* such as particle bombardment (gene gun) (Kuehnle and Sugii, 1992, Chia *et al*, 1994), protein electro injection, seed imbibation and pollen tube mediated DNA transfer (Nan & Kuehnle, 1995b). Among these, micro bombardment (Klein *et al*, 1987) was identified as highly suitable for *Dendrobium* based on high frequency recovery of transgenic plants. A co-cultivation method was standardized for transforming *Phalaenopsis* varieties *in vitro* with *Agrobacterium tumefaciens* using protocorm like bodies as explants (Hsieh *et al*, 1997). Yang *et al* (1999) and Yu *et al* (1999) exploited the protocorm like bodies (PLBs) to transform the orchid using particle bombardment in hybrids of *Cymbidium* and *Dendrobium*, respectively. Protocorm and PLB's of three genera *Brassica*, *Cattleya* and *Doritaenopsis* were genetically transformed via micro-projectile bombardment (Knapp *et al*, 2000).

Genetically transformed plants can be obtained by co-cultivation of *Agrobacterium tumefaciens* with cell clumps in *Phalaenopsis* orchid (Belarmino and Mii, 2000) and with their rhizome sections in *Cymbidium niveomarginatum* (Chen *et al*, 2002). A protocol was standardized to obtain a stable transgenic orchid (*Dendrobium nobile*) via Agrobacterium mediated transformation of PLBs (Men *et al*, 2003b). Transgenic orchid (*Phalaenopsis* cv. White Hikaru) plants were generated by inoculating a needle wounded PLB with *Agrobacterium tumefaciens* (Chai and Kim, 2004). The transformation of Phalaenopsis cv. Taisuco Crane protocorm like bodies (PLB's) by particle bombardment and *Agrobacterium tumefaciens* mediated transformation with beta-glucuronidase (GUS) fusion protein was studied by Chan *et al* (2003). Liau *et al* (2003) indicated that foreign DNA can successfully be integrated into the orchid genome and expressed transcriptionally and translationally in *Oncidium* orchid plants. Both genome breeding and molecular breeding approaches can be used concurrently in varietal development of *Dendrobium* as cut flowers and flowering potted plants. Chemical survey of *Dendrobium* species and hybrids has shows lavender cyanidin and peonidin to be the predominant anthocyanidin, and orange pelargonidin to be rare. Cloning and characterization of key anthocyanin biosynthetic genes such as of dihydro-flavanol 4-reductase enables more productive hybridization strategies to be implemented (Kuehnle *et al*, 2004).

6 Production of Quality Planting Materials

The lack of quality planting materials has become a major factor for limiting commercial cultivation of various orchids. There are six main techniques used for orchid propagation: division, Backbulbs, aerial cuttings, keiki, micropropagation and seed culture. Other techniques used in propagation are aerial shoots and tubers (Bhattacharjee and De, 2010).

6.1 Division

This is the easiest method of propagation used for sympodial orchids. In this case, the rhizomes are cut between pseudobulbs and the pieces are potted separately so that each part has at least three healthy pseudobulbs and one dormant bud for producing new growth. The best time for division of orchids is early spring. Division of an orchid encourages the plant to produce more vigorous shoots of a better quality. *Brassavola, Calanthe, Laelia, Miltonia, Odontoglossum, Oncidium, Cattleya, Dendrobium, Paphiopedilum* and *Cymbidium* can be multiplied through division.

6.2 Back Bulbs

These are previously flowered or unflowered back pseudobulbs. In this case, it may take up to three years to obtain a flowering size plant. A back bulb having roots is pulled or rhizomes are cut just beyond it and are inserted at one side of a pot filled with orchid compost or sharp sand or grit keeping the cut surface of the bulb nearest the edge of the pot. The shoots emerge from the bulbs within two or three months which can be potted in orchid compost. *Cymbidium, Cattleya* and *Coelogynes* are propagated through in this manner. In *Cymbidium*, it has been found that both saw dust and cocopeat are effective media for generation of plants through backbulbs. Prior to planting, treatment of the back bulbs of *Cymbidium* orchids with BA (200 ppm) or coconut water (1:5 or 1:10) is effective to enhance percentage of germination. By this method, both the media take 90-99 days in spring season and 42-48 days in the summer season to generate new plants (Tab. 6.1).

6.3 Cuttings

In monopodial orchids like *Vanda, Aerides, Arachnis,* and *Mokara*, the upper most parts of the stem are cut off just under the aerial roots and the removed part is planted in porous media for producing an individual plant, referred to as top cuttings. In sympodial orchids like *Dendrobium*, cuttings of 10-15 cm having 4-5 segments are taken from canes during spring or rainy season and planted in cocopeat or saw dust for rooting. Flower stalk cuttings are useful in genera like *Phalaenopsis, Phaius, Calanthe* and *Thunia*. Cutting are usually potted in propagation beds or directly in pots after treating the cut ends with fungicides like bavistin at 3 g/litre. The cutting of genera, like *Aerides, Arachnis, Vanda* are very hardy and are directly potted in pots, whereas those of *Dendrobium* and *Phalaenopsis* need special care to root and should be rooted in propagation beds. In *Phalaenopsis*, offshoots are produced by cutting out or mutilating the growing point, removing small leaves and treating the injured portion with fungicides.

Tab. 6.1: Generation of planting materials of Cymbidium through backbulbs

Name of hybrid	Type of Media	Season	Duration
Cym. 'H.C. Aurora"	Cocopeat	Spring	90 days
	Cocopeat	Summer	42 days
	Saw dust	Spring	93 days
	Saw dust	Summer	48 days
Cym. 'W.W.W.'	Cocopeat	Spring	99 days
	Cocopeat	Summer	42 days
	Saw dust	Spring	93 days
	Saw dust	Summer	48 days

6.4 Air layering

In this method, a cut is made through the stem 20 to 30 cm below the apex and most sphagnum moss is wrapped around the cut portion. The rooting media is kept moist and once the roots are formed, the layer is detached from the mother plant and potted in small-sized pots. *Vanda* and other monopodial orchids are easily multiplied by air-layering or marcotage.

6.5 Keiki

A keiki is a small plant which generally grows from one of the nodes along the stem instead of a branch. They occur through the accumulation of growth hormones at a specified point. Keikis grow in two forms: a regular and a basal keiki. The regular keiki is a small plant growing from one node along the flower stem, instead of a branch. This is induced by the accumulation of growth hormones at that point, either naturally or by the application of keiki paste, a cytokinin hormone which induces growth in the node of an orchid inflorescence. The basal keiki is a baby plant growing from the base of the mother plant. Sometimes keikis bloom while still attached to the mother plant. Keikies are used as propagules in *Dendrobium, Ascocenda, Phalaenopsis* and *Epidendrum.*

6.6 Aerial Shoots

Most of the Dendrobiums give aerial shoots or bulbs on old back bulbs devoid of leaves. They usually develop on the upper part of the back bulbs and grow out slowly. These aerial shoots take 90-120 days to develop roots. At this stage, they are detached along with the portion of back bulb and potted as an independent plant in orchid compost. In genera like *Goodyera,* the rhizome gives off special lateral branches which turn up and produce aerial shoots.

6.7 Tubers

In few genera, like *Peristylis* and *Nervillia,* the roots are produced from above the tubers, which are transformed into tubercles. These small tubers produce new plants the year after.

6.8 Tissue Culture

Tissue culture is one of the most rapid methods of multiplying vegetative plants. It develops new plants in an artificial medium under aseptic conditions from very small parts of plants, such as shoots tips, root tips and pollen grains. Thousands or even millions of identical plants can be produced from a small tissue in a relatively short time. Among these, meristem and shoot tip cultures are most popular for mass propagation of commercial species and hybrids. Axillary buds are a good source of explants in monopodials. Both liquid and solid media are used for culture of orchid tissues. The widely used media are Knudson's C medium, Vacin and Went's medium, Murashige and Skoog's medium. Additives like coconut water (15%) and banana pulp (10%) are found beneficial for promotion of shoots. The mineral salts, carbon source, vitamins and plant growth regulators are used in the media. Sucrose as a carbon source promotes organogenesis at suboptimal concentrations and protocorm formation at supra-optimal concentrations. Among vitamins, thiamine and growth regulators, auxin, cytokinins are used for callus formation. Protocorms of Cymbidium 'Soul Hunt-1' cultured on media incorporated with different levels of IBA & GA3 revealed that MS+AC+GA3 (0.5 mg/l) resulted in faster plb proliferation (18 days for 5th plb stage). A combination of both hormones on plb proliferation was found to have the best effect for MS+AC+IBA (0.5 mg/l)+GA3(1 mg/l). *In vitro* plants are hardened off *in vitro* only, before being transferred to the main field. Application of paclobutazol delays chlorophyll loss, reduces the activities of enzymes and delays senescence.

6.8.1 Shoot Tip Culture

In this method, shoot tips are extracted from the vegetative buds located on pseudobulbs. The necessary steps for propagation of micropropagated orchid planting materials are:

- Selection of healthy and disease free mother plant and establishment of mother blocks nursery,
- Indexing of viruses of mother plant in the nursery,
- Initiation of cultures,
- Proliferation of cultures,
- Primary hardening and rouging of undesirable plants,
- Secondary hardening and rouging of undesirable plants,
- Genetic fidelity testing and virus indexing at various stages of micropropagation.

6.8.2 Meristem –tip Culture

This method involves the use of the apical dome or shoot tip with a few leaf primordial of the size less than 1 mm in length as explants.

6.9 Seed Culture

Orchid sexual propagation is practised through seed embryo culture. Orchid seeds are very minute and do not have stored food for seed germination. However, during germination, fungi infect orchid seeds and help convert complex starch to simple sugars, which serve as an energy source. That fungi and orchids have symbiotic relationship in germination. Seedling orchids are grown with the objective of providing seedling plants and to breed new plants. This is an important field in orchid culture, where many hybrids and inter-generic crosses are being bred to exhibit new and different physical characteristics.

An F_1 Hybrid plant is produced from a seed obtained from cross pollination between two different species or two different varieties of the same species. Hybrid seeds can be produced the following ways:

- By hand pollination: This is usually practiced to produce new varieties for testing purposes. (Most new varieties are developed in this way).

- Production in seed plantations: In this method, two different varieties are interplanted or placed in an area and seeds are collected from those plants.
- Chance seedlings: In this case, seed is collected from plants i.e. the wild or elsewhere which are suspected to have cross pollinated with different varieties. The seedlings are grown on to flowering stage and then the best varieties are selected.
- Seed collected from F_1 Hybrid plants: The initial hybrid is produced by cross pollination (i.e: as above). This hybrid is grown until it produces seed, and the seed is then collected. This seed is second generation hybrid seed, and can be called as F_2 seed. New plants are grown as seedlings from this seed.

Orchid seed is not generally bought. It can be bred it or collected and can be sown as soon as possible. As the pods mature they change into yellow colour and then start to show signs of splitting along the placenta. The plants must be inspected carefully. At this stage, and not before, the seeds are harvested into a polythene or paper bag. Off one plant may have as many as a million or so viable seed, or as few as a dozen. (e.g.: one pod of a *Cymbidium tracyanum* contains nearly 3 million seeds).

Seeds should be shaken in chlorinated water (usually 1 in 20 with water) for 10 minutes before planting to kill any disease. Seeds are sown using droppers or needles in a sterile environment, into flasks. The flasks are sealed and placed in high humidity and warm environment (20-28°C). It normally takes 4 - 6 months to grow up into seedlings to the point where they can be transplanted. It can take 4 - 10 years for the seedlings to come into blooms.

Under artificial or laboratory conditions, a sterile artificial medium with sugar and other nutrients is necessary. Through research, an excellent medium for growing seeds without fungi can be developed: inside the bottle where orchid seedlings are grown is a miniature glasshouse, which protects seedlings from unfavorable environmental conditions. Using artificial media it has become possible to grow nearly all orchid seeds into mature plants.

6.9.1 Flasking and Reflasking of Protocorms

When orchid seed or embryo is planted in a culture bottle, numerous seedlings germinate in a very limited space with little available food. The first sign of successful germination is found when orchid seeds start to swell and turn green. As growth continues, the embryo becomes bigger and assumes a flattened top shape called 'protocorm'. A small amount of seed sown can produce hundreds of tiny protocorms growing in limited space. At this stage, they are transplanted into a fresh medium and kept for further development and rapid growth.

6.9.2 Composting and Repotting Seedlings

Orchid seedlings become ready for transplanting from culture bottles when roots and leaves are fully developed. Dendrobiums are potted after 4 to 6 months. *Vandas, Phalaenopsis* and *Cattleyas* in 6 to 8 months after reflasking before seedlings are ready for transplanting in pots. Seedlings should be potted only in sterile potting medium and pots to avoid damping-off diseases. Potting medium may consists of sterilized leaf mould, charcoal and chopped tree fern.

After removing seedlings from bottles, all agar is washed off and the seedlings are treated in fungicide suspension. Excess moisture is drained out and seedlings are sorted according to size. Small seedlings are transferred to community pots, while the bigger ones are potted individually in small pots.

6.10 Production of Disease Free Planting Materials Through Micro-Propagation

The necessary steps needed for producing virus, pathogen and insect free planting materials of orchids are as follows:

- Virus diagnosis using visual and molecular detection techniques applied at mother plant selection, 2nd and 3rd subcultures and primary and secondary hardening stages.
- Elimination of viruses using molecular techniques like ELISA, PCR, RT-PCR etc. and other techniques like chemotherapy using anti-viral substances such as acyclic adenosine analogue, thermotherapy using heat treatment either '*in vivo*' or '*in vitro*' and cryo-therapy means storage of samples at ultra low temperature of liquid nitrogen (-196°C).
- Management of other diseases during secondary hardening.

7 Physiology of Temperate and Tropical Orchids

Orchids are bilaterally symmetric (zygomorphic), many resupinate, one petal (labellum) are always highly modified, stamens and carpels are fused, and the seeds are extremely small.

7.1 Leaves

Like most monocots, orchids generally have simple leaves with parallel veins, although some Vanilloideae have a reticulate venation. They may be ovate, lanceolate, or orbiculate and very variable in size. They are normally alternate on the stem, often plicate, and have no stipules. Orchid leaves often have siliceous bodies called stegmata in the vascular bundle sheaths (not present in the Orchidoideae) and are fibrous.

The structure of the leaves corresponds to the specific habitat of the plant. Species that grow on sunny sites or dry areas have thick, leathery leaves and the laminas are covered by a waxy cuticle to retain their necessary water supply. Shade species, on the other hand, have long, thin leaves.

The attractive mottle of the leaves of Lady's Slippers from tropical and subtropical Asia, (*Paphiopedilum*) is caused by uneven distribution of chlorophyll. Also *Phalaenopsis schilleriana* is a pastel pink orchid with leaves spotted dark green and light green. The Jewel Orchid (*Ludisia discolor*) is grown more for its colourful leaves than its fairly inconspicuous white flowers. The number of stomata per unit surface area is always higher in upper leaves on the same stem due to stronger light intensity on the upper leaves. Epiphytes generally have smaller stomata than terrestrial species. Epiphytic orchids are characterized by thick and succulent leaves with thick cell walls, cuticles and small sub-stomatal chamber whereas those of terrestrial species are thin. Usually mature leaves are photosynthetically active. Leaves are sites for reduction of transpiration, water storage organs, retention of rain or condensed water and absorption of water as liquid or vapour.

The hard leathery leaf type of orchids are drought tolerant with very thick cuticles and thick walled epidermis together with extensive lignification offer excellent protection against desiccation. Thick leaves have Crassulacean Acid Metabolism (CAM), a very important adaptation to water stress. All thin orchid leaves show C_3 photosynthesis. Small and narrow leaves are better adapted to exposure than larger ones because they lose heat more efficiently by convection. Leaf hair may help conserve water by increasing the boundary layer thickness of air around the leaf and lengthening the diffusion pathway. Deciduousness occurring in sympodial orchids helps to avoid water stress during the dry season by shedding their leaves and entering a dormancy period. Water may be stored in pseudobulbs or other parts of the plant.

7.2 Stem and Roots

All orchids are perennial herbs and grow according to two patterns.

7.2.1 Monopodial

The stems grow from a single bud, leaves are added from the apex each year and the stem grows longer accordingly e.g., Vanda and Vanilla.

7.2.2 Sympodial

The plant produces a series of adjacent shoots which grow to a certain size, bloom and then stop growing, to be then replaced. The base of the stem of sympodial epiphytes, or in some species essentially the entire stem, may be thickened to form what is called a pseudobulb that contains nutrients and water for drier periods, e.g., *Cymbidium*, *Cattleya*.

7.2.3 Terrestrial orchids

Terrestrial orchids may be rhizomatous or form corms or tubers which contain reserve carbohydrates. Glucomannan is the major constituent of reserve carbohydrate in tubers. The free mannose, glucose, manobiose and maltose formed from the reserve polysaccharides are transformed to sucrose and transported to new tubers and to the newly formed upper part of plant.

In warm and humid climates, many terrestrial orchids do not need pseudobulbs. Epiphytic orchids have modified aerial roots that are sometimes a few meters long. In the older parts of the roots, a modified spongy epidermis called velamen has the function to absorb humidity. It is made of dead cells and can have a silvery-grey, white or brown appearance. The cells of the root epidermis grow at a right angle to the axis of the root to allow them to get a firm grasp on their support. Nutrients mainly come from animal droppings and other organic detritus on their supporting surface.

7.3 Flower

Orchidaceae are popular for their structural variations in their flowers. Some orchids have single flowers but most have a racemose inflorescence, sometimes with a large number of flowers. The flowering stem can be basal (that is produced from the base of the tuber) like in *Cymbidium*, apical (meaning it grows from the apex of the main stem) like in *Cattleya*, or axillary (from the leaf axil) as in *Vanda*.

The orchid flower, like most flowers of monocots, has two whorls of sterile elements. The outer whorl has three sepals and the inner whorl has three petals. The sepals are usually very similar to the petals but may be completely distinct.

The upper medial petal, called the labellum or lip is always modified and enlarged. The inferior ovary or the pedicel usually rotates 180 degrees, so that the labellum, goes on the lower part of the flower, thus becoming suitable to form a platform for pollinators. This characteristic, called resupination occurs primitively in the family and is considered apomorphic. Some orchids have secondarily lost this resupination, e.g. *Zygopetalum* and *Epidendrum secundum.*

A pollinium is a waxy mass of pollen grains held together by the glue-like alkaloid viscin, containing both cellulosic stands and mucopolysaccharides. Each pollinium is connected to a filament which can take the form of a caudicle, like in Dactylorhiza or Habenaria or a stipe, like in Vanda. Caudicles or stipes hold the pollinia to the viscidium, a sticky pad which sticks the pollinia to the body of pollinators. At the upper edge of the stigma of single-anthered orchids, in front of the anther cap, there is the rostellum, a slender extension involved in the complex pollination mechanism.

In Cattleya cut flowers, the respiration rate decreases with age. Tight buds have the highest rates of respiration, which declines after the opening of bud. All young flowers respire at higher rates than the older ones.

7.4 Pollination

Orchids have developed highly specialized pollination systems. Orchid flowers usually remain receptive for very long periods and most orchids deliver pollen in a single mass; each time pollination succeeds thousands of ovules can be fertilized.

Pollinators are often visually attracted by the shape and colours of the labellum. Some orchids mainly or totally rely on self-pollination, especially in colder regions where pollinators are particularly rare. The caudicles may dry up if the flower hasn't been visited by any pollinator and the pollinia then fall directly on the stigma. Otherwise the anther may rotate and then enter the stigma cavity of the flower (as in *Holcoglossum amesianum*).

In some extremely specialized orchids, like the Eurasian genus *Ophrys*, the labellum is adapted to have a colour, shape and odour which attracts male insects via mimicry of a receptive female. Pollination happens as the insect attempts to mate with flowers.

Many neotropical orchids are pollinated by male orchid bees, which visit the flowers to gather volatile chemicals they require to synthesize pheromonal attractants. Each type of orchid places the pollinia on a different body part of a different species of bee, so as to enforce proper cross-pollination.

A marked increase in respiration following pollination has been observed in orchids. Activities of a number of enzymes like catalase, peroxidase, polyphenol oxidase, ascorbic acid oxidase and glycolic acid oxidase increased following pollination. Activity of polyphenol oxidase in orchids is reportedly highest in the columns followed by aerial roots, tips, petals and leaves. A dramatic increase in catalase activity is observed in columns and petals of *Cymbidium lowianum* and *Dendrobium nobile* after pollination. The sharp rise in peroxidase activity in aging orchid flowers is caused by an increase in ethylene production during senescence.

7.5 Fruits and Seeds

The ovary typically develops into a capsule that is dehiscent by 3 or 6 longitudinal slits, while remaining closed at both ends. The ripening of a capsule can take 2 to 18 months.

The seeds are generally almost microscopic and very numerous; in some species over a million per capsule. After ripening they blow off like dust particles or spores. They lack endosperm and must enter a symbiotic relationship with various mycorrhizal basidiomyceteous fungi that provide them the necessary nutrients to germinate, so that all orchid species are mycoheterotrophic during germination and reliant upon fungi to complete their lifecycle.

7.6 Photosynthesis

7.6.1 C_3-Photosynthesis

All thin leaved orchids fix CO_2 via C_3 pathway. The thin leaved orchids have fewer layers of smaller mesophyll cells and a larger number of stomata than thick leaved species. They have high CO_2 points, prominent post-illumination CO_2 outbursts and active glycolic acid activity all of which are characteristics of plants with high photorespiration e.g., *Habenaria platyphylla, Arundina graminifolia, Coelogyne masangeana, Cymbidium cynense, Oncidium spp, Vanda tessellata, Eulophia keithii,* and *Spathoglottis plicata.*

7.6.2 C_4-Photosynthesis:

PEP is the initial C-acceptor and the product is oxaloacetate, which is readily converted to malate or aspartate. The malate is then decarboxylated to yield CO_2 which is refixed by RUBP carboxylase. C4 pathways occurs primarily in plants of tropical origin growing under high light and high temperature conditions e.g, *Arundina graminifolia.*

7.6.3 Crassulacean Acid Metabolism (CAM)

The majority of plants in the Orchid family use Crassulacean Acid Metabolism or CAM photosynthesis to fixate carbon dioxide. In these plants, the carboxylating enzyme for dark fixation is phosphoenolpyruvate carboxylase (PEPCase). PEPCase has a high affinity for the CO_2 molecule. Plants open their stomata during the cooler and more humid night-time hours, permitting the uptake of carbon dioxide with minimum water loss. During the day, they close their stomata and concentrate CO_2 around the enzyme RuBisCO increasing its efficiency, e.g., *Vanilla, Cattleya, Thunia marshiliana, Coelogyne cristata, Laelia spp, Dendrobium, Calanthe vestita, Bulbophyllum, Aerides odoratum, Phalaenopsis, Aranda,* and *Aranthera.*

7.7 CO_2 Enrichment and Orchid Growth

CO_2 enrichment generally causes plants to develop more extensive root systems to exploit additional pockets of water and nutrients and to enhance the activity of bacteria and other organisms that break nutrients out of the soil, which the plants can then exploit.

It is generally accepted that orchids have either C_3 or Crassulacean Acid Metabolism (CAM) mode of photosynthesis, and these are usually associated with thin or thick leaves (Arditti, 1992; Hew and Yong, 1997). In C_3 photosynthesis, the carboxylating enzyme Rubisco has a relatively low affinity for CO_2 molecule and therefore an increase in CO_2 concentration will increase the rate of CO_2 fixation. An increase in CO_2 concentration will also inhibit the rate of photorespiration. The net effect of these two events is an increase in net photosynthesis (Drake *et al.*, 1997; Hew and Yong, 1997).

An orchid leaf will have greater rates of photosynthesis at higher levels of atmospheric CO_2 concentration. This in turn will generate more carbohydrate available for growth and development.

7.7.1 Practical Aspects of CO_2 Enrichment

Carbon dioxide is generally introduced by one of three ways:
- Burning a hydrocarbon such as propane or kerosene.
- Placing containers of dry ice in the greenhouse or growth cabinet/room.
- Using pure carbon dioxide from a pressurized container.

The third option is the preferred one because pure CO_2 contains fewer growth limiting pollutants. For C_3 orchids (thin-leaved orchids like *Oncidium* 'Goldiana', *Spathoglottis plicata*), CO_2 enrichment should commence at sunrise or when photoperiod begins, and refrain during darkness hours. The average CO_2 level that is recommended is 700 to 1500 ppm. For CAM orchids (thick-leaved orchids, like *Dendrobium* and *Phalaenopsis*), CO_2 enrichment should commence at three to four hours before sunset, continue through darkness hours and stop when photoperiod begins.

7.8 Growth Physiology Stages in *Cymbidium* Orchids

The process from flower bud induction in the new growth to blooming can be divided into three stages.
Stage I: Flower Bud Initiation in the New Growth
Stage II: Flower Spike Initiation
Stage III: Spike Elongation and Blooming

7.8.1 Stage I: Flower Bud Initiation in the New Growth

This stage does not require a temperature as low as Stage II and III.

Normally, flowering of a cymbidium is initiated within the new growing pseudobulb. Under suitable growing conditions, where night temperatures are below the plant's required maximum night temperature, their pseudobulbs will be bigger and healthier with larger and thicker leaves.

Commonly, lower temperature at night reduces plant respiration; therefore plants deplete less stored energy. As a result, more energy is accumulated in their storage organs. Eventually, those pseudobulbs have a higher potential to produce better quality flowers with a higher flower count.

If the required night temperatures are not fulfilled, pseudobulbs will tend to produce poorer quality flowers and a lower flower count per stem. In severe cases, they may not produce any blooms at all. It is a general rule that the large-flowered types need greater energy storage in their pseudobulbs than the smaller-flowered types do. In other words, the large-flowered types require lower night temperatures than the smaller-flowered types do.

Traditionally, all commercial large-flowered cymbidiums have been developed from large-flowered species that originated from the foothills of the Himalayan Mountains, starting from northern India, Nepal, Bhutan, northern Burma, and southwestern China and throughout many smaller ranges in Vietnam, Laos and northern Thailand. In these original habitats, their climate is divided into wet and dry cycles.

- The wet season is the monsoon season of Asia. During this season, these areas receive ample rains with higher daytime temperatures. However, the nighttime temperatures drop drastically by 10 to 15°C and become cool. This is a common climatic occurrence at higher elevations. The monsoon season is also the season of active vegetative growth.
- The dry season is affected by the cold air-mass from northern Asia. Temperatures and humidity fall down. The monsoon rains completely stop. This is the time when the large-flowered cymbidium species stop vegetative growth, instead they enter into a reproductive cycle; i.e. the season of blooming.

The Mediterranean climate zones, *cool summer /mild winter*, such as the southwestern coast of the US, southern coast of Europe and South Africa, southern and southwestern Australia and New Zealand are considered ideal for most of the commercial large-flowered hybrids, which came from the Himalayan species. Because of their relatively temperate summers, the summer nighttime temperatures drop markedly. And, there is rarely a freezing period prolonged enough to damage plants or developing flowers. All traditional cymbidium hybrids perform well under these conditions.

For areas of humid subtropical climates, with *hot summer and cool-cold winter*, such as most of the eastern coasts of various continents between latitude 25-40; e.g. southeastern & southern USA, southern & eastern China, most of Japan, the eastern coast of Australia, southeastern Brazil to northern Argentina, northern India and northern Vietnam, there is often a prolonged period of hot & humid summer weather, with high night temperatures. These conditions cause stress to conventional cymbidiums, and result in lower energy storage in their pseudobulbs.

Such stress has a direct negative effect on new pseudobulb growth during the summer. Those affected pseudobulbs may end up smaller than their genetic potential. As a result, those stunted pseudobulbs will produce inferior quality flowers and usually fewer spikes. In the worst scenario, those sub-standard pseudobulbs may not produce any flowers at all.

Nowadays, growers can overcome these kinds of problems by growing varieties with a mixed background of tropical lowland species, namely Heat Tolerant Cymbidiums (HTC) and Warmth Tolerant Cymbidiums (WTC). Both HTC and WTC do not require night temperatures as low as that of conventional or standard cymbidiums. In climates with a prolonged hot summer, the new growths of HTC and WTC will still initiate flower buds in their growths. Therefore, non-flowering or reduced-flowering growths can be overcome.

7.8.2 Stage II: Flower Spike Initiation

This stage requires maximum night temperature lower than Stage I but higher than Stage III.

After the nearly-mature or fully-mature pseudobulbs with flower buds inside have been exposed to lower night temperatures for a while, spikes appear, emerging as cone-like nubs (similar to new growths but rounder) from the base of pseudobulbs, or within the lowest leaf axils. These enlarging flower buds developing into healthy new growths eventually become the inflorescences.

During this period, growers can manipulate spike maturation and alter how long it will take for them to bloom. In a protected environment where temperature, light, water and fertilizers can be fully controlled, the timing of blooming is partly controllable.

The same varieties, when grown in different conditions, can be made to bloom over an extended period. The group that received lower night temperature earlier will initiate flower spikes before other plants of the same clone. This can give an advantage over nurseries located in warmer places, with nurseries located at higher latitudes or at higher elevations experiencing earlier blooming.

Because both HTC and WTC have temperature trigger points of flower spike initiation higher than those of conventional/standard cymbidiums, they do not have to wait for the night temperatures to drop as low as the conventional/standard cymbidiums require. If HTC and WTC are grown alongside with conventional/standard cymbidiums, HTC will initiate their spikes first, followed by WTC and concluding with the conventional/standard cymbidiums. This indicates that nurseries located in warmer places or at lower elevations that grow HTC and WTC can have blooms no later than those that grow conventional/standard cymbidiums in cooler places or at higher elevations.

7.8.3 Stage III: Spike Elongation and Blooming

This stage is the most critical and requires maximum night temperatures lower than both Stage I and II.

When night temperatures keeps dropping continuously and the days get milder or cooler, the flower spikes elongate. Each flower bud enlarges and finally blooms. Depending on the overall temperature profiles, it normally takes at least 30-60 days from when the new flower spikes reach to the full bloom. In cases where autumn cooling is not stable, and especially if the night temperatures are not low enough, problems in spike development become obvious. As a result, the elongation may actually slow down, the lateral sepals become deformed, colours lose their intensity, pollen does not atain maturity and may darken, or at worst, the whole spike turns yellow and aborts.

Normally, cymbidiums with larger flowers and taller upright spikes require lower temperatures. They are more susceptible to bud drop due to heat stress than the varieties with arching or pendulous spikes and smaller flowers. Simply, the cut-flower varieties with the largest flowers and tallest spikes require cooler conditions during spike maturation than other varieties.

HTC and WTC help in reducing the problem of bud drop if the night chill is not stable as well as at higher temperatures that plants frequently face during transportation and display in the city markets.

7.9 Conclusions & Extensions

Cymbidium nurseries that most quickly complete the three stages have the blooms first and do not encounter problems with barren pseudobulbs.

At every stage, the cool-growing/conventional cymbidiums require lower temperatures than do HTC and WTC.

However, all cool-growing/conventional cymbidium hybrids do not require any temperature as low as 10°C at any growth stage. In reality, many cymbidium nurseries, especially those that produce cut flowers, are located in climates where external winter temperature drop to lower than 10°C at night throughout the whole of stage III. In addition, almost all cut-flower nurseries are located in a greenhouse or controlled environment because this is necessary for providing warmth and the stable temperatures desirable during stage III.

Colombia and Ecuador are the two countries in equatorial zones that currently successfully grow cut-flower cymbidiums because they have many suitable growing area at higher elevations. Cut-flower varieties are grown at 2600-2800 metres in Ecuador and above 1600 metres in Colombia. In these equatorial regions, the crops are not seasonal but grown round the year.

There are also enthusiasts of orchids and some new commercial nurseries growing cut-flower cymbidiums in northern India (Sikkim, elev. 1200-1400 metres) and China (Yunnan, elev. 1900 metres). These two places are located at lower elevations than Ecuador and Colombia because both Sikkim and Yunnan are located further away from the equator. In Sikkim, they experience the cold air mass flowing down from the snow-capped Himalaya, which provides a significant night temperature drop.

7.10 Orchid Pseudobulbs – A Genuine Importance in Orchid Growth and Survival

Most orchids have conspicuous storage organs. Corms, rhizomes, or tuberoids are common in terrestrial orchids while storage organs in epiphytic orchids are enlarged stems called pseudobulbs. Pseudobulbs are also found in some terrestrial orchids like *Cymbidium*, *Eulophia* and *Spathoglottis*.

Orchid pseudobulbs are of two types: heteroblastic or homoblastic. Heteroblastic pseudobulbs consists of only one internode, e.g. *Oncidium*, *Cattleya* and *Miltonia*. Homoblastic pseudobulbs consist of two or more internodes, e.g. *Eria* and *Dendrobium* (Arditti, 1992). A number of aspects have been studied in orchids, e.g. mineral nutrition (Hew and Ng, 1996), respiration (Hew, 1987), photosynthesis (Hew *et al*, 1989, 1996, 1997, Hew and Yong, 1994), flowering (Gow *et al*, 1982), flower physiology (Avadhani *et al*, 1994) and more recently, photo-assimilate partitioning (Yong and Hew, 1995a, Yong and Hew, 1995 b, Yong and Hew, 1995c, Ng and Hew, 1996).

7.10.1 Pseudobulbs – As Water Storage Organs

Orchid pseudobulbs serve as important water storage organs. The epiphytic biotope is characterized by frequent periods of water and nutrient shortage. The presence of fleshy organs in roots, stems or leaves confers epiphytic orchids with the ability to survive and grow in adverse climates. Pseudobulbs of *Oncidium* 'Goldiana' maintain relatively high water contents of 90-95% throughout development. In *Stanhopea* and *Pleione*, pseudobulbs are made up of an abundance of water-storing cells (Arditti, 1992). In addition, most orchid pseudobulbs possess a thick cuticle that is totally impervious to water and gases. In *Cymbidium sinense*, pseudobulbs are able to retain about 64% of their water content after 42 days of water stress conditions (Zengh *et al*, 1992).

7.10.2 Pseudobulbs – As Mineral Storage Organs

Epiphytic orchids face frequent periods of nutrient scarcity. They can tolerate low substrate fertility, being totally dependent on stem flow for nutrients. The low fertility tolerance of orchids is closely associated with the development of the pseudobulb. Tissue analyses of *Laeliocattleya* Culminant have shown that there is a net accumulation of nitrogen and phosphorus with age. In contrast, potassium content decreases with age, indicating that potassium is remobilised to support the growth requirements of new developing tissues (Davidson, 1960).

In *Oncidium* 'Goldiana', the highest uptake of nitrate is reported during the formation of new pseudobulbs. In addition, it is observed that mineral allocation to pseudobulbs within connected shoots of *Oncidium* 'Goldiana' is most active during formation and development of a new pseudobulb (Hew and Ng, 1996). There are remarkable reductions in the mineral content of mature pseudobulbs of connected shoots during the development of a new shoot. The remobilisation of stored mineral nutrients from older pseudobulbs coupled with the high rates of nutrient uptake is indicative of the demand for mineral nutrients by developing pseudobulbs. As such, it is important to keep connected back shoots intact during the propagation of sympodial orchids. The active accumulation of mineral nutrients during the period of pseudobulb development constitutes an important source of reserve for the subsequent development of the inflorescence and new shoots.

7.10.3 Pseudobulb Photosynthesis

Photosynthesis is the process by which carbon dioxide from the atmosphere is fixed into sugars in green plant organs. Leaves are the main photosynthetic organs in most plants. In addition to leaves, several other non-foliar organs of orchids possess chlorophyll and are capable of fixing carbon dioxide. Experimental evidences suggests that non-foliar green organs of orchids do contribute positively to whole plant carbon economy by refixing the carbon which would otherwise be lost through respiration. Most orchid pseudobulbs are impervious to water and gases due to the presence of a thick cuticle. The pseudobulb, a massive organ, therefore represents a substantial cost in terms of carbon for maintenance.

Although impervious to water and gases, pseudobulbs of *Oncidium* Goldiana, nevertheless are capable of photosynthesis. Pseudobulb photosynthesis in *Oncidium* functions, essentially, for the refixation of respiratory carbon produced by the underlying massive parenchyma (Hew and Yong, 1994). Enzymes within the tissue of the pseudobulb for carbon fixation are ribulose-1,5-bisphosphate carboxylase/oxygenase and phosphoenolpyruvate carboxylase. While most orchids are impervious to the external environment, gas exchange with the ambient atmosphere is mediated by a cavity rich in stomata on top of the pseudobulb in *Bulbophyllum minustissimum*. This is especially important for those orchids with rudimentary leaves (Winter *et al*, 1983).

In the CAM orchid, *Laelia anceps*, photosynthesis of leaves is largely affected by irradiance of the pseudobulb (Ando and Ogawa, 1987). Exposure of the pseudobulb to light is necessary for leaves to conduct daily gas exchange with the atmosphere. It has been proposed that the organic acid fixed during the night is transported to the pseudobulb and decarboxylated the next day, and that the transport of organic acid is enhanced by exposure of the pseudobulb to light. It appears that the pseudobulb can regulate the capacity for CAM in leaves of *Laelia anceps*, although evidence in CAM orchids for the basipetal transport of organic acids from leaves to pseudobulb is lacking. Presently, it is still not clear whether pseudobulbs of C_3 and CAM orchids have a regulatory role in leaf photosynthesis.

7.10.4 Pseudobulbs – as Carbohydrate Storage Organs

The ability of orchid pseudobulbs to photosynthesise points to their importance as carbon sources for the plant. Studies on both *Catasetum viridiflavum* (Zimmerman, 1990) and *Oncidium* 'Goldiana' (Hew and Ng, 1996) have shown that carbohydrate reserves in orchid pseudobulbs are important in the initiation of new growth. The pseudobulb of *Oncidium* accumulates massive amounts of carbohydrates during vegetative development. These carbohydrate reserves are subsequently remobilised to support new shoot and inflorescence development.

Storage carbohydrate of the pseudobulb is derived mainly from the import of currently assimilated carbon from the leaves (Yong and Hew, 1995a) and in part from its own regenerative photosynthesis (Hew and Yong, 1994). The carbohydrate reserves of connected back shoots also contribute to new shoot and inflorescence development (Yong and Hew, 1995c, Hew and Ng, 1996). While leaves are the main sources of currently assimilated carbon, pseudobulbs represent an important supplementary source of carbohydrate that is utilized to meet the increased demand for carbon during inflorescence and new shoot development. This observation explains the need for at least two connected back shoots for optimal inflorescence development (Yong and Hew, 1995b, 1995c).

7.10.5 The Absence of a 'Flag' Leaf' – an Apparent Anomaly Due to the Pseudobulb

A 'flag' leaf is the main leaf responsible for supplying carbon to the organ of economic importance. This is usually the leaf subtending the economically important organ. Based on gas exchange studies on *Oncidium* 'Goldiana', Hew and Yong(1994), showed that the rate of carbon dioxide uptake for the leaf subtending the inflorescence increased 1.4 fold during inflorescence development while the rate of carbon dioxide uptake for other mature leaves remained unchanged. This indicates that the leaf subtending the inflorescence is the 'flag' leaf. However, radioactive carbon tracer studies have shown an absence of a 'flag' leaf in *Oncidium* 'Goldiana'. All mature leaves within a single shoot supply similar amounts of carbon to the inflorescence (Yong and Hew, 1996). This apparent anomaly between gas exchange studies and radioactive tracer studies is interesting.

Radioactive tracer studies have shown that carbon produced in the leaves is transported to the pseudobulb in the first instance (Yong and Hew, 1995a) before being transported to the inflorescence. Tissue analyses of pseudobulb carbohydrate content showed that there is no net accumulation of carbohydrate during inflorescence development (Hew and Ng, 1996). Taken together, these results indicate that there is substantial mobilisation of carbohydrate to the inflorescence via the pseudobulb. It is likely that there is mixing of different carbohydrate pools (currently assimilated carbon from leaves with storage carbohydrate within the pseudobulb) during the transport of carbon from leaves to the inflorescence.

The pseudobulb is envisaged as central to the distribution of carbon within a single shoot of *Oncidium* Goldiana. Although the leaves are main sources that supply carbon for inflorescence development, the pseudobulb is responsible for the ultimate re-distribution of assimilated carbon from the leaves. This could account for the apparent absence of a 'flag' leaf in *Oncidium* based on radioactive tracer studies. Further research works needs to be done to substantiate the possible regulatory role of pseudobulbs in partitioning of assimilates in orchids.

7.10.6 Pseudobulbs and Myrmecophily

Ants are in frequent contact with epiphytic plants. Association between ants and orchids can be broadly classified into two categories: (1) ant-house and (2) ant-garden (Davidson and Epstein, 1989).

Ant-house orchids are characterized by the presence of a permanent dormatia in which ants take up residence while ant-gardens are nests of earthen material (called 'carton') constructed by ants on which the epiphyte grows (Beattie, 1985). Species classified as ant-house orchids include *Caulathron*, *Dimeranda* and *Schomburgkia* while *Vanilla planifolia* has been reported to be an ant-garden orchid. There is evidence to suggest that in both ant-house and ant-garden epiphytes the ant-epiphyte association is mutualistic. In the CAM orchid *Schomburgkia humboldtiana*, leaves grow from a large hollow pseudobulb and contain ant-nests (Griffiths *et al*, 1989). However, it appears that the hollow pseudobulb of *Schomburgkia humboldtiana* forms spontaneously without excavation by ants. This is likely the result of co-evolution, although the actual relationship remains to be unequivocally determined.

The occurrence of an ant-house in pseudobulbs of *Schomburgkia* provides an interesting material for the physiological role of the pseudobulb. It is unlikely that pseudobulbs of *Schomburgkia* are important in water storage like other orchids. However, it is possible for the pseudobulb of *Schomburgkia* to contribute to whole plant mineral and carbon economy through its association with ants. It is possible that the provision of an ant-house in the hollow pseudobulb constitutes an additional food source in the form of ant faeces and refuse. In addition, the fixation of respiratory carbon from ants may have a positive contribution to whole plant carbon economy.

Cattleya is a epiphytic plant generally found growing on trees of moist and wet forests from sea level to 4,900 feet (1,500 m) in elevation. Several published scientific studies have shown that flowering of *Cattleya* species and hybrids is promoted by exposure to short daylengths and cool temperatures. For example, in *Cattleya warscewiczii, Cattleya gaskelliana* and *Cattleya mossiae,* flower induction occurred only when plants were placed under photoperiods of nine hours (nine hours of light per day) at 13°C, while flowering was inhibited under 16 hours of light per day at 55°F (Rotor, 1959).

Dendrobium is one of the largest genera and is native to tropical and subtropical Asia, Australia and various Pacific Islands. The optimum temperature for flower induction consequently differs among *Dendrobium* selections due to its wide geographical distribution. In *Dendrobium nobile,* plants exposed to a constant 13°C produced flowers regardless of the daylength, whereas plants placed at 18°C remained vegetative and did not flower (Goh and Arditti, 1985). In contrast, *Dendrobium phalaenopsis* requires short daylengths and warmer temperatures for flowering. For example, flower-bud development and flowering of plants placed under nine-hour day lengths at 18°C are accelerated by six weeks compared to plants placed under longer day lengths at the same temperature. A similar response is observed at 13°C, but flower bud development is slower due to the cooler temperature.

Most *Phalaenopsis* species and hybrids require a period of exposure to relatively cool temperatures less than 28°C to trigger the elongation of the spike (Lee and Lin, 1984, Wang, 1995). Uniform spiking can be obtained when plants are grown at day/night temperatures of either 25/20°C or (20/15°C) for four to five weeks. When induced plants are placed at high temperatures (greater than 28°C), a spike can form a vegetative air plantlet (known as a keiki) instead of flower buds, or buds may abort. A few experimental studies have reported that short days enhance spiking and long days promote vegetative growth or the development of keikis in *Phalaenopsis* (Rotor, 1952; Griesbach, 1985). However, this short-day enhancement is thought to be a result of the extension of cool-night temperatures and not the daylength itself. Thus, it appears that photoperiod does not influence flowering of *Phalaenopsis* (Baker and Baker, 1991).

What the Future Holds

In recent years, orchids have become the second most valuable potted flowering plant in the United States, with a wholesale value of US$ 127 million in 2004. More than 12.7 million orchids were sold in the United States last year, with *Phalaenopsis* accounting for more than 75 percent of sales. Why are so many *Phalaenopsis* being sold and purchased when there are well over 25,000 described species of orchids from which to choose? One reason is that we understand how to regulate the flowering process. As mentioned earlier, growers can prevent flowering by maintaining the day and night temperatures above 82°F (28°C). To induce flowering, plants need to be grown at cooler temperatures. Unfortunately, there is virtually no adequate information available on the flowering of many other orchids, such as *Miltonia, Oncidium, Vanda* and *Zygopetalum*. As a result, growers cannot reliably flower an orchid such as *Zygopetalum* for a holiday such as Valentine's Day or Mother's Day, which is when consumer demand is greatest. We do not know if we can manipulate temperature, light or perhaps some other factor to control flowering. Without this information, growers are not able to produce a flowering crop when demand – and likely profit – is greatest.

8 Climate Change and its Impact on Orchid Productivity

The global population is expected to rise by about 9 billion people in 2050. Food, shelter, energy, employment, and health care etc. have to depend upon the bio-resources of the Planet Earth. This uncertainty is due to climate change, agriculture pattern, urbanization etc. which are closely linked with biodiversity, economy and the welfare of humanity. In 2050, 1.6 billion people of our country will need adequate, nutritious, safe and healthy food and adequate fiber within the available natural resources.
Some of the major challenges that we have come across are:
- Climate change due to rise in temperature, erratic precipitation and sea level rise
- Greenhouse gas emission
- Fragmented ecosystems
- Loss in biodiversity
- Trade and competitiveness

Before the early 1800's, the atmosphere contained about 290 ppm CO_2, in 1995 it was 360 ppm and according to the World Health Organization CO_2 reached 379 ppm in 2006. It is estimated that CO_2 will increase to 800 ppm by the end of this century and bring serious consequences to plants. A 1°C increase in temperature may reduce yields of major food crops by 3-7%. It is predicted that greater losses will occur at prolonged higher temperatures. A projected loss of 10-40% is expected in crop production by 2100.

8.1 Impact of Climate Change (Mitchell and Tanner, 2006)

8.1.1 Agriculture

- Shifts in food growing areas
- Changes in crop yields
- Increased irrigation demands
- Increased crop pests and diseases in warmer areas

8.1.2 Water Resources

- Changes in water supply
- Decrease in water quality
- Increased drought
- Increased flooding

8.1.3 Forests

- Changes in forest compositions and locations
- Disappearance of some forests
- Increased fires from drying of forest trees and grasses
- Loss of wild habitat and species

8.1.4 Biodiversity

- Extinction of some animal and plant species
- Loss of habitats
- Disruption of aquatic life

8.1.5 Weather Extremes

- Prolonged heat waves and droughts
- Increased flooding
- More intense hurricanes, typhoons, tornadoes and violent storms

8.1.6 Sea Levels and Coastal Areas

- Rising sea levels
- Flooding of low lying islands and coastal cities
- Flooding of coastal estuaries, wetlands and coral reefs
- Beach erosion
- Disruption of coastal fisheries
- Contamination of coastal aquifers with salt water

8.1.7 Human Population

- Increased deaths
- More environmental refugees
- Increased migration

8.1.8 Human Health

- Increased deaths from heat and epidemic diseases
- Disruption of food and water supplies

- Spread of tropical diseases to temperate areas
- Increased water pollution from coastal flooding

8.2 Actions

Community-centred development projects can be incorporated for adaptation to climate change through the following activities:

- Monitor species and the amount of vegetation to assess the impacts of climate change. Asking community members to report on invasive species and changes in growing patterns has been shown to effectively promote climate change awareness.
- Seasonal climate change projections can be reviewed during workshops with farmers, and decisions can be taken based on responses.
- Keep an updated climate change scenario on file and refer to this at each stage to make sure activities are not increasing vulnerability to climate change. The scenarios can also be used as an advocacy tool.
- Participatory appraisal techniques can be used to assess the impacts of climate variability and change in livelihoods and production. Simple cost/benefit analysis of different adaptation options can also be included.
- Fire management techniques and training need to be considered.
- Early warning systems are especially relevant for agriculture. Regional and local seasonal predictions are currently in use and being developed using weather forecasting tools.
- Developmental Agencies need to raise awareness among partners and make their programmes more resilient to climate change impacts.
- Development NGOs will need to plan their adaptation activities carefully to ensure consistent poverty reduction policies, plans and program. This may require strategic assessment of the role of adaptation in their program, as well as supporting mainstreaming in developing countries.
- NGO can link climate change to other related areas such as disaster risk management and sustainable development.
- Vulnerable groups must work on climate change adaptation from the starting point of current variability. Integrating the impacts of future changes into vulnerability reduction remains a significant challenge at the policy and strategic levels, as well as in communities and households.
- To develop a GHG Inventory Management System (GHG-IMS) through Black Carbon Research Initiative-National Carbonaceous Aerosols Programme (BCRI-NCAP), Long-term Ecological Research Observatory (LTERO) for climate change, Co-ordinated Studies (CS) in the North-Eastern Region on Climate Change (NECC) (CS-NECC) and Climate Change Assessment Studies (CCAS).
- In order to ensure sustainable management of biodiversity vis-a-vis climate change, adaptation is a key aspect in terms of chalking out a future strategy such as: to identify and conserve biodiversity components that are especially sensit-

ive to climate change; preserve intact habitats so as to facilitate the long-term adaptation of biodiversity; improve our understanding of the climate change – biodiversity linkages, and fully integrate biodiversity considerations into climate change mitigation and adaptation plans.
- India needs to undertake the following strategies at various scales (global, regional, national and local): increased intensity to promote sustainable management, conservation and enhancement of sinks and reservoir; take climate change consideration into account to minimizing adverse effects on the economy and on the quality of environment; promote and cooperate with scientific research; education training and public awareness; and the exchange of scientific information.

8.3 Challenges

8.3.1 Biodiversity and Conservation

Nearly 12.5% of the global vascular flora are facing extinction and therefore, conservation of rare and threatened plants are of international importance. Two thirds of orchid species are epiphytes and lithophytes, with terrestrial species comprising the remaining third (almost half) of the extinct species as per IUCN, 1999. In India, 250 species of orchids are under the threat in various categories.

- Conservation of biota in fragmented landscapes, protecting and increasing the habitat, improving habitat quality, increasing connectivity, managing disturbance processes in the wider landscape, planning for the long term, and learning from conservation actions undertaken.
- To maintain the tropical biodiversity, there is no substitute for primary forests, there is a need to increase the forest area under protected area network.
- The value and importance of indigenous peoples' and local communities' customary sustainable use and traditional knowledge in conserving and upholding biodiversity, land- and seascapes, and protected areas should be acknowledged. Incentives may be needed to entice people to participate in conservation and recovery programs.
- Implementation of community-based projects on biodiversity conservation provides opportunities to actively engage and involve local and indigenous people.
- There is an urgent need to develop the Biodiversity Profile of India so that we have adequate knowledge on existing species, ecosystems and genetic resources and threats to them in order to monitor and report on biodiversity (e.g., extinction rates, biodiversity loss). The main causes for a lack of knowledge on biodiversity loss include limited number of scientific experts, national indicators, research, finance and available technology and lack of biodiversity specific educational programs.
- More biosphere reserves, sanctuaries and germplasm banks need to be established.

- Promoting education and awareness about plant diversity conservation and sustainable utilization and biodiversity conservation at the local level are to be encouraged.
- An integrated orchid conservation approach including conservation genetics, mycorrhizal associations, pollinators interactions, *in situ* conservations (Biosphere Reserves, National Parks, Sacred Grooves, Gene Sanctuary and Individual Trees) and *ex situ* conservations (Field Gene Banks, Botanical Garden, Herbal Garden, *In-vitro*-conservation, Cryo-preservation and DNA Bank) will be taken up.

8.3.2 Genetic Improvement

- Genera and species wise cataloguing of all germplasm collection using IPGRI descriptors is required.
- The rich diversity of orchids in the country requires a well thought out Network Approach mode. The NRC for Orchids have to work on a network mode and also as a National Active Germplasms site with the various active centres working on specific groups of orchids. In view of the IPR regulations, it is of paramount importance to protect our germplasms using modern tools of bar coding. A network project involving groups with identical interest between universities and ICAR. These germplasm should be conserved with the duplicate sets grown in at least two locations, properly catalogued and characterized with national number obtained from NBPGR avoiding duplication. Cryopreservation to conserve germplasm can be taken up in collaboration with NBPGR.
- At present, orchid trade is solely based on the hybrids derived from varietals, and interspecies and inter-generic crosses. Building up a strong crop improvement programme based on sound breeding methodologies will yield the development of hybrids/varieties of internationally acceptable quality traits. It is essential to develop our own hybrids suitable for varied agroclimate for our country to fulfill the basic requirements of market demands.
- Evaluation of newly evolved genotypes to suit specific agro-ecological conditions is essential.
- Locating sources of resistance for biotic an abiotic stresses using conventional and biotechnological tools and developing varieties with high yield, quality and specific traits are suggested.

8.3.3 Frontier Science Technologies In Orchid Research and Development

- It is essential to use the available hybrids and segregating populations to develop Association mapping. Hence the facilities available at IIHR and NRCO may be used to develop genome assistant or marker assistant selections.
- The lead obtained in GIS with the help of facilities of ISSR for *Cymbidium* to cover other species which aid in location specific as well as species specific surveys will be effective.

- Characterization of rhizosphere microbial community structure and effect of engineered nanoparticles on microorganisms in the rhizosphere and phyllosphere.
- Commercialization of orchids through bioreactors covering micropropagation technology to industry in network mode.

8.3.4 Management of Natural Resources

- Cost effective agro-climatic management through optimization of a number of factors like light, temperature, humidity, water, air, growing media and nutrition for quality flower production. The standardization of growing media using cheap and indigenous materials such as leaf ferns, leaf moulds, green moss etc. may be explored and this must be used in combination with cropping systems to develop an orchid based farming systems.
- Development and popularization of cost effective agricultural practices (INM/IPM) for increasing productivity.
- Quantification of water use efficiency and water requirements in orchids based on growth habit.
- Carbon sequestration potential in orchid based cropping systems.

8.3.5 Post Harvest and Value Addition

- Development of location specific complete protocols starting from pre-harvest, harvesting, post-harvest techniques up to domestic and international markets for each genus of commercial orchids.
- Developing a comprehensive approach on value added products from wild orchids including species trade, drying, flower arrangements, herbal medicines, edible products and other aesthetic and aromatic products.
- Bio-prospecting using bioinformatics tools

8.3.6 Bio-Risk Management

- Surveillance, identification and characterization of new invasive insects pests and pathogens
- Pest-risk analysis
- Development of rapid and reliable diagnostics kits against pests and pathogens including invasive species
- Management alert and control of new invasive insect pests and pathogens

8.3.7 Policies

- Commercialization of the new upgraded technologies
- Genetic finger printing of rare, endangered and threatened species and their registration
- Finger printing and registration of newly released varieties or hybrids
- Patenting technologies related to orchids
- Confirmation and Documentation of ITK's

8.3.8 Transfer of Technology

- Constraint analysis and impact assessment of new technologies
- Production of quality planting materials, distribution and commercialization
- Large scale demonstration of proven technologies through training and FLD's
- Establishing agro-technology information centre like ITMU, AKMU
- Participatory planting material production of commercial orchids

8.4 Impact Assessment

Climate change due to global warming interacts with habitat loss and fragmentation, introduced and invasive species and population growths. This means that many ecosystems are likely to undergo severe modification. In Asia, climate change is projected to compound the pressures on natural resources and the environment associated with rapid urbanization, industrialization and economic development. Semi-arid vegetation will tend to be replaced by arid vegetation. Levels of precipitation are likely to change radically in many areas of the world. Increasing temperature may result in vegetational zones gradually moving vertically up mountain sides, both permitting tropical species to subtropical areas, subtropical species to temperate areas and eliminating the species in the highest zones (Liu *et al*, 2010).

Epiphytic orchids may be affected in various ways by changes in the availability of light, nutrients and moisture. Climate change is a major threat to pollination services and there is a need to conserve plant communities in which orchids live. The combination of higher temperatures and lower rainfall may make forests more susceptible to fire and it may lead to extinction of local species. During 1984, at the World Orchid Conference held in Miami, it was proposed that the orchid community should start banking orchid seed as an insurance against possible losses of species from their habitats in the wild. The majority of orchid species are capable of tolerating dry storage for many decades when stored at -20°C. Liquid nitrogen storage may further extend the life spans of orchid seeds. Living collections are recently underutilized as a conservation tool and there is a need to do more to induce members of the wider orchid community. A number of strategies can be followed for conservation, multiplication, production, improvement and protection of valuable orchid species and varieties (Tab. 8.1)

Tab. 8.1: Strategies for conservation, multiplication, production, improvement and protection of orchids

Goal	Approach	Performance Measure
Conservation of genetic resources	Integrated orchid conservation approach including GIS survey and remote sensing	• Conservation genetics with molecular methods and phylogenetic studies • Epiphytic (70%) and terrestrial (20%) mycorrhizal associations with analysis of DNA sequences • All Pollinators interactions with population genetics and phylogenetic analysis of orchids and pollinators • *In-situ* conservations (Biosphere Reserves, National Parks, Sacred Grooves, Gene Sanctuary and Individual Trees) of all available species) • *Ex- situ* conservations (Field gene banks, Botanical garden, Herbal Garden, *In vitro*-conservation, Cryo-preservation and DNA conservation) of more than 100 species
Evaluation, valuation and improvement of genetic resources effectively to meet the challenges of biotic and biotic stresses to sustain the impact of climate change in addition to quality	Character specific collection of exotic and indigenous germplasm, locating resistance source and evolving high yielding and disease resistant lines through selection, mutation, inter-generic and inter-specific crosses, polyploidy breeding and biotechnological tools for orchid improvement	• Specific collection of 850 indigenous germplasm from NEH region, 288 species of North Western Himalayas, 379 species of Peninsular India and 115 species of Andaman and Nicobar Islands. Exotic germplasm from Thailand, Britain, Singapore, Australia, Hawaii andNew Zealand will be attempted to enrich basic genetic materials. • Genera and species wise cataloguing of all 1350 germplasm of India collections using IPGRI descriptors. • Identification of genes contributing resistance to biotic and abiotic stress factors and quality characteristics of major 10 commercial orchid genera. • Improvement of *Cymbidium* to develop hybrids or varieties with less pre-blooming period and resistance to viruses. • Varietal evaluation of newly developed genotypes of commercial orchid genera to suit specific agro-ecological conditions. • Development of highly adaptive and tolerant genotypes to mitigate climate change and water stress • Genetic engineering and transformation for early flowering and extending shelf-life in commercial orchids • Development of Bar codes for germplasm identification at species level.
Increasing productivity of orchids through quality planting materials production and protected cultivation	Targeting the production levels by propagating and distributing quality planting materials of improved hybrids for effective spread	• Production of nucleus planting materials through mass multiplication using standardized protocols of commercial orchids at least one lakh per annum. • Standardization of protocols for mass multiplication of endangered, rare and threatened orchid species • Participatory quality planting material production of commercial orchids in collaboration with line departments • Developing technologies suitable for protected cultivation of disease free planting materials

continued **Tab. 8.1:** Strategies for conservation, multiplication, production, improvement and protection of orchids

Goal	Approach	Performance Measure
Developing efficient system for management of climatic factors and nutrients to get maximum production and developing an effective model	Generation of eco-region specific technologies based on maximum productivity of available natural resources	• Cost effective nutrient and agro-climatic management through optimization for quality flower production • Development and popularization of cost effective agricultural practices (INM/IPM) of 10 commercial orchid genera for increasing productivity • Quantification of water use efficiency and water requirements in orchids • Carbon sequestration potential in orchid based cropping systems • Development of at least two to five profitable orchid based cropping systems with other high value flower crops. • Reducing pre-blooming period of *Cymbidium* orchids from five to two years using effective management practices for higher income.
Management of biotic stress to reduce production losses by developing diagnostics, forecasting models, bio-control systems and effective management strategies	Identification of new and effective bio-molecules for management of biotic stresses coupled with development of innovative diagnostic techniques for rapid, accurate and cost effective detection of high impact pests and diseases.	• Surveillance, identification and characterization of new invasive pests and pathogens, pest risk analysis • Characterization of rhizosphere microbial community structure and effect of engineered nanoparticles on microorganisms in the rhizosphere and non-rhizosphere • Development of rapid and reliable diagnostics against pests and pathogens including invasive species • Management of new invasive insect pests and pathogens
Post-harvest and value addition	Development of post-harvest technologies to improve product quality and value addition, chemo profiling and identification of new flavour / bioactive principles.	• Development of pre-harvest, harvest and post-harvest technologies of major commercially grown orchids for specific target groups like domestic and export market and hybrid/variety specific technologies. • Development of packaging for marketing of commercially important fresh and dried produce using locally available materials. • Development of orchid based technologies for dry flowers and floral arrangement. • Use of orchid waste for production of phytochemicals including pigments, food, feed, herbal medicines and essential oils. • Patenting technologies related to species
Effective Transfer of Technology to the target groups	Participatory approach for effective transfer of technologies to empower stakeholders, analyzing feedback for further refinement	• Identification of various clusters of production, selection of beneficiaries and imparting training and technical inputs in order to produce orchids of international standards • Large scale demonstration of proven technologies through FLD's and technology dissemination using advanced tools • Upgrading of Knowledge, Know How techniques, Managerial Skills and Self employment among extension functionaries, farmers, school drop outs, young man and women generations by organizing Kisan Mela, exhibition, Brainstorming session, need based and focused Training Programme, Demonstration, Model training Courses etc. • Inter–institutional collaboration to facilitate popularization of effective technologies • Commercialization of technologies and patenting • Documentation of ITK's

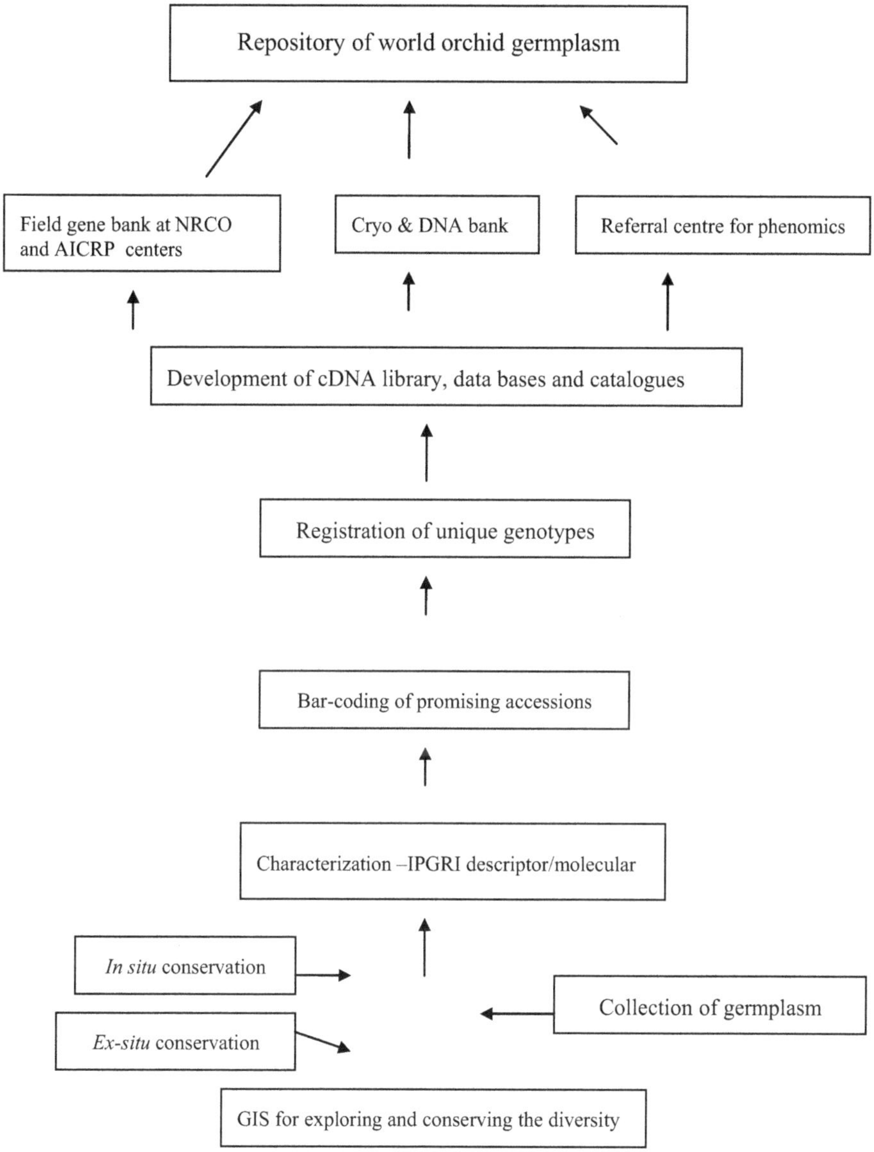

Fig. 8.1: Conservation of genetic resources

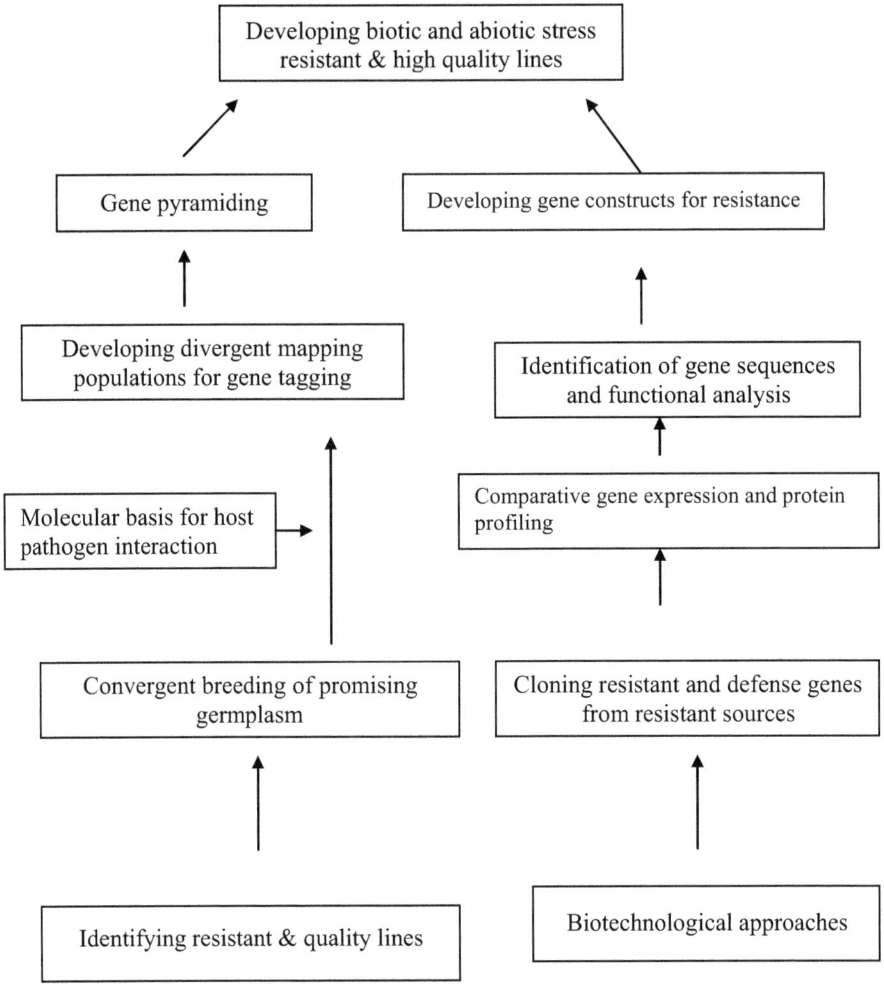

Fig. 8.2: Varietal development with resistance to biotic/abiotic stress

9 Production Technology of Commercial Epiphytic Orchids for Cut flowers

The term "Cut flowers" refers to blossoms or flower buds that are cut with branches, leaves and stems. Fresh cut flowers are highly perishable because they maintain only limited life supporting processes by absorbing water through their stems. They are used for bouquet preparation, decorations, vase arrangements; designs for weddings and funerals; gifts on occasions; informal display to beautify home and public places. Cut flowers are the most important floricultural products in the international flower trade. The most traded flower crops include roses, carnations, chrysanthemums, dahlia, gerbera, gladiolus, gypsophila, liatris, nerine, orchids, poinsettias, achillea, anthurium, tulips and lilies. These flowers are cultivated in open fields or protected structure.

Epiphytes grow on trees or other objects above ground level, but they do not act as parasites. Plants living epiphytically in a tropical rain forest face better light conditions than plants living on the ground level. On top of that, epiphytes are more easily pollinated by flying pollinators. The roots of an epiphyte are developed for a life above soil. For example, they need high air humidity since they often are forced to take most of their water from the air. Together with epiphytic ferns and bromelias, orchids grow and flower on tropical trees. Many orchids growing 'with' a tree have adapted very precisely to the tree's 'behaviour'. For example, when the tree drops its leaves to a period of rest, the orchid does the same thing.

9.1 Cymbidium (Orchidaceae)

9.1.1 Introduction

Cymbidiums are among the most popular winter and spring blooming semi-terrestrial orchids and originated from tropical and subtropical Asia. This includes North Eastern India, China, Japan, Malayasia, the Philippines, Borneo islands and North Australia, where it usually grows in cooler climates at high elevations. The important Cymbidium growing countries in the world are Australia, New Zealand, Japan, the Netherlands, the USA and England. Cymbidium consists of 70 semi-terrestrial and epiphytic orchid species of tropical and subtropical Asia.

9.1.2 Botanical Description

The plants are characterized by short and stout pseudobulbs ensheathed by encircling leaf bases. Leaves are long, ribbon shaped, leathery or soft and lanceolate. The flower spikes develop from the base of the pseudobulbs.

A plant has three types of bulbs. 1. Old Backbulbs without leaves, which act as a reserve food supply for emergencies. It is advisable to leave one of these on each divided plant. Back bulbs can make new plants but they may take years to flower. 2. Old bulbs with leaves, which support the new growth and may produce flowers for a number of years depending on the variety. When dividing, the plant must retain at least two old bulbs or have one back bulb attached to be able to reflower next year. 3. New leads or bulbs, which are the youngest bulbs on the plant that the flowers and most new growth come from. When dividing, at least 1 old bulb and one back bulb must be retained with this bulb to ensure that the plant may flower the following year.

The spikes are erect, arching or pendulous and arranged with 2 to 15 flowers. The individual florets are 1 cm to 12.5 cm across and are of various colours and shades. Cymbidiums are famous for their beautiful spikes derived from species and hybrids. Among the orchids, Cymbidium ranks first and in floricultural crops it accounts for 2.7% of the total cut flower production.

9.1.3 Classification

Cymbidium hybrids are classified into three groups-Standard, Intermediate and Miniature hybrids. Standard and Intermediate hybrids produce 90 to 120 cm long spikes with 8 to 15 flowers per spike. Miniature hybrids produce green, yellow or brown coloured flowers, 30 cm tall and each spike contains 30-40 flowers of 2.5 to 8.5 cm across. Novelty or Intermediate hybrids have been evolved by crosses between Miniature and Standard hybrids.

9.1.4 World Scenario

Cymbidium has been considered the top commercial orchids in Europe for many years. They fetch the highest price in the international markets, with the Asian countries of Singapore and Japan and the Netherlands being the major markets. Cymbidiums imported from the Netherlands fetched as much as US$ 11.18 per stem in Singapore, and those imported by Japan from New Zealand fetched US$ 3.33 per stem. As fars as the Dutch Auction market is concerned, the cymbidiums fetched the highest value, averaging Euro cents 331 per stem during 2003-2007 period (Tab. 9.1). East Sikkim has been declared an Agri Export Zone exclusively for production of cymbidium orchids. In Sikkim, more than 250 hybrids of cymbidium orchids are commercially cultivated in an around 25 ha of land and about 5 lakhs spikes are produced annually.

Tab. 9.1: Average Annual Prices at Netherlands Auction (2003-2007) (Euro cents /stem)

Orchids	2003	2005	2007	Average price
Phalaenopsis	38	46	37	40
Cymbidium (Big bud)	330	334	329	331
Cymbidium (Small bud)	138	148	140	142
Paphiopedilum	58	52	63	58

Source: CBI Market Survey, The Cut Flowers and Foliage Market in the EU

9.1.5 Importance and Uses

Cymbidiums are highly valued for their potential as genetic resources, cut flowers, hanging baskets, potted plants and herbal medicines (Tab. 9.2) (De and Medhi, 2012).

Tab. 9.2: Genetic resources for hybridization

Name of species	Uses
Cymbidium iridioides, C. eburneum, C. hookerianum, C. sanderae, C. lowianum, C. tracyanum, C. insigne, C. erythrostylum	Development of standard cymbidium hybrids
Cymbidium ensifolium, C. devonianum, C. tigrinum	Development of miniature cymbidium hybrids
Cymbidium atropurpureum, C. finalaysonianum	Development of cascading cymbidium hybrids

9.1.5.1 Cut flowers

Both standard and novelty hybrids are used as cut flowers.

9.1.5.2 Potted Plants

Species like *C. ensifolium, C. aloifolium, C. devonianum, C. eburneum, C. lancifolium, C. lowianum* and *C. mastersii* are used as potted orchids. Other miniature orchid hybrids that can be used as potted plants are Autumn Beacon Geyserland, One Tree Hill Solstice Gold B/CSA, Wakakusu Delight, Minneken Khobai, Kusada Fantasy Carioca, Phar Lap, Gladys Whitesell, Summer Love.

9.1.5.3 Hanging Baskets

Cascading hybrids are ideal for hanging baskets: e.g., Sarah Jean Ice Cascade Ad/CSA, Nicoles Valentine Geyserland HCC/AOS, Dorothy Stockstill Forgotten Fruit.

9.1.5.4 Herbal Medicines

In India, orchids have been used in the indigenous system of medicine since the Vedic period. In general, orchids through out the world are used for curing rheumatism, malaria, tuberculosis, cuts, wounds, burn injuries, asthma, bronchitis and several other ailments.

Tab. 9.3: Cymbidium orchids used in Indian medicine system

Botanical name	Parts used	Therapeutic uses
Cymbidium aloifolium	Rhizomes	Salep; used as nutrient and demulcent; as emetic and purgative
Cym. ensifolium	Rhizomes & Flowers	Eye sores
Cym. longifolium	Pseudobulb	As emetic and demulcent
Cym. giganteum	Leaf juice	Blood clotting

9.1.5.5 Stylish Packaging

These are specially designed to add value to the cut flowers.
Ciliandra: A gift of glass flute containing miniature cymbidium.
Oriental diamond: A fine packaging method for cut cymbidium is designed to add value to delicate flowers and used for special occasions.
Aura of Luxury: A new packaging method for cymbidium for luxury and style.

9.1.6 Varieties (De, 2014)

White: Jungfrau 'Snow Queen', Jungfrau 'Dos Pueblos', Camalex, Showgirl 'Cooksbridge', Showgirl 'Marion Miller', Swallow var. 'Takarazuki'
Pink: Lilian Stewart 'Coronation', Lilian Stewart 'Party Dress', Orkney 'Pink Heather', Ensikhan, 'Alpha Orient', Pacific Rose 'Swansea', Soulhunt Series, Valley Paradise 'Shangriila', Rievaulx 'Cooksbridge', Rincon Fairy 'Pink Perfection'.
Yellow: Angelica 'December Gold', Highland Sunset 'Plumpton', Mini Sarah 'Artisan', Hawtescens, Gwen Sherman, 'Arthur Fetzer', San Francisco 'Mona Lisa', Valya Craig 'Sutherland', Luana 'Imperial', 'Pine Clash Moon Venus', 'Valley Legend Steff'
Green: Joyce Duncan 'Susan Hughes', 'R.D. Hughes', Miretta 'Mcbean', Lucense, Tricia Allen 'The Globe', Sparkle 'Late Green', Levis Duke 'Belle Vista', Sparkle 'Late

Green', Amsebury 'Frank Slattery', 'Valley Zenith 'Top Spot', 'Madrid Forest King', 'Winter Beach Sea Green'
Red: Chief Joseph 'Pathfinder', Sensation 'Chianti' 4N, Terama 'Robin,'Barushka 'Dos Pueblos', Khyber Pass 'Rowes Red', James Toya, 'Fire Storm Ruby', 'Fire Storm Blaze' 'Bob Marlin Lucky'

9.1.7 Growing Requirements

In India, its cultivation is limited to Sikkim and the surrounding region of West Bengal covering Kalimpong, Darjeeling and Mirik. Other North Eastern states like Nagaland and Arunachal Pradesh are also promoting this flower. Higher elevations of 1500-2000 m with cool summer night and monsoonal summer rain are ideal for cymbidium cultivation.

9.1.7.1 Light

A full morning sun or bright dappled afternoon shade during summer and full sun in winter is ideal. Mature plants need 50-55% shade during hot weather. During growing season they require up to 5000-6000 foot candle light, whereas in flowering season up to 2000-3000 foot candle light is needed. Foliages should be yellowish green in colour.

9.1.7.2 Temperature

In general, cymbidiums can tolerate temperatures as low as 7°C. In the vegetative stage, plantlets grow best at temperature of 18°C at night and 24-30°C during the day. A temperature of 10-15°C is required for initiation of flower spikes. During the winter season (Late October to late February) a temperature of 7-12°C at night and 18-24°C during the day is maintained. Miniature hybrids can withstand temperatures 5°C higher than standard cymbidiums.

9.1.7.3 Propagation

Cymibidiums are propagated sexually through seeds and asexually through division or backbulbs. Division means splitting the plants into two to three parts each with one new shoot, and each will produce an individual plant. Propagation through back bulb is a slow process which will take three to four years to give a flowering size plant.

Tissue culture is the only way to produce millions of disease free and true to the type plants in the shortest time. In this method, callus (amorphous masses of cells), meristems and organs (root, leaf, flower, embryo, ovary, fruit, seeds, etc.) are isolated and cultured aseptically in a laboratory and supplied with defined media containing sugars, inorganic salts, vitamins and growth regulators.

9.1.7.4 Watering

In Cymbidium, watering is required year round to keep the pseudobulbs green and smooth. The frequency of watering is given below.
Summer: 2-3 times per week
Autumn: Once or twice per week
Winter: Once per week
Spring: As Autumn season

9.1.7.5 Relative Humidity

An optimum range of relative humidity is 50-80% and important for good growth and flowering. During hot weather, misting down the plants and the surrounding floors and benches maintains humidity, which prevents crinkling of leaves.

9.1.7.6 Air Circulation

Fresh air and good circulation are essential for orchid production. Leaves should move gently in a light breeze.

9.1.7.7 Feeding

Cymbidium orchids should be fed about once every two weeks. A foliar feeding of 0.2-0.3% NPK mixture is effective and applied in the following manner.
Spring season: 30:10:10
Summer season: 19:19:19
Autumn season: 10:30:20
Pre-harvest foliar sprays with micronutrient mixture (0.05%), glucose (0.1%) and mustard cake (1 kg/50 litres of water) were found promising in terms of number of inflorescence/plant (4 to 6) and number of flowers/spike (8 to 10) followed by GA_3 (50 ppm) and cow urine (1:20). Do not feed your cymbidium once it has started flowering.

9.1.7.8 Potting and Repotting

Cymbidiums prefer a coarse compost containing bark, cocopeat, cocochips, leaf mould, sawdust, brick pieces for development of new shoots and good pseudobulb growth. The compost should be very light, slightly acidic in reaction with a pH range of 5.5-6 and with good drainage properties and water holding capacity. In *Cymbidium*, the potting mixture combinations, cocochips + cocopeat + brick pieces + chopped fern leaves (3:3:1:1) followed by cocochips + cocopeat + brick pieces + slow release fertilizer (3:3:1:1) were found effective for good pseudobulb growth and flowering. Repotting is required every 2-4 years as indicated by the bulbs filling the pot or breaking down of potting mixture and if a plant is not able to produce 1-2 new bulbs each year from each bulb that grew the previous year.

9.1.7.9 Planting

It is advisable to begin at the end of February or beginning of March which will create a congenial environment for early establishment of the plant and thus better growth and development. It is a healthy practice to culture Cymbidiums in pots of 10" diameter and 18" height. The pots should be placed at spacing of 60 cm x 60 cm, which will accommodate about 3 pots / m^2 (gross). The total number of plants will be about 30,000 / ha. These must be cultured under shelter by providing poly-cover to obtain quality blooms. Some shading (about 30%) is essential during the blooming time for proper colour development of flowers and during the young plant stage. Cymbidium pots are kept on 2 to 3 feet high beds.

9.1.7.10 Training

Cymbidiums may be trained with bamboo sticks and yoyo to keep the spikes in upright position.

9.1.7.11 Harvest

The orchid flowers should be harvested at the proper stage to achieve quality flowers and maximum vase life. Morning is the best time for harvesting. Flowers are cut sharply with a knife or secateur and dipped immediately in a bucket of water. In most of the commercial orchids, the optimum harvesting stage is fully opened and mature flowers. A matured healthy plant in a 10" pot under good management will produce about 4-6 flowers per year. A healthy well grown orchid will produce flowers spikes every year and a plant 10 years old can produce from 10 to 20 flower spikes.

9.1.8 Insects Pests and Diseases

9.1.8.1 Insect Pests

Mites, thrips, scale insects, aphids, mealy bugs, grass hoppers and shoot borers are common insect pests of Cymbidium. All active stages (nymph and adult) of mite feed on undersurface of leaves and flowers by sucking the cell sap from the epidermal layer, especially along with the midrib and the base. The loss of cell sap causes yellowing of leaves. Sulphur, 3% wettable or colloidal, satisfactorily controls the nymph and adults.

There are five predominant species of scale insects *viz.*, ti scale, *Pinnaspis buxi,* florida red scale, *Chrysomphalus aonidum;* lecanium scale, *Lecanium* sp; soft brown scale, *Coccus hesperidum* and boisduval scale, *Diaspis boisduvali* which cause damage on orchids round the year. Both the stages of scale insects suck the cell sap from leaves, pseudobulbs, flower buds and flowers causing yellowing of leaves, vigor loss and stunted new growth. In the case of heavy infestation, infected plants become deformed, and sticky honeydew is excreted which attracts sooty mould on which dust particles deposit resulting in a negative effect on the rate of photosynthesis. Chemicals like imidacloprid 17.8 SL @ 0.003%, acephate 75 SP or carbaryl 50WP or monocrotophos 36EC @ 0.03-0.05% can be used against aphids, scale insects and mealybug.

Two species of aphids like yellow aphid, *Macrosiphum luteus* and black aphid, *Toxoptera aurantii* mainly cause damage to orchids. The nymphs and adults suck the cell sap usually from new flower spike and foliage. They also excrete honeydew on which sooty mould develops and affects photosynthesis. Thrips, *Dichromothrips nakahari* suck the cell sap from tender portion of plants and on leaves, it become discolored and shrivels. Both young and adult mealy bugs (*Pseudococcus* sp**)** suck the cell sap from the leaves and petioles or any joint portion of plants and as a result of that plants become weakened. Grass hoppers (*Hieroglyphus banian*) feed on young leaves, un-opened flower buds and flowers by cutting irregular shape with their biting and chewing mouth parts, and ultimately flower quality is affected.

9.1.8.2 Diseases

Black rot, Crown rot or heart rot (*Pythium ultimum, P. splendens, Phytophthora palmivora and P. parasitica*): Water soaked small brown spot on the aerial parts of plants, which quickly turn black. Fungicides spray viz Captan @ 2 g/l or Zineb @ 2 g/l water controls it.

Anthracnose (*Colletotrichum gloeosporioides and C. orchidacearum*): Initial symptom appears as a small oblong to circular oval, sunken and reddish brown to dark brown or gray coloured spots. Die back of leaves are also observed if the leaf tip is attacked. Spraying of Carbendazim (Bavistin) @ 1 g/ liter in 10 days interval checks the disease.

Blossom Blight (*Botrytis cinerea*): The pathogens produce numerous small dark spots on petals, especially on older flowers. Sometimes shot hole effect is found on infected flower petals. Spraying with Bavistin @ 1g/l liter or indofil Z @ 2g/liter at 7 days intervals are effective.

Bacterial soft rot (*Erwinia carotovora pv. carotovora* and *E. chrysanthi*): Deep grayish grey lesions on leaves. It causes leaf spot, soft rot and stem rot with fishy smell. Treating of infected plants with streptomycin or Oxytetracycline solution before planting checks the infection.

Cymbidium mosaic virus (Cymbidium mosaic potexvirus): The virus produces variable symptoms on different hosts. It produces mild or severe mosaic symptoms followed by necrosis. Start with certified and virus free plant material, proper sterilization of tools used in cultural practices, proper distance among plants has to be maintained to avoid virus infection, proper sanitation. Keep growing area free from plant debris. Quarantine new plants and control insect vectors are remedial measures.

Odontoglossum ringspot virus (Tobamovirus): It produces ringspot on Odontoglossum grande and diamond mottle symptoms. Start with certified and virus free plant material, proper sterilization of tools used in cultural practices, proper distance among plants has to be maintained to avoid virus infection and maintain proper sanitation. Keep growing area free from plant debris. Quarantine of new plants and control of insect vectors are remedial measures.

9.1.9 Post-harvest Management

A good quality cut flower of an orchid should have the following characteristics
- Minimum eight standard blooms per stem
- Flowers must be cleaned, evenly coloured and free from physiological disorders
- Stem must have flowers evenly arranged and around the stem.
- Two third of the stem should be covered with the flowers.
- Flowers must have a firm texture and a luminescent sheen
- Stems must be firm when held up
- The minimum base diameter of the stem should be of 10 mm

9.1.9.1 Stage of Harvest

In Cymbidium, flowers having 75% bloom stage or two buds opened stage with the spike length of 60-90 cm are harvested.

9.1.9.2 Grading

Cut spikes of Cymbidium orchids are graded based upon their classes, number of flowers per spike and spike length (Tab. 9.3).

Tab. 9.3: Grading of Cymbidium cut flowers

Category	Grade	Flower Count	Spike length
Standard	AAA	12-15	1.25 m
	AA	8	90 cm
Miniature	XL	>15	65 cm +
	L	12-14	55-64 cm
	M	8-11	40-54 cm
	S	<5	30-39 cm

9.1.9.3 Packing

After harvest the flower stems, are bunched into 5 or 10, wrapped in a specialized polythene cover and at the base of the stem a slant cut is made with a sharp knife. The stem bottom is inserted in a plastic plug containing clean water. This will keep the flowers fresh during transportation. In absence of the plug, a moistened cotton wrapped with a piece of polythene can do the job for domestic market. After plugging or wrapping with moistened cotton, the flower stems are placed in corrugated boxes and readied for dispatch to the market.

9.1.9.4 Storage

The cut spikes can be stored at 0.5 to 5°C for 14 days under dry.

9.1.9.5 Floral Preservatives

In the *Cymbidium* hybrid 'Red Princess' pulsing with 5% sucrose increases vase life up to 56 days followed by sucrose at 8% (54.78 days). In *Cymbidium*, 1-MCP and AVG are superior to STS in prolonging the vase life of cut flowers. In Cymbidium hybrid, 'Red Princess', 75% open flowers with 200 ppm 8-HQS showed highest vase life along with 100% opening. In *Cymbidium* 'Ensikhan' and 'PCMV', 4% sucrose + 100 ppm salicylic acid and 4% sucrose + 100 ppm $Al_2 (SO_4)_3$ are used as bud opening chemicals. 2% sucrose + 200 ppm 8-HQS is also used as holding solution.

9.2 Dendrobium (Orchidaceae)

9.2.1 Introduction

Dendrobiums are popular flowering potted plants and cut flowers around the world due to their floriferousness, wide range in flower color, size and shape, year round availability and lengthy vase life. Hawaii, California and Florida are major potted Dendrobium growing regions in the United States. In the Netherlands, production of potted orchids is now 40 to 50 million units with *Dendrobium* increasing in popularity. Imports from Thailand, the world's largest exporter of tropical cut orchids and second largest supplier to the EU, accounted for 22% of supplies to the EU. Thailand holds a particularly strong position in Dendrobium orchids.

9.2.2 Botanical Descriptions

Dendrobium consists of 1600 species of sympodial epiphytic orchids. The genus is characterized by long pseudobulbs or canes with soft leaves on the entire length or in some species, pseudobulbs are short or swollen terminating in two coriaceous leaves. The pseudobulbs are of four types, cane woody (a), cane cylindric (b), cane clavate fleshy (c) and bulbous round (d) (Tab. 9.4) (De *et al*, 2013).

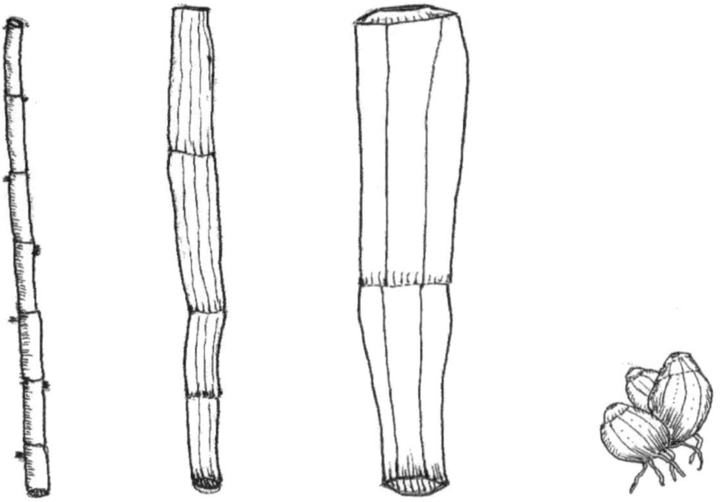

(a) (b) (c) (d)

Tab. 9.4: Nature of pseudobulb in Dendrobium

Characteristics	Example varieties/species
Cane woody	*Den. gibsonii, Den. bensoniae, Den. aphyllum, Den. ruckeri, Den. aduncum, Den. cathcartii*
Cane cylindric fleshy	---
Cane clavate fleshy	*Den. primulinum, Den. nobile, Den. parishii, Den. pendulum, Den. loddigesii,* Den. 'Bangkok Blue', Den. 'Big White Jumbo', 'Den. 'Thongchai Gold', Den. 'Erika', Den. 'Triple Pink', Den. 'Madam Pink'
Bulbous round	*Den. aggregatum, Den. chrysotoxum, Den. jenkensii,* Den. 'Julie'

The leaf size ranges from 2.5 cm to 40 cm, thick, and are deciduous or evergreen. In some groups, the flowers are joined in pairs or three on small peduncle on the entire length of the pseudobulbs, with caduceus leaves. In some species, with persistent leaves, the flowers are grouped in pairs or threes or alternately closely set, forming erect or pendent thyrses. In another group, flowers are generally solitary and small, arising from the axils of leaves. The inflorescences are terminal or subterminal and arranged with one to several dozens of flowers with extremely diverse dimensions, size and ranges of flower colour.

9.2.3 Importance and Uses

Dendrobiums are popular for cut flowers and for interioscaping. They are also valuable as pot plants or hanging baskets. Some species are hanged on the walls or on tree branches to cover the bare walls and branches. The dried stems of *Dendrobium nobile* are used for making herbal medicines. *Dendrobium* enhances salivation and is used for the treatment of dry mouth, dry coughs and severe thirst. Flowers could be used to cure eye ailments. The tonic made from *Dendrobium* nourishes stomach, lungs and kidneys. The plant is effective in treating pulmonary tuberculosis, impotence and anorexia. The pulp of pseudobulb is applied to boils and pimples, and the juice of the plant is used to relieve fever. Dendrobium blossoms are the most common genus used in cooking. In Thailand, these edible flowers are dipped in butter and deep fried while many European cooks garnish desserts and cakes with them. The starchy stems of *Den. speciosum* are roasted and eaten.

9.2.4 Hybrids (De, 2014)

White: 'Snow White', 'Pagoda White', 'Emma White', 'White Surprise', 'Jacquelyn Concert x Walter Oumae', 'Kasem White', 'Big White 4N', 'Big White Jumbo', 'White 5N'.
Blue: 'Vorawit Blue', 'Lee Chong Blue', 'Kultana Blue', 'Kiyoshi Izumi', 'Blue Fairy', 'Lee Chong Blue', 'Bangkok Blue'.
Pink: 'Chiengmai Pink', 'Ekapol Panda', 'Jisu's Star', 'Juree Red', 'Kiilani Stripe', 'Long Champ', 'Penang Sugar', 'Sagura Pink', 'Miss Singapore', 'Madam Pink', 'Sonia-16', 'Ear Sakul', 'Candy Stripe Pink', 'Sonia-17', 'Sonia-28', 'Dr. A. Abraham'.
Yellow: 'Sri Siam', 'Swan Lake', 'Thongchai Gold', 'Bonchoo Gold', 'Sarifa Fatima'.
Green: 'Daangsaard', 'Kanjana Green', 'Green Mist', 'Little Green Apples'.
Red: 'Meike Beauty', 'Pathum Red x Sabin', 'Little Lolita', 'Cleopatra', 'Diamond Star', 'Fireball', 'Kating Daang'.
Interspecific hybrids: 'Australian Lemon Pepper', 'Jiali Paradise', 'Womad', 'Green Elf', 'Memoria Dipper Nishi', 'Falcan', 'First Star', 'Falcan', 'Mini Snowflake', 'Scotts Valentine', 'Dounan Spicy', 'Black Gold', 'Winter Frost', 'Peng Seng', 'Silver Wings', 'Aminah Khatum', 'Bruce Gorden', Green Mist', 'Aussies Queen', 'Sweet Phurichaya', Big Alex', 'Carly Hera', 'Pink Glow', 'Molly's Angel', 'Go Secret'.
Inter-varietal hybrids: 'Candy Smile', 'Angel Moon', 'Million Gold', 'Liberty Girl', 'Asian Smile', 'Happy Holiday', 'Happy Smile', 'Pop Eye', 'Sunny Bird', 'Wonder

Rabbit', 'Sunny Bird', 'Peach Blossom', 'Long river Giant', 'Sally Fiesta', 'Rudkin', 'Ice Storm', 'Genting Melody', 'Hawaiian Twinkle', 'Sea Sky', 'Singa Beauty', 'Arthur Reserve', 'Green Wonder', 'Open Heart Leaf', 'Rising Star', 'Fairy Star', 'Nice Boy Wanda', 'Burbank Candy', ' Burnt Orange', 'Justine'.

Variety-species hybrids: 'Bright Angel', 'Blue Rain', 'Samson Toy', 'Margaret Thompson', 'Third Eyes Vision', 'Rods Eagle', 'Spider Lily', 'Sylvester', 'Paradise Fortune', 'Sky Mirror', 'Butter Fly Dawn', 'Island Snow', 'Australian Idol', 'Fine Ford', 'Half Moon Bay', 'Special Bride', 'Two Kings', 'Juliette Copper', 'Genting Lipstick', 'Singa Kagoshima', Jairuk Spin'.

9.2.5 Growth and Physiology

Like other flowering plants, orchids also have to attain a certain stage of growth and fulfill the energetic demand to initiate flowering. It may vary from 3 years to 7 years depending upon the type of species and hybrids. Orchid pseudobulbs are engaged in the control of physiological processes that are important for growth and survival. The ability to store water, minerals and carbohydrates in the pseudobulb has greater impact for survival in the harsh and nutrient limited epiphytic biotope. Pseudobulb photosynthesis recycles respiratory carbon that would contribute positively to whole plant carbon economy. There are significant variations in the content of carbohydrates in pseudobulbs and flowers and chlorophyll content in leaves among different hybrids. Reducing sugar analysis was carried out in the Dendrobium hybrid 'Thongchai Gold', where three stages of flowers, specifically opened flowers, half opened flowers and bud, were taken for analysis. It was found that the opened flower contains 29.00%, half opened flower contains 28.25% and buds contain 16.17% of reducing sugars (De *et al*, 2013).

9.2.6 Cultivation

9.2.6.1 Temperature

The cool growing Dendrobium orchid group thrives well in temperatures ranging between 10 and 24°C. The intermediate Dendrobium orchid prefers a temperature range of 14-26°C whereas the warm growing Dendrobium orchids prefer 16 to 30°C. The warmer group species like *Dendrobium phalaenopsis*, *Den. gouldii*, *Den. biggibum*, *Den. antennatum* and *Den. discolor* bloom at night temperatures above 16°C and the cool growing species such as *Den. lindleyi*, *Den. aggregatum*, *Den. parishii*, *Den. pierardii*, *Den. densiflorum*, *Den. chrysotoxum* and *Den. anosmum* perform well at night temperature of 10°C. Low day temperature causes leaf yellowing, defoliation and reduces vegetative growth, while higher temperatures delay flower bud development. Low temperature and short days could change the concentration of endogenous growth regulators leading to the induction of flowering in sympodial orchids.

9.2.6.2 Light

Most orchids generally prefer indirect or filtered light. Although it varies by species, growth habit and habitat, as the rule of thumb, 50% shading is advised for most of the commercial orchids. Under enough light, orchid plants have short, plump stems with bright green leathery leaves, while too much light causes yellowing, stunting and scorching of plants, and too much shade causes darker green, soft and succulent leaves with thin and spiny stems. All types of Dendrobium orchids require warm bright light (2500-3000 foot candles). They should get at least 12-14 hours of light each day year round.

9.2.6.3 Propagation

Conventionally, Dendrobiums are easy to propagate through keikis that produce along old canes or by division of pseudobulbs. 10-12 cm long cuttings also can be taken from a healthy, older and leafless cane keeping three nodes on each cuttings and placed in moist sphagnum moss for rootings. Commercially, Dendrobium hybrids are usually either seed-propagated or clonally propagated through tissue culture of apical and lateral buds that proliferate as protocorm like bodies.

For seed propagation, green capsules are surface sterilized and seeds are dropped on a basal salt medium containing 15% coconut water and 2% sucrose at pH 4.8 to 5.0 for germination. Three months after sowing, seedlings are transflasked with 75 to 100 plants per 500 ml flask in a salt medium containing 15% coconut water, 2% banana powder, and 1% sucrose at pH 4.8-5.0.

9.2.6.4 Atmosphere

Fresh air and good circulation are essential for orchid production. Full of continual light breezes make a good source carbon dioxide for photosynthesis.

9.2.6.5 Fertilization

Orchids are light feeders and require nitrogen from the beginning to two-thirds of their life cycle. During rest periods, they do not need any fertilizers. During flower initiation and inflorescence development, plants are fed with less nitrogen, more phosphorus and potassium. During the blooming time, a small level of nitrogen and phosphorus and high levels of potassium are maintained. In orchids, foliar feeding is found to be ideal. Frequent application of fertilizers in low concentrations is the best way of feeding orchids. A concentration of 0.2 to 0.3% of 30:10:10 (N:P:K) at vegetative stage and 10:20:20 (N:P:K) at blooming stage are applied for quality flower production. Sometimes, fresh coconut water, diluted cow urine is also useful as foliar sprays.

9.2.6.6 Potting Mixture

The potting medium of Dendrobium orchids should be loose, friable as well as well drained. A potting medium consisting of charcoal, brick pieces and coconut fibre in equal proportions is ideal for vegetative growth and flowering of epiphytic orchids like Aerides, Dendrobium etc. Under low humid conditions (30%), plastic pots with a mixture of bark/perlite/sphagnum moss or osmunda are used. Under average humidity (35-50%), it is advisable to use plastic pots with a mixture of bark and sphagnum moss. Under high humidity (55% and above), clay pots are used with bark, stone culture, charcoal or tree fern. In Dendrobium, among potting mixtures, cocochips + cocopeat + leaf mould + brick pieces (4:1:2:3) showed longevity on plants in Den. 'Ear Sakul'(71 days) followed by Den. 'Triple Pink' (70 days) and Den. 'Thongchai Gold' (69 days).

9.2.6.7 Watering and Humidity

Most orchids are damaged by overwatering rather than under watering. Overwatering leads to root rot and many other diseases. Most orchids prefer water of pH 5.0-6.5. Watering with lower or higher pH or with high levels of dissolved minerals can hamper nutrient uptake. Frequent watering is essential under high sunlight and high temperature conditions. Plants in small containers dry out more quickly than in large containers. Plants in earthen pots require more watering than plants in plastic pots. A hanging plant, since it has better aeration, needs more frequent watering than one in a pot. More frequent watering is also necessary for fresh potting materials. Watering should be practiced either in the morning or in the afternoon. Potting materials like coconut husk, tree fern etc., having more moisture retention capacity, need less water and less frequently. The single dominant factor that affects the cultivation of orchids is humidity, which should be as high as 50-75%. It varies species to species depending upon habit of growth, light, temperature and ecotypes. As rule of thumb, in high temperature, humidity should be kept high. Provisions of misting units or foggers or even humidifiers will ensure adequate humidity. Standing water beneath the benches may be kept to improve humidity.

9.2.6.8 Repotting

Orchid plants require repotting if there is no space left in the pots for new growths and if the substrate has decomposed or roots are rotting. Timing is the most important part of good repotting. The best time for repotting of an orchid is when new growth and new roots are just beginning to form, before those new roots reach even 1 cm long. In most of the orchids, it occurs right after flowering. It shows that repotting should be done between February and June.

9.2.6.9 Application of Growth Regulators

Experimental evidence has shown that photoperiod and low temperature modify the concentration of endogenous growth regulators. The combined application of GA_3 and BA improves inflorescence length and reduces percentage of abnormal flowers. In the *Dendrobium* hybrid 'Thongchai Gold' maximum flower spikes were found in drenching followed by morning and evening spray. Spike length was found to reach maximum in morning spray with GA_3 200 ppm (46 cm) while in evening spray with GA_3 100 ppm spike length was 43.6 cm and in morning spray with GA_3 100 ppm, spike length was minimum (42 cm). In *Dendrobium* hybrid, 'Emma White', treatment with NPK 20:20:20 with Ca, Mg and Mn along with BA 10 ppm and GA_3 100 ppm increased the number of leaves (20.06), pseudobulbs (2.73) and pseudo bulb girth (1.94 cm). Plant height was highest (59.79 cm) in a treatment with NPK 30:30:30 with Ca, Mg and Mn along with BA 25 ppm and GA_3 50 ppm. Treatment with NPK 20:20:20 with Ca, Mg and Mn along with BA 50 ppm and GA_3 100 ppm gave maximum number of spikes / plant (2) (De, *et al*, 2013).

9.2.7 Post-harvest Management

9.2.7.1 Harvest

Dendrobium orchid spikes possess the vase life of 2 to 3 weeks. Usually, 40-60 cm long floral spikes with 10-15 flowers are harvested at a stage when all flowers are open except the top bud. Harvested sprays should be immediately placed in clean buckets of water with the cut ends submerged about 2-3 inches. Then the sprays are taken into the cool shaded packing area. In *Dendrobium*, it has been reported that flowers harvested early in the morning, generally last longer than those harvested in the late morning.

9.2.7.2 Pre-cooling

This is the fast removal of field heat and it is an important operation in post-harvest handling and transport of cut flowers, wherever flowers are held dry pack. All flowers should be pre-cooled immediately after harvest by placing them in cold storage without packing or in open boxes until they reach the desired temperature. These temperatures variy with species and cultivar: *Cattleya* (7-10°C), *Cymbidium* and *Paphiopedilum* (0.5 to 4°C), *Dendrobium* (5-7°C). Pre-cooling lowers respiration rate and decreases the breakdown of nutritional and other stored materials in the stems, leaves and petals; and delays bud opening and flower senescence. It also prevents rapid water loss and decreases flower sensitivity to ethylene. Several pre-cooling techniques such as room cooling, forced air cooling, hydro-cooling, vacuum cooling and ice bar cooling etc. are available.

9.2.7.3 Pulsing

The absorption of chemical solutions containing sugars and germicides through the lower cut bases of flower stems is known as pulsing. Pulsing may be used by growers, wholesalers or retail florists in order to enhance the cut flowers subsequent vase life in water. Pulsing is employed with higher concentrations of sugar, mainly sucrose, the percentage of which varies with species and cultivars. Other chemicals used in the pulsing treatments are STS, $AgNO_3$, HQ, MH, AOA, $CaCl_2$, $CoCl_2$, nickel sulphate, aluminium sulphate and benzyladenine. Pulsing is found to be of great value in prolonging life, promoting opening and improving the colour and petal size of petals through osmo-regulation. In *Dendrobium* hybrid 'Pompadour'pulsing with 25 ppm $AgNO_3$ + 135 $Na_2S_2O_3$, $5H_2O$ for 30 minutes increases vase life of cut flowers. In *Dendrobium* cv. 'Sonia' pulsing with 4% sucrose + 400 ppm HQ recorded the highest vase life of 21.33 days.

9.2.7.4 Bud Opening

This is the procedure of harvesting flowers at a stage earlier than normally considered as the cutting stage and then opening the buds off the plant. Such types of post harvest handling may be applied by growers or wholesalers. Bud opening of flowers increases longevity of cut flowers by reducing the sensitivity of flowers to extreme temperatures, low humidity and ethylene, saving space during shipment and extending the useful storage life. The sugar concentration used is lower than the concentration of pulsing and the optimum temperature is kept lower. In the *Dendrobium* hybrid 'Thongchai Gold' opened flowers had 29%, half opened flowers had 28.25% and buds had 16.17% reducing sugars. In *Dendrobium* hybrids, HQS or $AgNO_3$ (50ppm) is effective for opening of tight bud cut flowers. It has been reported that a preservative solution containing 225 ppm HQS, 30 ppm $AgNO_3$ and 4% glucose increased bud opening and the time to wilting of the open florets of *Dendrobium* Cv. 'Ceasar'. In the *Dendrobium* hybrid 'Thongchai Gold', the maximum percent of fully opened buds (66%) was recorded with sucrose (4%) + $Ca(NO_3)_2$ (1%) followed by sucrose (4%) + acetyl acetic acid (100 ppm) (60%). Longest vase life (36 days) was found with sucrose (4%) + $Al_2(SO_4)_3$ (100 ppm) followed by sucrose (4%) + acetyl acetic acid (100 ppm) (33 days).

9.2.7.5 Preservatives

Preservatives are used in the holding solutions in the form of tablets containing a mixture of chemicals such as sugars, germicides, salts, growth regulators etc. In addition, the chemicals are employed during conditioning, pulsing and for making bud opening solutions to improve flower shape, size and opening and colour of the flowers.

Sugar, biocide, anti-ethylene compounds and hydrated compounds are used for conditioning. The sugar and biocide solutions are effective for opening of bud cut flowers.

The vase solution should contain sugars, an acidifying agent and a biocide. Citric acids are mainly used for acidifying agents and hydroxy quinoline as biocides. Metallic salts like silver nitrate, cobalt chloride, aluminium sulphate, zinc sulphate, calcium nitrate and nickel chloride have been found for prolonging post-harvest life of various cut flowers. A combination of biocide, sugar and hormone (8-HQC 100 ppm + sucrose 2% + BA 25 ppm) remarkably enhances the post-harvest life of the *Dendrobium* cut flowers. New chemicals that have been found promising as floral preservatives are ethylene inhibitors like amino-oxyacetic acid, 1-amino cyclopropane, aminotriazole, aminoethoxy vinyl glycine, alpha aminoisobutyric acid, diazocyclopentadiene and phenidone. Holding solutions for increased longevity of *Dendrobium* as reported by various workers are 8-HQC (200 ppm) + sucrose (2%), 0.5 mM AOA + 4% sucrose, AgNO3 (30 ppm) + 4% sucrose, 400 ppm HQ + 30ppm AgNO3 + 2% sucrose, 200 ppm 8-HQS + 50ppm AgNO3 + 8% sucrose (Bhattacharjee and De, 2005).

9.2.7.6 Grading

The export quality orchids are graded to maintain high standards of excellence. Sprays are graded according to length, colour, flower size etc. The grading is done in four standard sizes, based on the quality of the stalk and spike length for each grade (Tab. 9.5).

Tab. 9.5: Grading of Dendrobium cut flowers

Name of the grade	Length of the spike	No. of opened flowers
SMALL-S	30 cm	4-5
MEDIUM-M	40 cm	6-8
LARGE-L	45 cm	8-10
EXTRA LARGE-XL	50 cm	>10

9.2.7.7 Packaging

The flower spikes of Dendrobiums are first packed in polyethylene sleeves of standard thickness. The standard is to bunch around 5 spikes of the same grade and variety in a pack box. Each stem in the box should be placed in a tube containing water or preservative solution. During shipment, loss of water could be supplemented by employing flower tubes or vials, which could be filled with water or preservative solution. Instead of small tubes, cotton wrapping can also be used, in this case cotton pieces should be dipped in water or preservative solution. Then a piece of polythene can be used to cover the cotton and it should be tied with rubber band. And cushioning materials should be provided in the back side of the sleeve to avoid damage during transportation.

Graded flowers are then packed in suitable size of boxes. In order to check movement of spikes within the boxes during transit the base of the spikes should be tied to the base of the carton by adhesive tapes. *Dendrobium* flower spikes are normally packed in carton of different sizes. The length of the carton varies mainly based on the length of the flower spikes. The carton should be provided with sufficient numbers of holes or vents for aeration. The carton is exclusively designed to ensure flowers are better cared for and reach their destination in pristine conditions. In *Dendrobium* hybrid 'Sonia-17' a low gauge polyfilm of 100 gauge thickness the cotton dipped in 8-HQS (25 ppm) covering the base of the spike had maximum vase life and flower quality.

9.2.7.8 Storage

Storage of cut flower is an essential part of floriculture industry. Tropical orchids like *Dendrobium*, can be stored at 7-10°C and 90-95% RH. The longest vase life of 19 days has been observed in *Dendrobium* hybrid Sonia 28 followed by 15 days in Sonia 17 when stored at 10°C. The orchids stored below optimum temperature cause chilling injury characterized by darkening of the labellum. In extreme cases, the sepals and petal also get affected. Orchids are sensitive to ethylene and the storage environment should be free from ethylene which can be effectively accomplished by proper ventilation and placing ethylene scrubbers or absorbent containing potassium permanganate in the area.

9.2.8 Value addition

Value addition in flower crops can directly or indirectly influence the floral market to a great extent. Value addition in flower crops by employing techniques like colouring in white flowers, flower dehydration, flower processing, advances in flower arrangements etc. can add value up to 5 to 10 times

9.2.8.1 Tinting

This is one of the most important value addition techniques for imparting desired shades of colour to the flowers. It is a very useful technique in flower crops where pigments are absent or light and dull. Aesthetic beauty of the cut flowers and dry flowers were enhanced through tinting. Translocation, immersion and spraying are methods followed in tinting. Stopping irrigation two days before the harvest of flower improves the flower colour. It can be combined with pulsing solution. Edible dyes of 0.25 to 1% can be used along with pulsing solutions for 30 minutes to 3 hours. Different dyes and strains such as food colours, feulgen stain, bromocresol blue, bromocresol green, eosin yellow, ammonium purpurate and phenol red at varying concentrations can be used to get flowers with different shades of colours. Artificial colouring can be done by using edible dyes like Apple green, Kalakhatta, rose pink etc. Employing this technique, white flowers can be used to result in all shades of red, blue, green and yellow.

9.2.8.2 Preparation of Bouquet

Bouquets are prepared with different colours of *Dendrobium* cut flowers selected and combined according to colour and arranged with some foliage plants and packed in boxes.

9.3 Phalaenopsis (Orchidaceae)

9.3.1 Introduction

Phalaenopsis is also known as 'Moth Orchid' because when the *amabilis* species was first observed in its natural habitat, the long inflorescences of pendulous white flowers that festooned the jungle tree tops were thought to be clusters of moths. This orchid originated in the jungles of South and Southeast Asia, Indonesia, Malayasia and the Philippines.

9.3.2 Botanical Description

Phalaenopsis consists of 70 species of monopodial orchids ('Moth Orchids')distributed in Asia, the Philippines, Indonesia, Malayasia, Australia and New Guinea. The plants are pseudobulbless with short stems covered by clasping leaves. The leaves are leathery, and thick. The inflorescence arises from the axil of leaves, drooping or erect bearing spikes of 100 cm length. The flowers are spectacular, long lasting and white, pink, yellow or mottled. There are two types of *Phalaenopsis* . In the first type, leaves are thick and fleshy, elongate-elliptic and obtuse in apex. The flowers petals are broader than the sepals and the lip possesses two attractive centre lobes and appendages. The flowering stem is up to 60 cm long and bears 15 or more blooms. Species belonging this group are *Phalaenopsis parishi*, *P. aphrodite*, *P. stuartiana*, *P. schilleriana* and *P. sanderiana*. In the second type, plants are short stemmed bearing fewer blooms. The flowers are smaller with equal sizes of sepals and petals and without any appendages. Species belonging this group are *Phalaenopsis cormi-cervi*, *P. leuddemanniana*, *P. equestris* and *P. mannii.*

9.3.3 Importance and Uses

Phalaenopsis are commonly used as pot plants and cut flowers and are suitable in hotel arrangements, hanging arrangements, households, boutique stores, weddings, funerals, birthdays, etc. The purity and brightness of the *Phalaenopsis* make it especially appealling for weddings and corsages.

9.3.4 Hybrids (De, 2014)

Tab. 9.6: Common commercial Phalaenopsis hybrids

Colour	Name	Remarks
White	Taisuco Crane, Taisuco Kochdian, Cygnus, Yukimai, Sogo Musadian, White Dream, Florida Snow	Large flowered
Pink	Nobby's Pink Lady, Minho Valentine, Minho King Beauty, New Cinderella, Taisuco Firebird, Sogo Smith	Large flowered
Striped hybrids	Sogo Zebra, Chih Shang Stripes, Okay Seven	
Spotted hybrids	Carmela spots, Rousserole, Soroa Delight, Leopard Prince	Pretty
Yellow	Carol Campbell, Emil Giles, Brother Lawrence, Taipei Gold, Golden Bells, Sogo Managers, Brother Passat	5-10 flowered
Spotted & Barred yellow	Golden Amboin, Yellow Queen, Ching Her Buddha, Golden Sun	
Orange red to deep lavender	Cordova, Lonnie Morris, Sogo Pony, Sogo Rose, Sogo Grape	
Red spots on yellow or cream coloured background	Brother Purple, Sogo Prince, Golden Peoker	
Colour blending	Sweet Memory, Zuma Aussie Delight, Pago Pago, Brother Sara Gold	
White or Yellow base colour with black blotches	Ever Spring Light, Yu Pin Pearl, Yu Pin Panda, Bright Peacock	Mutants of Taiwan
Multiflora hybrids	Be Glad, Cassandra, Vilind, Carmelas Pixie, Zuma's Pixie, Timothy Christopher, Be Tris, Quevedo	Bred at California
Miniature hybrids	Micro Nova, Mini Mark, Anna-Larati Soekardi	Require lower day and night temperature and lower humidity

9.3.4.1 Bigeneric Hybrids

Aeridopsis = Phalaenopsis x Aerides
Arachnopsis = Phalaenopsis x Arachnis
Doritaenopsis = Phalaenopsis x Doritis
Phalanetia = Phalaenopsis x Neofinetia
Renanthropsis = Phalaenopsis x RenantheraVandaenopsis = Phalaenopsis x Vanda

9.3.4.2 Trigeneric Hybrids

Sappanara = Phalaenopsis x Arachnis x Renanthera
Trevorara = Phalaenopsis x Arachnis x Vanda
Lycockara = Phalaenopsis x Arachnis x Vandopsis
Rhyndoropsis = Phalaenopsis x Doritis x Rhyncostylis
Moirara = Phalaenopsis x Renanthera x Vanda
Yapara = Phalaenopsis x Rhyncostylis x Vanda

9.3.4.3 Tetrageneric Hybrids

Bogardora = Ascocentrum x Phalaenopsis x Vanda x Vandopsis
Bokchoonara = Arachnis x Ascocentrum x Phalaenopsis x Vanda
Edeara = Arachnis x Phalaenopsis x Renanthera x Vandopsis

9.3.4.4 Pentageneric Hybrids

Sutingara = Arachnis x Ascocentrum x Phalaenopsis x Vanda x Vandopsis
Macekara = Arachnis x Phalaenopsis x Renanthera x Vanda x Vandopsis
Paulara = Ascocentrum x Doritis x Phalaenopsis x Renanthera x Vanda

9.3.4.5 Natural Hybrids

Phal. Amphitrita, Phal. intermedia, Phal. leucorrhoda, Phal. rothschildiana, Phal. Singulifera

9.3.4.6 Inter-specific Hybrids

'Borneo Belle', 'Cecile', 'Fuscabell', 'Anna', 'T.H. Pearl', 'renchy's Plastic Yellow', 'Herman Sweet', 'Gold Veins', 'Amabell', 'Algicora', 'Smiling Tiger'.

9.3.4.7 Inter-varietal Hybrids

'Free Gold', 'Durga Ko Dil', 'Popp Queen', 'Red Hot Girl', 'White Ghost', 'White Galaxy', 'Dotted Perfection', 'Hamana Gold', 'Triastar Diamond', 'Tiraster Golden Lip', 'Archie

Goodwin', 'Taida Sun Smile', 'Pink Pixie', 'Exotics Plum Red', 'Woodson's Dalmation', 'Walnut Valley Peachy'.

9.3.4.8 Variety-species Hybrids

'Hemlata and Chris', 'Hama Snow', 'Berry Blossoms', 'Good Time Charlie', 'Stone Trail', 'Taisuco Imperator', 'Taisuco Sumatra', 'Thirty Eight Special'.

9.3.5 Climatic Requirements

9.3.5.1 Light

Phalaenopsis orchids are not like Cymbidium, Cattleya, regarding their light needs, and do well in indirect light. For this reason, healthy *Phalaenopsis* can be grown indoors in windowsills, sun rooms, and shaded greenhouses and under artificial lights in most temperate climates. The requirement of light is 1000-1500 foot candles for winter and 800-1200 foot candles in summer. *Phalaenopsis* can be grown under artificial light. *Phalaenopsis* can be grown 9-12 inches under fluorescent lights or 4 to 6 feet under 400 watts high intensity discharge lights or high pressure sodium lights. If *Phalaenopsis* does not have a flower spike growing by February, move it where it will receive more light. The indication that the correct amount of light has been received is that the foliage will appear yellow-green, not dark green. Dark green leaves or the new leaf growing longer and narrower than the old leaf indicates the light is too low. Too much light causes white, dried, burned areas on the leaf, and will have short flower spike. Inadequate light results in succulent, floppy, dark green foliage with no flowers. In cloudy winter conditions artificial light should be supplemented, whereas direct sunlight will damages the plant, therefore placing the plant near a bright window is good. In green houses, 750-1500 foot candle light is judicious. If grown indoors, *Phalaenopsis* should be placed in a North East window to avoid hot and direct sun. Higher intensity light is recommended for boosting growth at the vegetative stage, particularly for growing leaves and roots and flower induction. At flowering, *Phalaenopsis* can tolerate low light intensity (100 foot candle).

9.3.5.2 Temperature

The *Phalaenopsis* is a tropical plant, and consequently temperatures lower than 15°C and above 32°C should be avoided. For an appropriate growth endeavors should be made to maintain an average temperature of 26-27°C during the growing phase and 19-21°C during the phase of flowering. During winter the temperature should be maintained between 18 to 20°C. A temperature of 18°C is particularly necessary in the event that the induction of buds needs to be enhanced in conditions of inadequate light or high daytime temperatures. Care should be taken to ensure a minimum night time temperature of 15°C -20°C for the rest of winter. Although they are warm growers, above 30°C, they tend to stop growing, so if this is likely take steps to reduce the temperature by improving ventilation or misting the floor etc. Continued periods of exposure to low temperature below 10°C or a rapid drop in temperature can cause chilling injury which ultimately results in poor growth and yellow sunken spots on the leaves. Higher temperature coupled with higher humidity will not affect the plant. *Phalaenopsis* can tolerate higher temperature (30°C to 35°C) for a few hours if sufficient water is in pot. They need cool temperature for a month at the time of bud development. Higher temperature delays the flowering and aborts the bud at 28°C and more during flower bud development.

9.3.5.3 Humidity

Phalaenopsis can perform better with the humidity of 50% or higher. At sufficient humidity plants grow lushly and leave looks healthy. Insufficient humidity cause stunting of plant, accelerated premature falling of buds, dehydrated and shriveled leaves and papery texture of flower edges. In greenhouses, the humidity can be easily controlled by watering the foot path and benches or by humidifier. Humidity can be improved sufficiently in the growing area. Keeping of water in plastic tray below the benches is a good option to maintain the humidity in green house. Misting can also be done to increase humidity for a limited period, but it causes leaf spot diseases if there is not good air circulation. To increase the humidity level, companion plants like ferns, bromeliads, and other foliage plants can be placed near the *Phalaenopsis* plant in the house.

9.3.5.4 Ventilation and Air Circulation

It is necessary to provide enough space between plants and to allow air movement (slight breeze) to help dry the leaves. Using an electric fan to induce air movement inside the greenhouse is common. Plants should not be left in stagnant air, as they can become infected by bacteria and fungus spores. Fans should be used constantly to reduce temperature on hot days and to dry plants out for cold nights. In most of the areas where Phalaenopsis are found, gentle air blows. Air movement in growing environment ensures good growth and less infestation of diseases and pests. Ceiling and oscillating fans are effective for providing gentle air flow in hobby greenhouses or indoor growing areas. Both can cover large areas with a constantly changing air flow pattern without excessively drying the plants.

9.3.6 Growth and Flowering

The *Phalaenopsis* passes through two distinctive phases i.e. vegetative and flowering for production. Plants are grown in separate greenhouses with required temperatures during these two phases.

9.3.6.1 Vegetative Phase

The vegetative phase begins when plants are removed from tissue culture flasks and are transplanted into community trays. Forty to fifty plants from a flask can be accommodated into 1300 to 1500 cm^2 square or rectangular trays or in 128-cell plug tray. The vegetative phase is again divided in three phases. The first phase comprises pricking out from flask (25-30 weeks), while the second phase includes repotting to an intermediate pot (27-32 week) and the third phase is to planting in a bigger pot (10-12 week). During vegetative phase, the plants must be grown at 28°C or higher to avoid immature inflorescences. The higher temperature will promote rapid growth of leaves. *Phalaenopsis* can tolerate temperatures as high as 32°C, to 35°C for a few hours per day subject to sufficient moisture in the substrate and good aeration. In general, about 50 weeks are required to complete these phases. During the initial stage of growth, during the first 4 to 8 weeks after transplanting the recommended maximum light intensity is 75µ mol. $m^{-2}s^{-1}$ to acclimatize the plants to the greenhouse. A maximum of 300µ mol.$m^{-2}s^{-1}$ light intensity is maintained during the growth phase.

9.3.6.2 Flowering Phase

Phalaenopsis plants already developed 4 to 5 large leaves, can be exposed to cooler temperatures to induce the flowering process. Flower induction occurs when exposed to less than 26°C temperature during the day. Generally, growers maintain a 25°C/20°C day/night temperature regime for flower spike initiation. However, plants exposed to a constant 25°C temperature, produce spikes at a faster rate than those exposed to 26°C. After being exposed to this temperature for a period of 4 to 5 week, plants can be grown at 17°C to 26°C to time the flowering at specific date. Inflorescence and flower bud number are higher when *Phalaenopsis* is induced at 14 to 17°C compared to warmer temperatures. In warmer regions, the flowering can be regulated in air conditioned green houses for year round production.

At a favorable environmental condition (28°C) bud initiation occurs after the spikes attain 5 cm in length. However, if a plant with a young inflorescence (less than 10 cm) is subsequently grown at 28°C or higher spike can form 'keiki' instead of flower buds' or buds may abort. During the flowering phase, higher temperatures may increase spike but number and size of flower may decrease. If the light intensity is excessively low during flowering phase, flowering can be inhibited.

9.3.7 Propagation

9.3.7.1 Division/ Cuttings

This type of propagation can be done by using two types of techniques. *Phalaenopsis* can be propagated asexually by keikis forming on old flower stalks or on the sides of the main stem. Top cutting can be practiced on long drooping plants, after which, new plants will shoot from the remaining stem. Using lanolin paste with Benzyl adenine (BA) on buds in flower stalks to induce kiekies has given successful results. Kiekies can be planted in clay or plastic pots with chopped charcoal and coconut husk. Stem propagation sometimes also involves the application of keiki paste or a similar product to dormant nodes of the flower spike, an ideal method for non-commercial growers. When the keikis have 2-3 roots, it can be removed, by slicing between the stem and the keikis, or cutting the stem above and below keikis attachment point. The new plant can now be potted up and grown on. If more flowers are desired, cut the stem as above, but do not move the plant.

The *Phalaenopsis* mother plant is topped and it continues to grow vertically and discard its lower leaves that had served as a storage vessel of water and nutrients. New roots are produced above the leafless stem, as the plants continue growing vertically. The stem can be cut below the new roots. The top part, with leaves and roots, is repotted after proper care of the cut. The remaining stub can be left for a few days/ weeks. Soon, new little plants will be found growing out of the old stub. These keikis can be repotted in the media.

9.3.7.2 Tissue Culture

Phalaenopsis can be propagated by tissue culture technique using shoot tip, nodal flower-stalk, internodal section, leaf tip and root tip in Vacin and Went medium, Murashige-Skoog medium, REM, Knops Solution, Hyponex/Kyoto medium New Dogashima medium.

9.3.8 Cultivation

9.3.8.1 Plant Material

The plants are generally available in flasks containing 20 to 25 plants or in nursery trays containing 40-50 plants. As soon as they are received, it is necessary to acclimatize the plants for a few weeks to the local growing. The plants in nursery trays are better as there is less mortality and a reduction of cultivation period by approximately 5-7 month.

9.3.8.2 Potting

The mature plant with 2 to 3 leaves and healthy roots is generally shifted to a nursery tray from the flask. Once they harden and the leaf attains a span of 10-15 cm size in the nursery tray, they can be shifted to pots. The plants need to be sorted out in two grades i.e. bigger and smaller plants, before planting. Small plants often require 3-4 months more before they are shifted to pots than bigger plant. Smaller plants can be kept together in community pot. Seedlings can be planted 1 cm apart.

The potting material should be damped and should be packed firmly and it is required to water the seedling gently and thoroughly. The community pot needs to be covered with plastic and left slightly open. This will provide the seedlings with a humid environment and need to keep the plant in partial shade. In few months the seedlings will be ready for transplanting as like bigger plant.

9.3.8.3 Containers

The most common pot or container is a plastic or clay pot. In general, the orchid pot should have a greater number of holes and larger size, both at in the bottom and sides for better drainage. *Phalaenopsis* pots are shorter and shallow (10 cm) or have a larger diameter with a broader base for more stability. Most of the growers prefer plastic pots over clay pots because they are inexpensive, durable, have a lower accumulation rate of salts and are easy to clean. In *Phalaenopsis*, a healthy and more roots can be obtained in transparent pots as compared to dark coloured pots. Generally, white colored transparent pots of 12 to 20 cm sizes are preferred for *Phalaenopsis.*

9.3.8.4 Potting Media

Potting material must support the plant, drain out rapidly, retain moisture and should have durability. *Phalaenopsis* cannot withstand long dry spells because they do not have any water storage organs. A good potting mix contains coconut husk pieces, charcoal or broken pieces of roofing tile with coconut peat. 25-30% of the media should be charcoal or tile pieces to ensure adequate drainage. *Phalaenopsis* can be potted on clay or plastic pots, with charcoal and coconut husks, or mounted in wooden slabs. Currently, the use of tree fern roots is being discouraged for the conservation of our endangered giant tree ferns. Pre-soaked coconut husk is another good medium for *Phalaenopsis*. Another alternative medium is a 1:1 mixture of sand and coconut choir dust. For immature or miniature *Phalaenopsis* a fine mix consist of 3 parts coco-peat, 1 part fine charcoal and 1 part perlite and 1 part sphagnum moss is filled in 10 cm size pot. However, for standard size plant the growing media consist of 4 part coconut husk, 1 part medium charcoal, 1 part medium size perlite and 1 part medium sphagnum moss.

9.3.8.5 Repotting

Repotting should be done if there is wilting of foliage, the plant is loose in the pot, flowers do not last long, there are brown tips on the leaves etc. The plants prefer good aeration around the roots and this is the easiest way to check the condition of the root system. The *Phalaenopsis* are best repotted after flowering in the late spring or early summer. Repotting is done about once a year in an appropriate sized pot. During repotting, it is required to cut or break the base of the stem is carefully. Except main also other roots should be cut off and soak the plants and roots in a standard Mancozeb solution for 10 minutes then seal the bottom of the stem with a thick paste of Mancozeb or Thiram. Plants should be air dried before repotting. Growing media is the same as initial potting media.

9.3.8.6 Watering

Water needs are influenced by humidity, potting media, type of pot, size of pot, time of year and the orchid. Over watering is the biggest killer of orchids. *Phalaenopsis* do not like water as do most other types of orchid. The pot should not be watered until very nearly dry. *Phalaenopsis* do not prefer frequent watering, although younger plants require more watering than matured plants. This can be accomplished by misting the plants instead of water sprinkling. Watering is necessary twice per day during very hot and dry months, and only 2-3 times per week during wet months or none at all for matured plants. The quality of water is an important factor for governing the growth and flowering. Water must be free from chemical and visible contamination. Water for *Phalaenopsis* should not contain excessive harmful elements like sodium, chlorine and bicarbonate. In the absence of good quality of water, deionised water can be used as substitute. However, rainwater is best for all orchids.

9.3.8.7 Fertilizers:

Phalaenopsis plants need nutrition twice or once a week. Seedlings could be sprayed with very dilute fertilizer every day after watering. Plants need to be watered heavily once a week to flush excess salts and should be moist before applying of fertilizers. Use of balanced foliar fertilizer with trace elements is always beneficial for orchids. Weak fertilizer applied frequently is better than strong application.

Reduction of fertilizers is suggested once flower spikes form and under low light conditions. It is important to note that feeding depends upon some other environmental factors: the higher the light the more food is required and the more food is available at the roots, the more of water is necessary to prevent salt damage to the roots. When plants are grown under lower light and temperature, less water and nutrition is required. The pH of fertilizer solution needs to be maintained between 5.2 and 6.2. The EC of the nutrient solution should lie within the range of 0.8 and 1.2 mS/cm. Slow release fertilizers like Nutricote (13:13:13) or Osmocote (13:13:13) can be used at a level of 1 table spoon/plant while planting. The most common form of fertilizer used with *Phalaenopsis* is the water soluble type. The water soluble is a 19:19:19 or 20:20:20 NPK with micronutrients which are available in market and users friendly. Generally, 1g/l at weekly intervals is sufficient for growth. It is better to fertilize the plant frequently with diluted rate. The excess rate or higher concentrations will damage the plant. At the active growth stage the *Phalaenopsis* require more nutrition than in winter or inactive phase. In the dormant period, the fertilizers should be applied at lesser rate (50%). If salt accumulation occurred, it needs to clean with flushing of water.

9.3.8.8 Flower Production

Mature plant with 5 fully developed expanded leaves may be induced to flower. Generally, flower production starts 8 months after planting subject to proper management. It is important that plant should be healthy, large enough and the pots have a good root system to bear the flowering. *Phalaenopsis* will naturally induce spike when it is fully grown. Flower production can be manipulated by controlling light and temperature regimes. The spikes become ready to harvest when the spike has approximately three unopened buds. New and heavier branches could develop from the plant base after harvesting. The development of this new branch will take more time than the development of a branch from a bud. There should be at least 12 plants/sq.m and the complete growing period approximately 5 years. On average, the plant produces 2 branches /plant/year.

9.3.8.8.1 Premature Flowering

Sometimes flower spikes develop before the plant matures. The premature flowers will need to be removed to boost vegetative growth. At the early stage, the flower inflorescence, or stick, is soft enough and can be pinched off easily. During the growth phase, the maintenance of temperature at 27°C limits the premature flowering.

9.3.8.9 Training of Spike

Phalaenopsis is a marketable flower and consumers prefer straight flower spikes with better presentable form. The *Phalaenopsis* flower stems are supported with sticks made of plastic, bamboo or metal at the time when flower buds begin to swell. Tying of spike should be secure and firm with the stake. Use of twist tie, soft cotton string, small green cable ties and loop tape are better than wire, which could damage the stem. The first tie has to be placed on the lower part of the spike close to the first node. Another tie should be placed a few inches higher on the flower spike. If necessary, more ties can be placed at suitable distance. The spikes are attached to overhead wires with the help of strings to avoid bending. The flower sprays are supported with a small hook from the strings when the lower bud starts swelling.

9.3.8.10 Harvest and Yield

The 40-60 cm long spikes containing at least 8-10 flowers are harvested when all flowers are fully open. Average flower production is 6 to 7 stems / plant /year.

9.3.9 Post-harvest Management

9.3.9.1 Vase life

Normally, *Phalaenopsis* cut flowers last for 15 to 21 days.

9.3.9.2 Pulsing

Chemicals like Chrysal are added to water to enhance vase life. In *Phalaenopsis*, pulsing with 0.5 mM STS for 24 hours blocks the deleterious effect of ethylene.

9.3.9.3 Storage

The harvested flowers need to be kept at proper temperatures. In general, *Phalaenopsis* flowers can be stored at 7 to 10°C for two weeks. The low temperature helps to slow down respiration and depletion of reserved food materials of petals.

9.3.9.4 Packaging

The *Phalaenopsis*, cut flower are packed in single window gift boxes of 100 cm x 15 cm x 11.5 cm. 25 to 30 flowering stems are packed in a box depending on the number of flowers/ stem. During packing of flowers, boxes are cooled at 20°C. Pre-cooled boxes are more efficient than packing a box and then placing in a cooler place, a process that requires hours to bring flowers to the optimum temperature range. Cut stems are kept in cooler places prior to packing. It is best to pack in cooled rooms to reduce respiration and condensation build-up in the slips.

9.4 Cattleya (Orchidaceae)

9.4.1 Introduction

Cattleyas are considered by most people to be fascinating home garden orchids. Flowers are long lasting and possess a beautiful fragrance. Cattleya hybrids also produce the biggest orchid flowers. Named in the honour of William Cattley, a noted 19th century English Horticulturist, this genus falls into subtribe Laeliinae, which comprises many Cattleya like species. Of the many intergeneric hybrids, *Laelia, Sophrinitis* and *Brassovola* are the main genera used in hybridization programs. *Cattleya* orchids are also known as the 'Queen of Orchids'.

9.4.2 Botanical Description

Cattleya consists of 113 species from tropical America. The plants possess elongated pseudobulbs and may be unifoliate or bi-foliate. The leaves are thick and leathery. The present day hybrid cattleyas belong to the unifoliate group. The flowers are 5 to 15 cm in size and they occur in all colours, except true blue and black.

Unifoliate cattleyas bear up to 5 flowers per inflorescence whereas bifoliate possesses 2 to 25 flowers per inflorescence. They are from the tropical Americas, but can also be found in the West Indies and Mexico. *Cattleya* orchids are epiphytes and have well-developed water-storage organs (called pseudobulbs) and large, fleshy roots. *Cattleya* orchid plants are long-lived perennials and will usually flower annually. These orchid plants are naturally erect, without need of much staking, and of a medium olive-green color.

9.4.3 Importance and Uses

In addition to cut flowers and pot plants, Cattleya orchids are ideal components for flower decorations including bouquets and flower arrangements. In addition, this orchid is used for making clay flowers.

9.4.4 Hybrids (De, 2014)

9.4.4.1 Parents for Large Blue Cattleya Hybrids

Cattleya warneri var. *coerulea*, *Cattleya labiata* var. *coerulea*, *Cattleya mossiae* 'Reineckiana Blue Lip', *Cattleya gaskelliana* ' Blue Dragon', *Cattleya trianae* 'Blue Bird'

9.4.4.2 Bigeneric Hybrids

Brassocattleya = Cattleya x Brassolvola
Epicattleya = Cattleya x Epidendrum
Laeliocattleya = Cattleya x Laelia

9.4.4.3 Trigeneric Hybrids

Vaughnara = Cattleya x Brassovola x Epidendrum
Brassolaeliocattleya = Cattleya x Brassovola x Laelia
Rolfeara = Cattleya x Brassovola x Sophronitis
Osmentara = Cattleya x Broughtonia x Laeliopsis
Dialaeliocattleya = Cattleya x Diacrinum x Laelia
Epilaeliocattleya = Cattleya x Epidendrum x Laelia
Sophrolaeliocattleya = Cattleya x Laelia x Sophronitis

9.4.4.4 Tetrageneric Hybrids

Iwanagara = Cattleya x Brassovola x Diacrinum x Laelia
Yamadara = Cattleya x Brassovola x Laelia x Epidendrum
Potinara = Cattleya x Brassovola x Laelia x Sophronitis
Kirchara = Cattleya x Epidendrum x Laelia x Sophronitis

9.4.4.5 Intervarietal Hybrids

'Day Tripper', 'Spring Break', 'Beau's Apricot Gem', 'Dendis Angel', 'Dendis Bee', 'Sweet Peggy', 'Redland Icicle', 'Moonlight Glitter', 'French Cream', 'Mackay Drop', 'Nobiles Virgin', 'Summer Bouquet', 'Tokyo Life', 'Poo Hoon', 'Montking', 'Black Wonder', 'July Sun', 'Pink Jaguar', 'Memoria Madan Tamang'.

9.4.4.6 Inter- specific Hybrids

'Eva's Fabula Andena', 'Oleo Stella' 'Oro Verde Azul', 'Mareeba Freckles', 'Brazilian Jewel', 'Bird Flap', 'Bucks Fizz', 'Genes Dream', 'Little Lars', 'Comwhite', 'Amazon Blue', 'Mini Green', 'Jungle Nobility', Berlin Moon', 'Brazilian Ruby', Brazilian Fire', 'Brazilian Star'.

9.4.4.7 Variety Species Hybrids

'Maryglossa', 'Betty Blue', 'Koko Drop', 'My Partner', 'Pradit Spot', 'Summer Poole', 'Angaur' 'Baby Tata', 'Little Bells', 'Balkis Gem', 'Snow Field', 'Summer Pink', 'Berlin Star', 'Berlin Beauty', 'Memoria Bishnu Subba', 'Memoria Naren Rai', 'Leopard', 'Snow Tiger'.

9.4.4.8 Blue Cattleya Hybrids

Cattleya Alcmeda, *Cattleya* Bobby Howarth, *Cattleya* Intertexa, *Cattleya* Mrs. Myra Peters, *Cattleya* Veriflora

9.4.5 Cultivation

Cattleya orchids belong to a group of orchids which is the most colorful of all orchid species. Due to their relative ease of culture and adaptability, *Cattleya* orchids are among the most popular orchid genus grown because of their cross compatibility many genera, which have similar structures and a range of interesting and unique characteristics. Additionally, the *Cattleya* orchids belong to both the intermediate and warm climate groups of orchids under cultivation. *Cattleya* orchids lend themselves to many different types of cultivation and they will grow outdoors in trees in tropical landscaped gardens as well as in pots in shade houses and climate controlled green houses. *Cattleya* orchids are usually slow growers and can take up to three or even four years to flower. However, once they start flowering they will continue to do so all year round under their optimum growth conditions.

9.4.5.1 Temperature

The ideal temperature range of *Cattleya* orchid species is between 15 and 30°C. Depending upon the temperature, *Cattleya* orchids should be watered about once or twice a week. In winter, the *Cattleya* orchid plants become stressed if the temperature drops below 15°C. It is also essential to keep the orchid plant a little drier under those conditions because extended exposure to cold and damp simultaneously will result in rot. The *Cattleya* orchid goes into a state of dormancy and will wake up when the weather warms up again. Do not feed or water the *Cattleya* orchid plant when it is in a state of dormancy damage. During the summer, the *Cattleya* orchid plants should be watered and fed. *Cattleya* orchids should always be protected against frost in the cold winters.

9.4.5.2 Light

Cattleya orchids need medium to bright light with 2000 to 3000 foot candles. They thrive well under a 40% shade cloth. Dark green but limp foliage indicates that they received little light. The *Cattleya* and its allied inter-generic hybrid orchids are capable of withstanding higher light, but they may become stressed if exposed for too long.

9.4.5.3 Water and Humidity

Cattleya orchids should be watered about once or twice a week, and more frequently during the hotter months of the year. *Cattleya* orchids may be allowed to dry out between applications of water. It is advised to always water the *Cattleya* orchids early in the day, so that the foliage will be dry by nightfall. They require 40-60% relative humidity. During the summer months, it is necessary to feed and water *Cattleya* orchids to plump up the pseudobulbs and thus provide nutrients for storage purposes to be used during the winter rest period.

9.4.5.4 Feeding

Cattleya orchids are known as heavy feeders. This is especially true during their active growth period in spring. A high nitrogenous orchid fertilizer must be applied on a regular basis during spring. Application of dilute 20:10:10 (N:P:K) weekly fertilizer produces more flowers. Furthermore, well-rooted plants should be given regular feeds of nitrogen-based fertilizers. During summer, the fertilizer type may be changed to a good quality bloom booster that will harden the plant off and make it ready for flowering.

9.4.5.5 Potting Mix and Repotting

All *Cattleya* orchids, including all hybrid and orchid species that have been hybridized from the *Cattleya* orchid require a very open, friable potting mix and a very free draining compost containing at least 50 percent bark. They should be potted in a porous, free-draining medium. The most commonly used potting mixes are bark, shredded tree-fern fibre, various types of rock like granite chips, processed coconut fiber and lately, mixes based on peat moss and perlite and also osmunda.

Cattleya orchids should be potted on an annual basis when the orchid plant is young or in its active growth period. Once the orchid plant has matured and begun to flower, repotting may occur every alternate year. Repotting of *Cattleya* orchids is best practiced soon after flowering, just when the new growth appears at the base of the pseudobulbs, and just before any new roots may have begun.

9.4.6 Pests and Diseases

All excess humidity increases the risks of fungus and bacterial infections. This can further result in a loss of new growth and in severe cases causes even permanent spotting on the leaves.

The ant is a serious pest to *Cattleya orchids*. They relish the sugary substance that is produced by the new flowers. They are carriers of scale insects. To prevent this, a grower should remove old bracts and sheaths and groom the orchid plants regularly.

Virus infection on *Cattleya* orchids are usually manifested as white striations in the flowers. This phenomenon is referred to as color-break. This color-break soon develops further into a brown streak. Viral infections on *Cattleya* orchids can be controlled effectively by making use of clean secateurs, blades, etc.

9.4.7 Post-harvest Management

9.4.7.1 Stage of Harvest

Spikes are harvested with fully open flowers. There may be one or more flowers on a spike of 25-40 cm length depending on the type.

9.4.7.2 Storage

Cut flowers can be kept at 7-10°C for 14 days in dry storage and 10 to 13°C in wet storage.

9.4.7.3 Vase life

Cut flowers last for 10 to 15 days in normal water.

9.4.7.4 Preservatives

STS 1 mM + sucrose 1 percent.

9.5 Oncidium (Orchidaceae)

9.5.1 Introduction

Oncidium species, their hybrids and inter-generic hybrids are suited to intermediate and warmer climates. The plants are epiphytic using hosts such as cactus plant and trees for support. Hybridization of *Oncidium* orchids with Brassia orchids, Miltonia orchids and *Odontoglossum* orchids results in an *Oncidium* orchid that has some warmth tolerance and beautiful flowers. Hybrids and some commercial species are used for cut flowers, hanging baskets and potted plants.

9.5.2 Botanical Description

Oncidium consists of 750 species of sympodial epiphytic orchids from America, Mexico and Argentina. The plants have either pseudobulbs or a fan of very thin leaves. They bear numerous attractive blossoms in various size and forms and are commonly called 'Golden Showers' and 'Dancing Ladies'.

The pseudobulbs are topped by one or more leaves which are small, soft, pencil like or very large, leathery and thick. Usually, single inflorescence is produced from a single growth or in some cases, two inflorescences may be produced. Inflorescences develop from the base of the pseudobulbs or from the axil of leaves. The flower size varies from 1 cm to 12.5 cm across. The flower colours are mostly shades of yellow and brown, and in some cases they are red, pink, magenta, green or white. In general, all three sepals are alike in size, shape and colour, although in some cases these vary. The two lateral petals are similar in size and shape while dorsal sepals are larger. *Oncidium* species are characterized by (i) presence of coloumn wings (ii) presence of complicated callus on the lip (iii) pseudobulbs with one or three leaves (iv) several basal bracts at the base of pseudobulbs.

9.5.3 Species & Hybrids (De, 2014)

Tab. 9.7: Common Oncidium Species and Hybrids

Types	Species & Hybrids
Yellow Flowered	*Oncidium splendidum O. lanceanum, O.spacealatum*
Golden Shower type	Aloha Iwanga Dogasima, Goldiana, Gower Ramsey, Golden Shower, Sum Lai Who Jungle Queen, Taka H & R, Sharry Baby Sweet Fragrance AM/AOS
White coloured	*Oncidium variegatum* White
Red coloured	Popki Red, Irine Gleason Red, Vision Brownish Red, Catherine Wilson x New Calidonia Brownish Red
Pink Coloured	Robson Orchid Glad
Cream Coloured	*O. lowianum* hybrids

9.5.3.1 Bigeneric Hybrids

Aspasium = Oncidium x Aspasia
Brassidium = Oncidium x Brassia
Miltonidium = Oncidium x Miltonia
Odontocidium = Oncidium x Odontoglossum
Trichocidium = Oncidium x Trichocentrum
Oncidipilia = Oncidium x Trichopilia

9.5.3.2 Trigeneric Hybrids

Aliceara = Oncidium x Brassia x Miltonia
Wilsonara = Oncidium x Cochlioda x Odontoglossum
Colmanara = Oncidium x Miltonia x Odontoglossum

9.5.3.3 Tetrageneric Hybrids

Withnerara = Oncidium x Aspasia x Miltonia x Odontoglossum

9.5.3.4 Interspecific Hybrids

'Dark Tower', 'Ruby Frost', 'Sanddrinho', 'Barbara Ann', 'Himekogane', 'Peach Pie', 'Flamingo', 'Flower Fairy', 'Nutmeg Dancer', 'Karukera Beauty', 'Midnight Moon', 'Lemon Ice', 'Caribbean Stars', 'Ruby Jewell', Debonoir', 'Fragrance Fantasy', 'Red Fantacy'.

9.5.3.5 Inter-varietal Hybrids

'Volcano Gold', 'Dark Sun', 'Bright Night', 'Coral Gold', 'Golden Sun', 'Sundown', 'Sweet Sunset', 'Golden Bonanza', 'Dear Friend', 'Orchidom Happy', 'Kulmura Gold', 'Aka Cacao', 'Music Shower', 'Summer Glow', 'Baby Breath', ' Island Gold', 'Kona Boy', 'Space Baby', 'Millenium Gold', 'Golden Prince', 'Golden River', 'Sun Shade', 'Ash Hollow', 'Private Dancer', 'Green Valley', 'Fragrant Red Barry', 'Golden Sunray', 'Sungold', 'Sharry Baby Dancing Doll', 'Sharry Baby Pink Lip', 'Sharry Baby 'Tricolor'.

9.5.3.6 Variety-Species Hybrids

'Cameo Moonlight', 'Dusk', 'Jazzberry', 'Long Yellow', 'Orchidon Gold', 'Tokyo Fantacy', 'Panache Gold', 'Sarah Elizabeth Merritt', 'Barbie Doll', 'Elegant Dancer Pretty Lady', 'Yellow Canary', 'Butterfly Profusion', 'Pacific Perfume', 'Copper Hills', Annabel', 'Forbes Island', 'Sweet Sugar'

9.5.3.7 Colmanara Hybrids

Colm. 'Jungle Monarch', Colm. 'Wildcat', Colm. Wildcat 'Bobcat', Colm. Wildcat 'Cheetah', Colm. Wildcat 'Rainbow', Colm. Wildcat 'Red Star', Colm. Wildcat 'Tiger'.

9.5.3.8 Medicinal Oncidium

Oncidium cebolleta is reported to contain phenanthrene derivatives.

9.5.4 Cultivation

9.5.4.1 Temperature

The majority of species grow well in temperature between 25-30°C during daytime and 20°C during the night. A little bit of heat can be tolerated by the warmth tolerant *Oncidium* orchid hybrids if humidity and air movement are increased as the temperatures rise.

9.5.4.2 Light

Light needs can vary from bright to nearly full direct sun depending on the *Oncidium* orchid species and hybrids. Most *Oncidium* orchids will thrive with one to several hours of sun a day and a light intensity of 2500 foot candles. Generally, thicker-leaved plants, such as **Tolumnias** also known as "**mule-ear**" and "**equitant**" *Oncidium* orchids, can tolerate more light. *Oncidium* orchids in a greenhouse are best advised to make use of a shade cloth which can be anywhere from 30 to 50 percent shade, depending on the orchid plants. In the home, *Oncidium* orchids grow best in the east, south or west windows. Many types of *Oncidium* orchids will even grow under artificial light.

9.5.4.3 Water and Humidity

Most *Oncidium* orchid species and the *Oncidium* orchid hybrids prefer free compost that holds moisture rather than water. Water requirements vary with the type of orchid plant. Generally, *Oncidium* orchid plants with large fleshy roots or leaves require less-frequent watering than thin-leaved or thin-rooted plants. Watering should be thorough, and the medium should be allowed to dry at least halfway through the pot before watering again. This may be every 2 to 10 days depending on weather, pot size and material, type of orchid and type of potting medium. Plants not actively growing should be watered less. Humidity ranges should be between 50 and 60 percent. Many *Oncidium* orchids require less humidity than other orchids. Most greenhouses have adequate humidity and the *Oncidium* orchid can easily be kept well under those circumstances. Under indoor climate, *Oncidium* orchid plants survive better above moist pebbles in trays. If *Oncidium* orchids are grown in a shade house or garden outside, then they will require more frequent mistings to reduce the possibility of dehydration. Dehydration of *Oncidium* orchids is usually manifested as crinkled leaves.

9.5.4.4 Feeding

The *Oncidium* orchid is known to be a heavy feeder. The *Oncidium* orchid and its hybrids must be fed at half the recommended strength of most fertilizers once a week while the orchid plants are actively growing. A high nitrogenous fertilizer in spring is beneficial in terms of new growth and a balanced fertilizer during the summer and the winter keeps the *Oncidium* orchid in peak condition. During autumn, it is best to feed the *Oncidium* orchid with a high potash-based fertilizer to aid new flowering stems from the bulbs.

9.5.4.5 Potting Mix and Repotting

Potting of *Oncidium* orchids should be done when new growth is at about one-half mature stage. This is usually done in the spring. One can use fine-grade potting media with fine-rooted *Oncidium* orchid plants and coarser mixes with large-rooted orchid plants, and the standard size is medium grade. The *Oncidium* orchid should be positioned in the pot so that the newest growth is farthest away from the edge of the pot and allows the maximum number of new growths before crowding the pot. The roots of the orchid plant are spread over a cone of potting medium which is then filled in around the roots. The medium around the roots of the orchid plant is firmed with your fingers. The humidity is kept high and the potting medium kept dry until new roots form.

The *Tolumnias*, or *equitant* and *mule-ear Oncidium* orchids, as well as other fleshy-leaved or large-rooted orchid plants are easily grown on slabs of cork bark or tree fern or in pots filled with a coarse, well-drained medium such as charcoal, or even osmunda. This measure facilitates the necessary drying between water applications required by these *Oncidium* orchids; these orchids detest wet feet.

9.5.5 Pests and Diseases

The *Oncidium* orchid species is almost pest and disease free as fewer pests attack them. On rare occasions, *Oncidium* orchids get aphid infestation. This is observed usually with those orchids that are placed in open shade houses.

9.5.6 Post-harvest Management

9.5.6.1 Stage of Harvest
Spikes are harvested when more than 80 percent of the flowers on the spikes are fully open on a spike length of 60 cm with many florets.

9.5.6.2 Storage
Cut flowers can be stored wet at 8 to 12°C.

9.5.6.3 Vase life
Cut flowers last for 12 to 15 days in normal water.

9.5.6.4 Preservatives
8-HQC 200 ppm + 4% sucrose; $AlCl_3$ 100 ppm + boric acid 500 ppm + 4% sucrose + aspirin 100 ppm are commonly used preservatives for holding solutions.

9.6 Paphiopedilum (Orchidaceae)

9.6.1 Introduction

Paphiopedilums are most popular orchids because of their unique shape, colour and ease of cultivation. They are commonly called 'Slipper Orchids' and can be grown in windowsills, basements, Wardian cases, small greenhouses, in large commercial greenhouses, lathhouses and shade houses and can be grown in different climates. Paphiopedilum is a Greek word; 'Paphion' means an epithet for Aphrodite (the Goddess known as Venus to the Romans) and 'pedilon' meaning slipper. In India at least seven species have been reported so far. Paphiopedilums were originally introduced to England in the 19th century.

Paphiopedilum venustum was the first species to flower in England in 1819, and subsequently other species like *Paph. insigne, Paph. javanicum, Paph. purpuratum, Paph. barbatum, Paph. lowii* and *Paph. glanduliferum* were introduced. The first hybrid of Paphiopedilum was flowered by John Domany of Messrs. It was Veitch who developed a hybrid and named it *Paph. Harrisianum (Paph. villosum x Paph. barbatum)*. The second hybrid was named as *Paph. Vaxillarium (Paph. barbatum x Paph. fairrieanum)*. More than 13,000 hybrids have been registered that are very attractive in colors, size and shapes. These are grown as pot plants or cut flowers.

9.6.2 Botanical Description

Paphiopedilum consists of 50 species of terrestrial or lithophytic orchids distributed in the Himalayas, China, South East Asia, Indonesia and New Guinea. These orchids are stemless, pseudobulbless sympodial orchids with well developed leathery, elliptic – lanceolate leaves clasping at the base. The flowers are borne singly or in few flowered racemes on a short to elongate inflorescence. The inflorescence is 60 cm tall and often purplish brown. The dorsal sepals are distinct with markings. The two lateral sepals are fused to form a vertical sepal. The lateral sepals are narrow and long with wavy margins. The petals are at right angles to the sepals and sometimes curve forward towards the lip.

9.6.3 Hybrids

9.6.3.1 Natural Hybrids

Paph. harrisianum: *Paph. villosum* x *Paph. barbatum*
Paph. vexillarium: *Paph. barbatum* x *Paph. fairrieanum*

9.6.3.2 Bigeneric Hybrids

Phragmipaphiopedilum = *Phragmipedium* x *Paphiopedilum*

9.6.3.3 Parents for Primary Hybrids

Paph. viniferum, Paph. armeniacum, Paph. micranthum, Paph. malipoense

9.6.3.4 Parents for Mottle Leaf Hybrids

Paph. armeniacum, Paph. micranthum, Paph. bellatulum, Paph. concolor, Paph. superbiens, Paph. wardii

9.6.3.5 Parents for Maudiae Hybrids

Paph. sukhakulii

9.6.3.6 Parents for Smallest Paphiopedilums

Paph. bellatulum, Paph. concolor, Paph. godefroyae. Paph. niveum

9.6.3.7 Parents for Multi-stemmed Flower

Paph. charlesworthii, Paph. henryanum, Paph. spicerianum, Paph. tigrinum

9.6.3.8 Multi-floral or Strap Leafed Paphiopedilums

Paph. rothschildianum, Paph. sanderianum, Paph. stonei, Paph. philippense, 'Prince Edward of York', Paph. 'Michel Koopuwitz', Paph 'Saint Swithin', Paph 'Mount Toro'.

9.6.3.9 Fragrant Slipper Orchids

Paph. delentii, Paph. malipoense, Paph. micranthum, Paph. armeniacum, Paph. 'Joyce Hasegawa', 'Paph. 'Lynleigh Koopwitz', 'Paph. 'Magic Lantern', Paph. 'Harold Koopwitz'.

9.6.3.10 Hybrid Phragmipediums

Phrag. 'Sorcerer's Apprentice', Phrag. 'Grande', Phrag. 'Don Wimber', Phrag. 'Elizabeth March', Phrag. 'Hanne Popow', Phrag. 'Jason Fischer', Phrag. 'Living Fire'.

9.6.3.11 Interspecific hybrids

'AI Rutel', 'Alexej', 'Hung Sheng Eye', 'Dungu Roth', 'Georges Morel', Memoria 'David Hanson', 'Caroline Hands', 'Armeni White', 'Magic Lantern', 'Malvic', 'Aaron Poock', 'Chai-Ui Lady', 'Natasha Von Fox', 'Chiu Hua Dancer'.

9.6.3.12 Intervarietal Hybrids

'Big Tiger', 'Big Jaguar', Double Moon', 'Heavy Ken', 'Across World', 'Dark Shadow', 'Winter Vacation', 'World Records', 'Angel Dust', 'Tree of Crow', 'California Spirit', 'World Edition', 'Angel Lace', 'Red Delight', 'World Command', 'Mystic', 'Pulsar', 'Two Loves', 'Black Pepper'.

9.6.3.13 Species-varietal Hybrids

'Bubble Chocolate', 'Chunky Fairy', 'Arthur Bell', 'Angel Drop', 'Baby Angel', 'Black Beam', 'Doctor Mike Gothic Venture', 'Irish Lace', ' Night Leopard', 'Winter Passage', 'Red Glory', 'Power Spice', 'Professor Plum', 'Asian Sunshine', 'Vintage Venus'.

9.6.4 Cultivation

9.6.4.1 Light

Generally, the slipper orchids grow on shallow humus or on rocks in forest areas where conditions are cool and plenty of shade is available to protect from direct sun light. Some of the species like *Paph. villosum, Paph. parishii* and *Paph. lowii* grow as epiphytes and on tree trunks of forest trees. Most of the paphs require lower light conditions in comparison to other orchids, but some species like *P. randsii* and *P. phil-*

ippenense require higher light levels. High temperatures are detrimental for Paphiopedilums, and shade cloth roller blinds may be used in the glasshouse to control the temperature. The shade net can be drawn back during the cloudy season in summer and rolled up during winter season.

Paphiopedilum readily adapt to indoor conditions and can be grown successfully in artificial light. The optimum light for the growth of *Paphiopedilum* is between 6000-15000 lux (800-1200 foot candles). Direct sunlight is not ideal for slipper orchids.

9.6.4.2 Ventilation and Humidity

Paphiopedilums require good ventilation and high relative humidity particularly during summer months for the growth of leaves and roots and to minimize bacterial and fungal infection. The air movement helps to evaporate excessive water in the compost or potting medium. Therefore, the polyhouse should be equipped with exhaust fans of suitable size and misting systems to maintain proper humidity. The humidity can be maintained by fogging with a very fine spray, which allows tiny droplets of moisture to settle on the leaf surface thereby reducing the leaf temperature without saturating compost. In warm climates, evaporative coolers are essential for commercial glasshouses.

When the temperature decreases during the winter season, the humidity has to be decreased gradually. The optimum humidity for good growth of Paphiopedilums ranges between 65-75%. Excessive humidity may cause fungal and bacterial infection. During extreme winter when temperature falls, heaters are essential to maintain proper temperature and humidity in the polyhouse. Under low humidity the maturing buds will show yellowing and dryness.

9.6.4.3 Temperature

Paphiopedilums can survive in a range of temperatures. The best temperature in polyhouses for good growth is between 15°C to 26°C. However, some species may tolerate up to 36°C. Decrease in night temperature is required for initiation of flowering particularly in *P. rothschildianum* and its hybrids. Paphiopedilums will generally tolerate lower than ideal temperatures provided the medium is dry. During hot summer, glasshouse fans are used to increase air movement with maximum ventilation and constantly introduced humidity. Shading is also used to minimize the temperature in the polyhouse.

9.6.4.4 Containers and Compost

Paphiopedilums generally prefer light friable humus, and well drained crevices and cracks. Two types of pots are available: clay and plastic pots. Clay pots are expensive, heavy and dry out quickly whereas plastic pots are light and inexpensive and extra drainage holes can be made easily. Plastic pots can be reused after washing in a bleaching solution followed by washing with water and proper drying. They prevent the potting media from drying out too quickly and should be used with an open, friable and airy mix.

The potting media should consist of coconut husk, expended clay, charcoal, peat or Sphagnum moss, perlite, vermiculite etc. Coconut husks are economical and are broken down rapidly by heat and fertilizers. In warmer climates, *Sphagnum* moss is widely used in clay containers which allow the mix to dry out without leaving the roots saturated. Charcoal can be used as an additive, but it attracts excess salts which is harmful for emerging roots. The most widely used potting mix contains equal part of peat and coarse grade perlite mixed with three parts of medium sized bark. A small amount of dolomitic lime is also recommended as an additive over the compost. The epiphytic Paphiopedilums are also mounted on tree fern slabs or cork rafts.

9.6.4.5 Repotting

In general, the repotting should be done once or twice a year after flowering. Whenever a plant becomes pot bound and looks unwell, repotting is recommended. The plant should be removed from the pot and the roots inspected. Dead roots are removed and the plant is repotted in the same or larger pot and care must be taken to avoid damaging roots, especially with plants that have live yellow tips. For repotting, damp compost is recommended and irrigation is not given to the newly repotted plants for at least two weeks. Additional humidity and fogging are necessary for rooting in newly potted medium. New roots are encouraged to grow in fresh potting medium. At this stage plants should be watered normally.

9.6.4.6 Watering

Rain water is supposed to be the best for paphs orchids but it is not always available. For newly potted plants the media should be allowed to dry, followed by watering. Paphiopedilums require more frequent watering during summer. In general, good humidity is maintained by misting. One can judge the amount of watering required by checking the dampness of the pot medium with one's fingers. Watering should be done as early as possible in the morning to ensure that excess water will evaporate and not remain on the foliage and in the growing tip overnight. It is advised that once a month the plant should be watered thoroughly so that excess mineral salts are leached out of the compost. It is also very important that the compost should not be allowed dry completely. If it is, misting should be done to maintain proper wetness.

9.6.4.7 Fertilization

Paphiopedilums lack pseudobulbs and do not have any reserve food materials for drawing nutrition intermittently. The media is supplemented with several granulated fertilizers at the time of potting. Fertilizers are required during the active vegetative growth stage of the plant. The basic fertilizers are highly diluted to a quarter or no more than half the strength suggested by the supplier.

During spring, nitrogen based fertilizer 30-10-10 (N:P:K) is applied to encourage growth followed by a general fertilizer (20-20-20) in the summer and then potash based fertilizers are applied (10-30-20) for flowering and hardening of the plants. After this application, water is applied thoroughly once a month to leach out any excess mineral salts, which would be detrimental to the health of the plant. During winter, a weak solution of general feed is applied.

9.6.4.8 Propagation

Vegetative propagation is the common method for multiplication of Paphiopedilums. It is necessary to ensure that the plant is divided equally, because inadequate division may be detrimental and allow attack by fungal and bacterial pathogens. Some species, particularly green strap-leaf species and succulent brachypetalums, do not survive after division, whereas the mottled leaf types can survive after division. Plants that have already flowered will often send out a new shoot when potted. Another method for multiplication of Paphiopedilums is selective breeding, as there is no proven technique of mericloning.

Paphiopedilum is also propagated by seeds in flasks by many growers. One flask contains about 25-30 seedlings, but growing these plants is a difficult task for small growers. Only those flasks that are vigorous, uniform with healthy roots and no fungal infection may be selected. The choice of flask and selection of stock depends on the experience, expertise and skill of the growers.

After the discovery of tissue cultures techniques during the 1960s, it was revolution in mass propagation of orchids and other crops. Today, all the ingredients of tissue culture are easily available from laboratory suppliers and the media can be prepared and sterilized within a very short time. Seeds are sterilized in bleaching solution and are then spread over solidified agar gels in the flasks. After few weeks to months whitish yellow bodies are formed which are called protocorm. Subsequently the flasks are transferred to growth chamber where temperature is maintained at about 21°C. These flasks soon develop into tiny plantlets and are kept in the chamber for six to nine months until the plantlets are large enough to be removed.

To grow Paphiopedilum seedlings by seed culture is a costly affair for the small grower. However, a number of commercial laboratories offer flasking service. The best time to deflask the young plants is during spring or autumn.

9.6.5 Pest and Diseases

Among the orchids, Paphiopedilums are least affected by insect pests. Maintaining good hygiene in the polyhouse is the first step to manage the diseases and pests.

Regular removal of the weeds and dead leaves plays an important role in minimizing disease and pest infestations of orchid polyhouses. Keeping clean floors free from algae and mosses will prevent snails and slugs. Plant benches may be cleaned and sprayed with liquid preventatives and slug pellets. Ants play an important role in bringing aphids, mealybugs and scale insects around plants. Therefore, by controlling ants, other pest problems can be minimized. Many pests can be removed manually; application of insecticides should be the last option. Before introducing the new plants into the glass house, they are thoroughly checked for pests and diseases, and proper quarantine procedures must be followed.

Spider mites and false spider mites cause damage to the leaves by causing pitting or silvery spots, depressions or punctures on the leaves. Heavy infestations are accompanied by white webbing on the underside of the leaves. By maintaining high humidity, spider mites can be controlled. The plant should be watered thoroughly before spraying miticides.

In Paphiopedilums, the bacterial diseases caused by *Erwinia cypripedii* and *Pseudomonas cypripedii* are most common. The infected area turns black or brown as the rot spreads. To control the disease, the infected plant should be separated immediately and the infected part may be removed by cutting, followed by treatment of the cut with sulphur. If the infection is extreme, the plant should be removed from the pot and the compost should be discarded, then the entire plant should be allowed to soak in Physan solution before repotting.

Sometimes crinkling and pleating of the leaves are reported, which are not very harmful. These symptoms are due to a lack of humidity at some stage of development of leaves. The pleating may arise due to genetic problems or from an incompatible breeding. Brown leaf tip or leaf tip burn usually indicates an excess of unleached mineral salts in the compost. The decomposing acidic, soggy compost is responsible for root tip burn, rotting and poor growth of roots.

9.6.6 Post-Harvest Management

Stage of harvest: 3 to 4 days after opening of flower
Storage: 0.5 to 3°C for 20 days
Vase life: 25 to 30 days
Preservatives: 8-HQC 200 ppm + sucrose 2%; $AgNO_3$ 300 ppm + sucrose 2%

9.7 Vanda (Orchidaceae)

9.7.1 Introduction

The Vanda orchids are completely tropical and are easy to grow. Some species can be exposed to full sun. They are cross compatible with other genera like *Ascocentrum*,

Aerides, Rhyncostylis, Neofinetia, Renanthera and even *Phalaenopsis*. All orchids having the same characteristics as the *Vanda* genus, are called vandaceous orchids and they are grouped together to form the *Vanda Alliance*. Genera like *Vandas, Aerides, Ascocentrum, Renanthera, Rhyncostylis, Aranda, Mokara, Kagawara* are included in the *Vanda Alliance*. Many of orchids under this group are called as 'Scorpion Orchids' or 'Spider Orchids'. They are ideal for hanging baskets, pots or tree logs. Leaf juice of *Vanda coerulea* is used against diarrhea, dysentery and external application for skin diseases. Leaf paste of *Vanda teres* is applied to reduce temperature in fever. Leaves of *Vanda cristata* are used as tonic and expectorant.

In *Vanda tessellata*, a paste made from the leaves is effective against fever. It is an ingredient of Rasna Panchaka Quatha. In Ayurvedic medicine, it is used in the treatment of arthritis and rheumatism. The leaf juice extract is used in the treatment of otitis media. The root is an effective against scorpion sting and bronchitis.

9.7.2 Botanical Description

Vanda consists of more than 70 species of monopodial epiphytic orchids distributed in India, China, The Himalayas, Sri Lanka, Philippines and throughout South East Asia. These commercial orchids are grown in Thailand, Singapore, Malaysia and Hawai.

They are diverse in vegetative and reproductive growth. Based on leaf characters they are grouped into four categories: strap shaped, terete, semi-terete and channeled (Tab. 9.8). The lip of strap shaped leaves is very irregular while the terete leaves are of pencil thickness. The inflorescence arises from the axil of the leaves in strap leaved orchids, whereas in the case of terete leaved orchids inflorescence emerge from the side of the stem of opposite leaf. The inflorescence is axillary, erect, and simple. The flowers are small to large, few to many, fleshy, heavy textured, long lasting and yellow, brown, purple, magenta, blue, and lavender in colour. The flower size varies from 2.5 to 10 cm.

Tab. 9.8: Leaf Type in Vanda

Leaf types	Example species/varieties
Terete	*Vanda teres*, V. 'John Clubb' V. 'Miss Joaquim'
Semi-terete	V. Emma van Deventer, V. Ruby
Channelled	*Vanda sanderiana, Vanda lamellata*
Strap	*Vanda coerulea, Vanda cristata, Vanda parviflora, Vanda coerulescens, Vanda tassellata, V. stangeana, Vanda alpina*, V. 'KS.SD', V. 'Prao Sky Blue', V. 'Pures Wax', V. 'RBSD Black', V. 'PAT D', V. 'Sansai Blue', V. 'Motes Indigo', V. 'Pakchong Blue', V. 'Roberts Delight Blue'

9.7.3 Hybrids (De, 2014)

9.7.3.1 Natural Hybrids
Vanda boumaniae, *Vanda charlesworthii*, *Vanda* 'Miss Joaquim'

9.7.3.2 Bigeneric Hybrids
Aeridovanda = Aerides x Vanda
Aranda = Arachnis x Vanda
Ascocenda = Ascocentrum x Vanda
Opsisanda = Vanda x Vandopsis
Renantanda = Renanthera x Vanda
Rhynchovanda = Rhyncostylis x Vanda
Trichovanda = Trichoglottis x anda
Vandoritis = Doritis x Vanda

9.7.3.3 Trigeneric Hybrids
Aeridovanisia = Aerides x Luisia x Vanda
Burkillara = Aerides x Arachnis x Vanda
Fujiora = Ascocentrum x Trichoglottis x Vanda
Goffara = Luisia x Rhyncostylis x Vanda
Holttumara = Arachnis x Renanthera x Vanda
Joannara = Renanthera x Rhyncostylis x Vanda
Kagawara = Ascocentrum x Renanthera x Vanda
Leeara = Arachnis x Vanda x Vandopsis
Moirara = Phalaenopsis x Renanthera x Vanda
Mokara = Arachnis x Ascocentrum x Vanda
Renanda = Arachnis x Renanthera x Vanda
Trevorara = Arachnis x Phalaenopsis x Vanda
Wilkinsara = Ascocentrum x Vanda x Vandopsis
Yapara = Phalaenopsis x Rhyncostylis x Vanda
Vascostylis = Ascocentrum x Rhyncostylis x Vanda

9.7.3.4 Tetrageneric Hybrids
Alphonsoara = Arachnis x Ascocentrum x Vanda x Vandopsis
Bokchoonara = Arachnis x Ascocentrum x Phalaenopsis x Vanda
Himoriara = Ascocentrum x Phalaenopsis x Rhyncostylis x Vanda
Isaora = Aerides x Ascocentrum x Phalaenopsis x Vanda
Knappara = Ascocentrum x Rhyncostylis x Vanda x Vandopsis
Okaara = Ascocentrum x Renanthera x Rhyncostylis x Vanda

Robinara = Aerides x Ascocentrum x Renanthera x Vanda
Yusofara= Arachnis x Ascocentrum x Renanthera x Vanda

9.7.3.5 Pentageneric Hybrids

Knudsonara = Ascocentrum x Neofinetia x Renanthera x Rhyncostylis x Vanda
Macekara = Arachnis x Phalaenopsis x Renanthera x Vanda x Vandopsis
Paulara = Ascocentrum x Doritis x Phalaenopsis x Renanthera x Vanda
Sutingara = Arachnis x Ascocentrum x Phalaenopsis x Vanda x Vandopsis

9.7.3.6 Vanda Hybrids

'Roberts Delight', 'Dr. Anek', 'Pakchong Blue', 'Miss Joaquim', 'Fuch's Delight', 'Lumpini Red', 'Motes Indogo Blue', 'Pat Delight', 'Rasri Gold', 'Samsai Blue', 'Adisak Smile', 'Robert Sorenson', 'Kasems Delight', 'Richard Brandon', Kultana Delight', 'Brigton's Pride', 'Hope', 'Mark Lewis', 'Dr. Mahathis', 'Kultana Miami White', 'Kultana Violet', 'Grand Lady', 'Fuch's Midnight', 'Motes Nut Brown', 'Prayoon Delight'.

9.7.3.7 Aranda Hybrids

'City of Singapore', 'Hilda Galistan', 'Urmila Nandey', 'Christine', 'Thailand Sunspot', 'Millenium Dawn', 'Broga Giant', 'Salaya Red', 'Propine White', 'Propin Spot', 'Lueng Cholburi', 'Ishbel Manisaki', 'Baytown', 'Chao Praya Blue', 'Chao Praya Dot Com', 'Chao Praya Beauty', ' Ethan Pride', 'Taksari Chandrabir'.

9.7.3.8 Ascocenda Hybrids

'Carol Belk', 'Renuka Angle', 'Joyce Bevins', 'Adisak Blue', 'Renu Gold', 'Tipi Blue Boy', 'Bobs Fortune', 'Rubychai', Shah Rukh Khan', Yang Sophia Firuz', 'Abdul Ghani Othman', 'Chunika', 'Fuch's Star', 'Kultana White', 'Golden Peace', 'Lani Beauty', 'Mahogany Gem', 'Copper Pure', 'Pokey Dot'.

9.7.3.9 Mokara Hybrids

'Madam Panne', 'Bangkok Gold', 'Chao Praya Gold', 'Chark Kuan Orange', 'Chark Kuan Pink', 'Kelvin Red', ' Kelvin Orange', 'Walter Ouame', 'Jitti Orange', 'Happy Beauty', 'Margaret Thatcher', 'Sarita Gold', 'Laura Bush', 'Winny Burang', 'Gladys Oumae', 'Denis Child', 'Ratchaburi Blue', 'Lion's Gold', 'Chao Praya Sunrise', 'Pak-Kred', 'Golden Rooster', 'Jasso's Pride', 'Chao Praya Classic', 'Chao Praya Spots', ' Calypso', 'Bota Gold'.

9.7.3.10 Kagawara Hybrids

'Red Lava Orange', 'Mist', 'Christie Low Redland', 'Broga Cemerlang', 'Chao Praya Fantasy', 'Curtis Lutchman', 'Emily Kavita Rajah', 'Diinesh Gold', 'Lion's Flame', 'Irene Hew', 'Bukit Timah', 'Sandy Gold'.

9.7.3.11 Vascostylis Hybrids

'Ique Pink', 'Kruewan', 'Banjong Jasmine', 'Mishima Lime', 'Vanity Fair', 'Firuz', 'Pine Rivers', 'Banjong Elite', 'Pichtawit Gold', 'Chao Praya Lime', 'Lanna Rosy', 'Jeans Delight', 'Bay Sapphire', 'Spring Hill'.

9.7.3.12 Aeridovanda Hybrids

'Bensiri', 'Noreen', 'Early Bird', 'Shiv Sidhu', 'New Dawn', 'Harrison Luke Somsri Sunlight'.

9.7.4 Cultivation

9.7.4.1 Light

All Vanda orchids require high light levels. Strap leaf Vandaceous orchids require 60-65% shade whereas terete or semi-terete vandas need more light and require 50% shade. On average they require a light intensity of 4000 foot candles.

9.7.4.2 Temperature

The Vanda orchid prefers warm temperatures in summer and winter. They are highly susceptible to cold and frost. In cooler regions, the orchid plants will go into a dormant state during winter. Cylindrical Vandas require a minimum temperature of 16-17°C at night during winter, and a maximum of 30-32°C in the summer. Flat leaved vandas need a winter night temperature of 10-12°C and a summer day time temperature of 22-25°C They are capable of tolerating higher and lower temperatures for short periods only.

9.7.4.3 Propagation

Easily propagated by cuttings of the shoots along with roots or air layerings. Rapid large scale multiplication is possible through tissue culture.

9.7.4.4 Watering

Vandaceous plants love good quality water. Being a monopodial, the plants cannot retain water for any great length of time. It is advisable at all times to water these plants as early in the morning as possible, so that the foliage of the plant is dry by noon. Orchids do not appreciate being wet and cold, so water your plants in the winter time only on bright and clear mornings as necessary. Adult plants are watered once a week in the winter time and every day during summer.

9.7.4.5 Fertilization

All Vandaceous orchids are heavy feeders. During the growing period from the beginning of September to the end of March, the adult plants are fed once a week as follows: For two weeks we use a well balanced fertilizer (30:10:10) and the following week we use a low nitrogen fertilizer (10:30:20). The young seedlings are fed once a week; one can use a well balanced fertilizer (30:10:10) and the following week using a low nitrogen fertilizer (10: 30: 20).

During the dormant period when the weather starts getting cooler and the daylight hours shorten, the food to plants is reduced. Adult plants are fed only once a fortnight with a low nitrogen fertilizer in order to keep the velamen on the root system plump. The young seedlings are fed only once a fortnight, rotating a balanced fertilizer with a low nitrogen fertilizer as mentioned above.

9.7.4.6 Potting Media

Freshly deflasked seedlings are potted into a fine mix in community pots, and the larger plants (from flask) are potted directly into 50 mm tubes using the same mix. When the plants have outgrown their 50 mm tubes, they are either potted into 10-12 cm black plastic baskets or 15 cm clay pots which have extra holes in the sides.

The mix used in 15 cm pots is 50% good quality treated bark, 30% larger chunks of cocochips and 20% brick pieces. In the baskets or clay pots, one can put extra coarse bark and cocochips to support the plants until the roots have gripped the container. This assists Vandaceous orchids by providing plenty of good air circulation around the root system. They do not like to have wet feet especially during prolonged rainy or cold weather when they can get bacterial soft rot.

9.7.5 Insect -pests and Diseases

Fungal and bacterial diseases cause collapse of the plant's tissues, frequently with a water soaked appearance. Bacterial infection may attack the leaves, the stem or the root system. These diseases require high humidity to survive. Some occur at high humidity and low temperatures, while other types are not activated until the temperatures are relatively high. These diseases can be discouraged by watering orchids as early as possible in the morning so when the temperature reaches its peak, the plants dry up and will remain so as the temperature falls at night.

Fungal and bacterial problems can be kept to a minimum with good sanitation, good air circulation around the plants, and regular spraying with a preventative fungicide to protect the plants against infection.

Mancozeb or Dithane (0.2%) is also used periodically if required, for instance after prolonged rainy weather. Before the onset of winter (beginning of May) a mixture of Mancozeb and Spraytech Oil is sprayed on the plants. Infections such as bacteria can enter the plants through injury to the root system and then the plant is susceptible to Fusarium Wilt. The affected plants start to drop the lower leaves. Spraying with Mancozeb (0.2%) or Zineb (0.2%) controls the disease. Vandaceous orchids are almost free from pests and insects. Occasionally scale or cockroaches may attack, causing a widespread infection in the shade-house. If required, spray with a mixture of Diazinon and White Oil for scale or Diazinon to prevent cockroaches.

9.7.6 Post-harvest Management (Bhattacharjee and De, 2005)

9.7.6.1 Stage of harvest:

Spikes are harvested when all flowers are already open, depending on types 10 to 16 flowers on a stem length of 50-60 cm.

9.7.6.2 Storage

Cut flowers are wet stored at 8 to 13°C.

9.7.6.3 Vase life

Cut flowers last for 14 to 15 days in normal water.

9.7.6.4 Preservatives

$AgNO_3$ 30 ppm + sucrose 1.5% is commonly used preservatives in holding solutions for enhanced vase life of cut flowers.

10 Production Technology of Commercial Terrestrial Orchids for Cut Flowers

10.1 Introduction

Terrestrial orchids grow on the ground level in the soil. Most of these can be found in North America, Europe, and cooler regions of Asia. There are a great number of species even among the terrestrial orchids. They are either found on the forest floor under shade (for example *Calanthe, Acanthephippium, Eulophia macrostachya, Tainia, Phaius* etc.) or in open grasslands and meadows like *Habenaria, Peristylus, Pachystoma, Ipsea, Eulophia nuda* etc. These orchids are rich in starch and alkaloids and medicinally important.

Based upon growth habit, terrestrial orchids are either creeping or erect type and can be divided into two types: the solitary type and those growing in clumps or tufts. Within the solitary type group of terrestrial orchids there are two kinds of growth habit; one with plants producing single, erect, leafy stems arising from underground tubers or corms which are terminated by an inflorescence. They are deciduous after fruiting and seeding are completed. The tubers or corms continue their life cycle after completion of an annual period of dormancy. Orchids with the second type of solitary growth habit produce a flowering shoot and leafy shoot from separate buds on the rhizome. Genera such as *Nervilia, Eulophia* and *Pachystoma* are examples of this type. The erect terrestrials are usually evergreen, retaining their leaves for more than a year. Each new growth starts from the base of the leafy pseudobulbs as in the genera *Liparis, Acanthephippium, Calanthe* and *Phaius*.

The creeping terrestrial orchids have mostly slender, smooth, fleshy rhizomes with elongated internodes and several short or long roots at the nodes. Axillary buds, formed below the apex of the rhizomes produce and ascending leafy and flowering shoot. New annual growth is repeated at the end of flowering, fruiting and, in most cases, after leaf fall. The mother ascending shoot gradually becomes prostrate and becomes an additional part of the rhizome. Examples of orchid genera in New Guinea with this type of growth habit are *Erythrodes, Eurycentrum, Eucosia, Hetaeria, Macodes, Vrydragzynea* and *Zeuxine*.

Terrestrial orchids generally prefer:

- A potting medium that retains moisture
- Even watering
- Moderate humidity
- Less air movement
- Less fertilizer
- Cooler temperatures
- Lower light levels

10.2 Calanthe

Calanthe is a genus of about 200 terrestrial species that are widespread throughout all tropical areas but are highly concentrated in Asia. The first man-made orchid hybrid created was a *Calanthe* Dominii, back in 1853, and *Calanthe* were very popular hothouse plants during the Victorian Era.

There are two types of *Calanthe*, the deciduous ones which generally have large, silvery pseudobulbs, drop their leaves in winter, and require less water during winter; and the evergreen *Calanthe* which have either no pseudobulbs or very inconspicuous ones, usually keep their leaves for several seasons, and require even moisture year round. The scape is axillary, terminal or lateral from a leafy pseudobulb. The racemes are long, bearing few to 20 flowers, which are subtended by leafy bracts. The flowers are basically white, red or mauve, medium sized and attractive.

10.2.1 Genetic Resources (De, 2011)

Calanthe triplicata: This species is native to the Philippines, Pacific Islands and Australia. An evergreen species with hairy long inflorescence. The flowers are white coloured with reddish and yellow tinge turning black with age and produced during summer and autumn.
Calanthe masuca: An evergreen species of Sikkim Himalaya and Western Ghat with elliptic–ovate to lanceolate leaves. Inflorescence is 90 cm tall and crowded with 12 to 15 small blue violet flowers.
Calanthe Regnieri: A deciduous species from Vietnam. The inflorescence is 45 cm long, pubescent with 8-10 flowers. The flowers are white tinged with pink and produced during winter season.
Calanthe rosea: A deciduous species from Burma. The pseudobulbs are spindle shaped. The leaves are broadly lanceolate. The inflorescence is raceme and many flowered. The flowers are long lasting and rose pink to white or dark rose in colour and produced in autumn and winter.
Calanthe veratrifolia: This species is native to India and Australia. The leaves are elliptic and distinctly folded light green in colour. The inflorescence is 150 cm tall, erect and many flowered. The flowers are long lasting, white and produced during May.
Calanthe vestita: A deciduous species from India, Malay Peninsula, Burma and Thailand. The pseudobulbs are conical, egg shaped carrying 3-4 leaves. The inflorescence is 150 cm tall, arching and arranged with large white flowers.
Calanthe veitchii: It is a hybrid of *Calanthe rosea* and *Calanthe vestita*. The pseudobulbs are elongated with deciduous leaves. The inflorescence is 75 cm tall with rich rose flowers. The flowers are long lasting and suitable as cut flower and produced in January –February.

Calanthe sylvatica: A terrestrial tropical and subtropical orchid from North East India. Flowers are large lilac to deep purple with a prominent spur and produced in August-September.
Calanthe alasmifiolia: Native to Bhutan, India, Sikkim, West Bengal, Arunachal Pradesh and Nepal. The inflorescence is corymb. The flowers are white and produced in May-June.
Calanthe biloba: Native to Bhutan, India and Nepal. Pseudobulbs are elongated. The flowers are yellow-ochre spotted with violet or purple brown and produced in September-November.
Calanthe brevicornu: Native to Bhutan, India and Nepal. Pseudobulbs are round. The inflorescence is terminal and arranged with 8 to 15 flowers. Flowers are brick red to purple red colour striped with white or yellow and produced in May-June.
Calanthe chlorleuca: Native to Bhutan and India. Pseudobulbs are conical cylindric. The flowers are yellow green, fragrant and produced in April-May.
Calanthe puberula: Native to India, Bhutan and Nepal. The pseudobulbs are ovoid-conical. The flowers are loosely arranged with inflorescence. The flowers are pale-lavender and produced in June to August.
Calanthe discolor: An evergreen species. Flowers are pure white.
Calanthe aristulifera: A rarest endemic species of Japan. An evergreen species, 45 cm tall spikes bearing light pink flowers.

10.2.2 Hybrids

During the 1850's, Mr. Dominy developed first hybrid, *Calanthe dominyi*, which is a cross bettwen *Calanthe masuca* and *Calanthe furcata*.

10.2.2.1 Inter-specific Hybrids

Hizen = *Calanthe discolor x Calanthe aristulifera*
Ghita Norby = *Calanthe triplicata x Calanthe madagascariensis*
Egg Drop = *Calanthe arisamensis x Calanthe striata*
Tydares Sieboca = *Calanthe striata x Calanthe masuca*
Calanthe Sieboldii 'Takane' hybrids, Calanthe Kozu 'Spice' hybrids

10.2.2.2 Bigeneric Hybrids

Phaiocalanthe = *Phaius x Calanthe*
Gastrocalanthe = *Gastrorchis x Calanthe*

10.2.2.3 Inter-varietal Hybrids

'Elegance', 'Tsiku Flamingo', 'Tsiku Honolulu', 'Narita', 'Urayasu', 'Pink Champagne', 'Mont Nicholle', 'Kitayama', 'Koriyama', 'Ranyu', 'Bouvet', 'Mac's Gold', 'Mont Isaac', 'Mont Couchon', 'Mont Remon'.

10.2.3 Uses

These species are used as cut flowers. Calanthe orchids are rich in linalool, methyl benzoate, methyl salicylate, carvone and cinnamic aldehyde. *Calanthe masuca* is used for the treatment of acnes and inflammatory sebaceous cysts.

10.2.4 Cultivation

Basically they grow terrestrially in deep patches of forest humus, on rotting logs, and among rocks with crevices filled with leaf litter. In cultivation one should practice in pots in order to get the best growth and blooming from the plants. The plants grow well in a temperature range of 15-25°C, light intensity of 1800-2500 foot candles and relative humidity of 50-70%. They require ventilated, well lighted and airy conditions for proper growth and flowering. The compost mixture should consist of charcoal, tree bark, sphagnum moss, pea nut shells, cow dung, perlite and sandy soil. Dilute liquid manure is to be applied at monthly intervals. Propagated by the division of bulbs.

10.3 Eulophia

Eulophia consists of 300 species of sympodial terrestrial orchids from Africa, Madagascar, Malayasia, Sri Lanka, China, America and India. The pseudobulbs are subterranean topped by 3 to 5 linear-lanceolate leaves. The leaves are leathery or soft, 1.8 m tall and 10-12.5 cm wide. The inflorescence is simple or branched and bears few to many attractive flowers which are long lasting. The flower spikes are 1-2 m tall and arranged with diverse types of flowers, which are yellowish or greenish in colour with some brown and purplish markings.

10.3.1 Genetic Resources

Eulophia alata: Native to South Florida, Brazil and West Indies. The pseudobulbs are corm like, subterranean and borne in a long creeping series and 7.5 cm across. Leaves are dark green and arise from pseudobulbs, 1.2 m tall and 7.5 cm across. The inflorescence is 1.5 m tall, raceme, and bearing attractive flowers. The flowers are 4 cm in

diameter, scented, long lasting, greenish bronze yellow or dull yellow, green purple or maroon in colour with cup shaped maroon purple lip and produced in autumn and winter.
Eulophia andamanensis: Native to Andaman Islands. The stems are tuberous at base.
Eulophia ovalis: Native to tropical Africa. Plants are 70 cm tall with purple and white or cream to lilac in colour.
Eulophia explanata: A tall robust species from Garhwal, Kerala, Chhotanagpur and Kumaon with a short pseudostem covered by sheathing bases. The inflorescence is 60 cm tall with lemon yellow flowers. The flowers are produced during June-July.
Eulophia guineensis: Native to West tropical Africa. The pseudobulbs are clustered, furrowed, and irregular with stalked narrowly elliptic leaves. The inflorescence is raceme like, erect, 90 cm tall and has 10-15 flowers. The flowers are fragrant, long lasting, waxy, greenish brown with a white lip.
Eulophia mackinnonil: A tropical and subtropical species from Western Himalayas. Pseudobulbs are flat, triangular with acuminate and lanceolate leaves. The inflorescence is a raceme and 8 to 12 flowered. The flowers are yellow to reddish brown and produced in July-August.
Eulophia spectabilis: A terrestrial species from Bhutan, India and Nepal. Pseudobulbs are spherical and subterranean with 3 to 4 pleated leaves. The inflorescence is 5-15 flowered. Flowers are purple brown to pure green in colour and produced in April-July.
Eulophia nuda: A tuberous plant with smooth spherical leaves. Leaves are elliptic. Flowers are lax-raceme, 9-20 flowered and rose-pink in colour.
Eulophia parviflora: Native to Eastern Cape and grown in grassland and rocky slopes. Flowers are pleasantly scented with brown and yellow flowers.
Eulophia aculeata: Native to southwestern Cape to Mpumalanga. Plants are rhizomatous, 50 cm tall with ivory to greenish, pink or purple flowers.
Eulophia clavicornis: Native to tropical Africa. Plants are rhizomatous, 80 cm tall with white to purple or yellow petals.
Eulophia foliosa: Native to South Africa. Plants are 60 cm tall with dense elongated clusters of lime green flowers.

10.3.2 Hybrids

Eulophia x burundiensis (*Eulophia cucullata x Eulophia flavopurpurea*), an inter-specific hybrid is native to South Africa. Flowers are large and bright yellow in colour. Popular as pot plants. Another hybrid available in this genera is 'John Davison'.

10.3.3 Medicinal Eulophia

Tubers of *Eulophia ochreata* are used for the treatment of earache and toothache. In India, the tubers of *Eulophia nuda* are used to treat tumours, scrofulous glands, blood disorders, bronchitis, tuberculosis and as a vermifuge. Raw tubers are eaten in rheumoid arthritis. *Eulophia speciosa*, a native to Africa is used as food sources, for medicinal purposes and in superstition. The attractive flowers are used on letter stamps.

10.3.4 Cultivation

Eulophias prefer humid and well lighted location, exposure to direct sun should be avoided. The sympodial terrestrial species prefer a temperature range of 10 to 15°C, relative humidity of 40-60% and light intensity of 3000-5000 foot candles depending upon the species. The pot mixture comprising of leaf mould, sandy soil, tree bark, sphagnum moss and perlite mixture is good. Application of dilute liquid form of well rotted cowdung manure at monthly intervals is beneficial. Epiphytic species should be grown on osmunda fibre or tree fern block with sphagnum moss.

10.4 Phaius

Phaius consists of 30 species of terrestrial orchids distributed in East Africa to Tropical Asia, Pacific Islands, Himalayas, New Caledonia, Indonesia and Fiji Islands. The pseudobulbs are stocky and thickened and arranged with 2 to 8 large, thin, deeply grooved, long and lance shaped leaves. The individual leaf is 1.2 m long and 20-25 cm wide. The inflorescence is 90 to 120 cm long, arises from the rhizome base between the point of attachment of two leaves. The flowers are large, showy, 10 cm across, long lasting and of various colours.

10.4.1 Genetic Resources

Phaius tankervilliae: Native to tropical Asia, Australia and Pacific Islands. The pseudobulbs are lightly clustered, dull green with very large folded and heavy leaves. Inflorescence is stout longer than leaves, tipped by a spike of 15-20 flowers. The flowers are 11 cm across, long lasting, sweet smelling. The fragrant flowers are silvery on the outside and have yellowish throats. Commonly known as 'Nun's Orchid'. Flowers are produced during December-March.
Phaius humboltii: Native to Madagascar. The leaves are 50 cm long and broadly lanceolate. Inflorescence is 15 to 20 flowered, erect and thick. The flowers are 6.5 cm across and rose coloured. Flowers are produced during June-July.
Phaius flavus: Flowers are 45 cm tall and are usually produced during April-June. The species is native to Japan and Asia. Commonly known as 'Rock Orchid'. Flowers are yellow with a reddish brown banded lip. Pseudobulbs are conical, topped with 50-60 cm long plicate variegated leaves.

10.4.2 Commercial Hybrids

'Masako', 'Morningstar', 'Spring Fever'

10.4.3 Herbal Medicines

Paste of pseudobulbs of *Phaius tankervilliae* is used to heal swellings of hand and legs, and in poultices to soothe pain of abscess.

10.4.4 Cultivation

Phaius should be grown in a temperature range of 16°C to 24°C night to day. However, the plants will tolerate lower temperatures to 10°C or higher to 32°C. They enjoy bright light or light shade. Plants are grown in pots with a mixture of 2 parts coarse peat moss, 2 part sandy loam and 1 part each perlite and fine bark and watering with good drainage. Fertilize the plants at every third watering with a balanced Orchid fertilizer, such as 18-18-18, diluted to half the strength recommended. Repotting of phaius orchids is required every two or three years. They are propagated by flower stalk cuttings or divisions.

10.5 Pleione

Pleione consists of 20 species of cool growing terrestrial or lithophytic orchids distributed in China, Formosa, the Himalayas and South East Asia. These deciduous orchids are known as 'Indian Crocus'. The psedubulbs are angular, one clustered, small and sometimes mottled with black. These pseudobulbs are topped by a solitary folded leaf. Pseudobulbs arise from the base of each pseudobulb, one or two flowered and attractive. The flowers are delicate and frilled.

10.5.1 Genetic Resources

Pleione humilis: This species is native to Sikkim and grown in high altitude. The pseudobulbs are egg shaped and clustered with solitary, folded and deciduous leaves. The flowers are 12.5 cm across, fragrant, white and produced during September-October.
Pleione maculata: A dwarf orchid from India. The pseudobulbs are flask shaped or almost conical, compressed and tuberculate at base. The leaves are folded, two and deciduous. The flowers are solitary, fragrant, 5 cm across, long lasting, white and produced during October-November.

Pleione praecox: Distributed in South China, India, Burma and Himalayas. The pseudobulbs are egg shaped, lightly clustered, bottle shaped and depressed at the apex. The leaves are 2, folded, deciduos and elliptic to lanceolate. The flowers are solitary, fragrant, long lasting, 12.5 cm across, magenta red in colour and produced during December.

10.5.2 Hybrids

10.5.2.1 Inter-specific Hybrids

'Io', 'Mandrill', 'Gelada', 'Kima', 'Langur', 'Polto', 'Dr. Mo Weatherhead', 'Bonobo', 'Mariemonda', 'Lucey', 'Muriel', 'Orangutan', 'Nozomi', 'Spring', 'Renate', 'Lilac Wonder', 'Umpqua Jan Shan', 'Yeti', 'Zottel', 'Sinope', 'Kyoto', 'Confirmation'.

10.5.2.2 Inter-varietal Hybrids

'Darwin', 'Gorilla', 'Wahakari', 'Jake Butterfield', Helen Brownsword', 'Vicky', 'Michael Butterfield', 'Snow Monkey', 'Adabra', 'St. Helens', 'Badaling', 'Fan Jiang', 'Kiku', 'Mikado', 'Trask', 'Ashmore', 'Calypso', 'Titan'.

10.5.2.3 Variety – species Hybrids

'Aye-Aye', 'Austice Harris', 'Sifaca', 'Tai Pan', 'Kong', 'Mardin', 'Leda', 'Bo Pan', Little Goose Pagoda', 'Yokata', 'Sakura', Floor Shan', Hazlebury'.

10.5.3 Medicinal Pleione

Pseudobulbs of *Pleione maculata* are used in liver complaints and stomachache. Pseudobulbs of *Pleione formosana* are rich in dihydrophenanthrene, pleioanthrenin, bibenzyls, pleiobibenzynin and cyclomargenyl -p-coumarate. The pseudobulbs are used to treat tumors.

10.5.4 Cultivation

Pleiones prefer bright filtered light with the intensity of 1000-1500 foot candles. Most Pleiones require cool to intermediate temperatures. Maximum day temperatures should be between 22-25°C, and night minimums between 15-18°C. During the winter months, the plants benefit from a night minimum between 10-15°C.

Plants are grown in shallow clay pots or pans. The potting mixture should consist of tree bark, peat and sphagnum moss. Plenty of fresh air movement and watering are essential. Plants are propagated by divisions of bulbs.

11 Orchids for Pot Culture/ Hanging Baskets/ Tree Mounting

Orchids are one of the most distinctive plants of nature and highly priced in the international flower trade due to their incredible range of diversity in size, colour, shape, forms, appearance and long lasting qualities of flowers. They are considered to be highly valued cut flowers as well as potted plants. Out of 1331 species of orchids of India, 856 species are found in North Eastern Hill Region due to the congenial climatic conditions, diversified topography and altitudinal variation. Among flower crops, orchids contribute a 10% share in international trade. In addition, the Sikkim Himalayan region is the centre of origin of an important species like *Cymbidium*. Other valuable genera of commercial importance available in the region are *Coelogyne, Dendrobium, Paphiopedilum, Pleione, Rhyncostylis* and *Vanda* etc. The varied agroclimatic conditions of our country are highly favourable for cultivation of commercial orchids like *Cattleyas, Vanda, Dendrobium, Arachnis* and *Aerides* in tropical areas, Cymbidium hybrids, Paphiopedilum, Dendrobium hybrids and Phalaenopsis in subtropical areas and Cypripediums, Cymbidiums etc in temperate areas. Among orchids, *Cymbidium, Dendrobiums, Phalaenopsis, Cattleyas, Vandas, Arandas, Arachnis* etc are used as cut flowers whereas *Aerides, Arachnanthe, Bulbophyllum, Calanthe, Coelogyne, Eria, Phaius, Phalaenopsis, Pleione,* and *Rhyncostylis* are used as potted orchids.

11.1 Agronomic Management

11.1.1 Humidity

The single dominant factor which affects the cultivation of orchids is humidity, which should be as high as 75-85%. It varies species to species depending upon habit of growth, light, temperature and ecotypes. Monopodial orchids require higher humidity than sympodial ones. As rule of thumb, in high temperature, humidity should be kept high. The provision of misting units or foggers or even humidifiers will ensure adequate humidity, and standing water may be kept beneath the benches to improve humidity.
Humidity ranges of some orchid genera are given below:
Dendrobium: 50-70% *Paphiopedilum, Cupripedium*: 40-70%
Cattleya, Laelia: 40-55% *Rhyncostylis, Aerides, Phalaenopsis, Vanda:* 70-75%

11.1.2 Light

Most orchids generally prefer indirect or filtered light. Although it varies species to species, growth habit and habitat, as the rule of the thumb, 50% shading is always advised for most of the commercial orchids. Under enough light, orchid plants have short, plump stems with bright green leathery leaves. Yellowing, stunting and scorching of plants occur under too much light, while under too much shade plants have darker green, soft and succulent leaves with thin and spiny stems. Low light orchids are classified as those requiring 1200-2000 foot candles of light intensity, and originated from dense forests e.g. *Phalaenopsis, Calanthe* etc. Medium light orchids grow at the top of the tree canopies and prefer 2000-3000 foot candles of light intensity. *Cattleya* and related genera and hybrids fall into this category.

High light orchids grow in full sun habitats and need 3000 foot candles or more light intensity, which includes *Cymbidium* and vandaceous plants. Terete vandas like *Vanda teres, V. hookeriana* require plenty of sun light and can grow under direct sun whereas strap leaved vanda such as *Vanda coerulea, V. tricolour, V. sanderiana* can not tolerate direct sun and need 50-60% shade during summer.

11.1.3 Temperature

Orchids are classified into three categories based on temperature. Warm orchids like *Aerides, Vanda, Rhyncostylis* and some *Dendrobium* species grow at day temperature up to 32.2°C and 15.5°C for a minimum night temperature; intermediate species like *Cattleya, Laelia, Brassovola, Oncidium, Miltonia* etc prefer 26.6°C days and 12.8°C nights. Cool species such as *Cymbidium, Odontoglossum, Cypripedium* etc. prefer 24°C days and 10°C nights. Most orchids require a lower night temperature for both robust growth and to initiate blooms. A night temperature of 10-13°C is ideal for initiation of flowering in *Cymbidium*. In *Phalaenopsis*, day and night temperatures of 24°C and 17°C, respectively are essential for optimal flower production. Day temperatures above 21°C and night temperatures in the range of 15.5 to 21°C are favourable for *Vanda*. Small and young orchids require 5°C more temperature than mature plants.

11.1.4 Atmosphere

Fresh air and good circulation are essential for orchid production, full of continual light breezes, make a good source carbon dioxide for photosynthesis.

11.1.5 Propagation

Orchids are highly valued for long lasting cut flowers. Both monopodial (Single stemmed growth) and sympodial (Multistemmed growth) are preferred for commercial cultivation. Orchid genera like *Aerides, Arachnis, Epidendrum, Renanthera, Phalaenopsis, Vanda* and *Dendrobium* are propagated by cutting. Flower stalk cuttings are useful in genera like *Phalaenopsis, Phaius, Calanthe* and *Thunia*. Air layering is effective in *Vanda*. Genera of sympodial orchids like *Cattleya, Dendrobium, Cymbidium, Epidendrum* are multiplied by division. Off –shoots or keikis emerge from main stem are used as propagules in *Dendrobium, Ascocenda* and *Phalaenopsis*. The older canes of sympodial orchids like *Cymbidium* (called 'backbulbs') are forced to root in a moist medium. Seed culture involves germinating seeds and growing seedlings under aseptic condition in an agar medium supplemented with nutrients in Knudson C media. Various plant parts like shoot tips or meristem, leaf and leaf segments, stem segments, floral parts, aerial parts have been used for tissue culture of orchids. Among these, meristem and shoot tip culture are most popular for mass propagation of commercial species and hybrids, while axillary buds are good sources of explants in monopodials. Both liquid and solid media are used for culture of orchid tissues, with Knudson's C medium, Vacin and Went's medium, Murashige and Skoog's medium being the most widely used. Additives like coconut water (15%) and banana pulp (10%) are beneficial for the promotion of shoots, and mineral salts, carbon source, vitamins, plant growth regulators are used in the media. Sucrose is used as a carbon source to promote organogenesis at suboptimal concentrations and protocorm formation at supra-optimal concentrations.

Among vitamins, thiamine and growth regulators, auxin, cytokinins are used for callus formation. *In vitro* plants are hardened off *in vitro* only, before transferring to a main field. The application of paclobutazol delays chlorophyll loss, reduces the activities of enzymes and delays senescence. Nitsch media with carbon sources of 40 g/l is suitable for the conservation of the *Cymbidium* hybrid 'Sleeping nymph'. It was also found that MS media with activated charcoal and sucrose at 20 & 30 g/l is suitable for fast multiplication of pseudobulbs.

The combination effect of BAP and NAA was found to be suitable for faster multiplication of pseudobulbs in *Cymbidium* hybrid 'Pine Clash Moon Venus' (MS +AC 1.5 g/l + BAP 0.05 mg /l + NAA (0.25 mg/l).

11.1.6 Pots and Potting Media

Terrestrial and semi-terrestrial plants like *Paphiopedilum* and *Cymbidium* perform better in deep pots. Orchid plants as a rule grow to be near one another to aid in creating a microclimate with higher humidity. Basket culture is useful for those orchids like *Vanda, Rhyncostylis, Arachnis* with pendent flower spikes and long dangling roots. Clay pots are best suitable for terrestrial orchids while plastic pots are used for epiphytes. Slabs or logs of tree fern are effective for cool growing orchids.

A potting media of terrestrial orchids should have equal parts of leaf mould, soil and sand. Clay soil, bonemeal, sawdust, charcoal dust, manure, wood savings etc. are also used in various proportions for satisfactory growth of terrestrial orchids.

A potting media consisting of charcoal, brick pieces and coconut fibre in equal proportions is ideal for vegetative growth and flowering of epiphytic orchids like *Aerides*,

Dendrobium etc. Under low humid conditions (30%), plastic pots with a mixture of bark/perlite/sphagnum moss or osmunda are used. Under average humidity (35-50%), it is advisable to use plastic pots with a mixture of bark and sphagnum moss. Under high humidity (55% and above), clay pots are used with bark, stone culture, charcoal or tree fern.

11.1.7 Watering

Most orchids are damaged by overwatering rather than under watering; overwatering leads to root rot and many other diseases. Most orchids prefer water of pH 5.0-6.5. Watering at lower or higher pH or with high levels of dissolved minerals can hamper nutrient uptake. Frequent watering is essential under high sunlight and high temperature conditions; plants in small containers dry out more quickly than in large containers. Plants in earthen pots require more watering than plants in plastic pots, while a hanging plant needs even more frequent watering as it has better aeration than one in a pot. More frequent watering is necessary for fresh potting materials and watering should be practiced either in the morning or in afternoon time. Potting materials having more moisture retention capacity (like coconut husk, tree fern etc) need less water less frequently, and vice versa.

11.1.8 Nutrition

Orchids are light feeders and require nitrogen from the first two-thirds of their life cycle. During rest period they do not need any fertilizers, but during flower initiation and inflorescence development, plants are fed with less nitrogen, more phosphorus and potassium. During blooming, a small level of nitrogen and phosphorus and high levels of potassium are maintained. In orchids, foliar feeding is found to be ideal. Frequent application of fertilizers at low concentrations is the best way of feeding orchids: a concentration of 0.2 to 0.3% of 30:10:10 (N:P:K) at the vegetative stage and 10:20:20 (N:P:K) at blooming stage are applied for quality flower production. Sometimes, fresh coconut water and diluted cow urine are also useful as foliar sprays. In *Cymbidium*, foliar application of 0.3% 19:19:19 of N, P, K at 7 days interval increased the plant height and leaf length. In *Aerides multiflorum* and *Dendrobium moschatum*, treatment with 1000ppm N and 500 ppm each of P and K enhances vegetative growth and flowering. In the *Dendrobium* hybrid, 'Sonia 17' application of 0.2% NPK (30:10:10) along with 200 ppm BA increases number of spikes per plant and number of flowers per spike.

11.1.9 Repotting

Orchid plants require repotting if there is no space left in the pots for new growths and if the substrate has decomposed or roots are rotting. Timing is the most important part of good repotting. The best time for repotting of an orchid is when new growth and new roots are just beginning to form (before those new roots reach even 1 cm long). In most orchids, this occurs right after flowering, indicating that repotting should be done between February and June.

11.2 World Production Scenario of Potted Orchids

The production of potted orchids for the mass market has gone beyond the United States and has global economic importance. The largest exporters of potted orchids include Taiwan, Thailand, the United Kingdom, Italy, Japan, New Zealand and Brazil, while the largest importer of potted orchids was the United States during 2002. The production of finished potted orchids in Europe is often for domestic consumption. In 2002, the total number of potted orchids produced in China and Japan were 4 million and 28 million, respectively. At flower auctions in Holland, *Phalaenopsis* orchids were the most valuable potted plant sold with a wholesale value of €173.7 in 2006. Potted *Dendrobium* orchids are produced in China, Taiwan, Thailand, Philippines, USA, Japan and Germany, while potted *Vanda, Mokara, Oncidium, Cattleya, Ascocenda, Oncidium, Renanthera* and *Arachnis* are produced in large scale in Thailand.

11.3 Description and Culture

Description and culture of orchids for pot culture, hanging baskets and tree mounting such as *Aerides, Anoectochilus, Ascocentrum, Bulbophyllum, Coelogyne, Doritis, Epidendrum, Gastrochilus, Laelia, Miltonia, Odontoglossum, Renanthera, Rhyncostylis* and *Zygopetalum* are given below (De, 2011; De, 2014).

11.3.1 Aerides

Aerides consists of more than 60 species. These are monopodial epiphytic orchids grown on trunks or branches of trees. The stems are round, leafy and branched with aerial roots. The strap leaves are thick, leathery and bilobed. The inflorescence is auxillary and drooping and up to 60 cm long in some species. The flowers are many and 2 cm in diameter, closely set on the inflorescence, and generally fragrant.

11.3.1.1 Genetic Resources

Aerides falcatum: It is distributed in the NEH Region, Thailand and Burma. The plants are 150 cm tall with narrow and thick leaves and pendulus or arching, 50 cm long inflorescence. The flowers are available in the month of March to May, white, tinted rose and fragrant.

Aerides Lawrenceae: This species is distributed in Philippines. The plant is 150 cm tall with thin leaves and pendent inflorescence. Flowers are large, white, purple spotted and available in the autumn season.
Aerides multiflorum: This species is distributed in India, Burma, Thailand, Vietnam and the Himalayas. The plants are short stemmed with leathery leaves and densely flowered inflorescence. Flowers are fragrant, waxy, white with rose purple shades and available in June-July.
Aerides odoratum: This species is distributed in India, China, Himalayas, Burma, Thailand, Philippines and Indonesia. The plants are 150 cm tall with branched stems and leathery, bilobed leaves. The inflorescence is cylindrical, many flowered, pendulous, waxy white with touch of amethyl purple and fragrant. It is available in June-July.
Aerides album: A native of Philippines, 150 cm tall, branched and glossy green bilobed leaves. The inflorescence is pendent, arching and densely many flowered. The flowers are long-lasting, white with purple tipped petals and sepals and available in August-October.

11.3.1.2 Intergeneric Hybrids

Aredocentrum = Aerides x Ascocentrum
Aredefinetia = Aerides x Neofinetia
Aeridopsis = Aerides x Phalaenopsis
Aeridovanda = Aerides x Vanda
Renades = Aerides x Renanthera
Rhynchorides = Aerides x Rhyncostylis

11.3.1.3 Common Commercial Hybrids

'Edward', 'Brighton's Sparkle', 'Brighton Pink', 'Cagayan', 'Hollyhock', May Woo', 'Jueile', 'Pink Flush', 'Sandorata', 'Lokelani', 'Cressida'.

11.3.1.4 Uses

Aerides is ideal for pots and slat baskets. They make spectacular specimen plants in baskets or mounted on trees with long flower stems. Scented species are utilized for the extraction of essential oils, and *Aerides guttalum* is used as an emollient drug. Tubers of *Aerides crispum* are rich in aeridin and phenanthropyran and anti-bacterial.

11.3.1.5 Cultivation

Aerides prefers a bright light of 2400-3600 foot candles, not direct sun. These plants perform well under warm humid conditions with plenty of free moving air, and should not be subjected to temperatures below 15°C. A porous potting media consisting of shredded tree fern fibres or coarse bark along with gravels is ideal. These orchids grow throughout the year and require a good amount of water at all times. Fortnightly or monthly feeding with ½ or ¼ strength of balanced fertilizers of N:P:K (20:20:20) is desirable. Aerides are easily propagated by keikis divisions.

11.3.2 Anoectochilus

Anoectochilus consists of about 25 species, which are commonly known as 'Jewel Orchids'. These are pseudobulbless terrestrial orchids with creeping rhizomes and petioled green, brown, copper colour veined and striated, rose, yellow or white leaves. Flowers appear in glandular pubescent spikes.

11.3.2.1 Genetic Resources

Anoectochilus albolineatus: This species is found in Burma and Malaya Peninsula and produce velvety dark purplish leaf with red veins.
Anoectochilus brevilabris: It is distributed in Sikkim, Arunachal Pradesh, Nepal and Bhutan. The plants are 10-30 cm tall having beautiful leaves patterned with pale gold to copper veins on velvety maroon brown leaves.
Anoectochilus roxburghii: A native to the Himalayas. The leaf is deep velvety green with lighter centre, veined and reticulated golden yellow, reddish margins.
Anoectochilus argyroneurus: This species is native to Java. Leaves are brilliant green, silvery white veined and green spotted.
Anoectochilus regalis: This species is from Sri Lanka. The leaves are dull green and bronze shaded veined and reticulated golden yellow.
Anoectochilus yunngianus: Leaves are brown green and covered with a network of gold and copper veins.

11.3.2.2 Uses

- Jewel orchids are grown for attractive foliages.
- Medicinal Anoectochilus:
  - *Anoectochilus formosanus*: An anti-inflammatory agent, anti-febrigoagent, anti-depressing agent and against the virus influenza A.
  - *Anoectochilus roxburghii*: Distributed in the Himalayas and Vietnam. It is rich in novel flavonoid glucoside and has potential in fighting cancer. Popular varieties are 'TE', 'Alba', 'Variegated', 'Variegated Alba'.

11.3.2.3 Cultivation

Plants grow well in intermediate to warm conditions under shade environment. They can be placed in a wet interior with one to two hours sunlight everyday. A well drained potting mixture consisting of sand, bark, perlite and soil is ideal. Plants are propagated by seeds or rhizomes.

11.3.3 Arachnis

Arachnis consists of more than 20 species and is commonly known as 'Scorpion Orchid.' These are monopodial epiphytic plants, 30 cm to 4.5 m tall with many leathery leaves. Leaves are strap shaped, clasping the stem at their base. Flowers are showy, large borne on simple racemes or branched panicles and the inflorescence is 50 to 65 cm long arise from the stem just above the clasping leaves. Flowers are green, yellow, brown and purple in colour.

11.3.3.1 Genetic Resources

Arachnis cathcartii: This species is distributed in Sikkim and Bhutan. Plants are 30-45 cm tall with bilobed coriaceous, lorate and leathery leaves. The inflorescence is stout, pendulous or horizontal. The flowers are fleshy, waxy, long lasting and pale yellow with red brown bands, and are available in March-April.
Arachnis clarkei: Distributed in Sikkim and Bhutan. Plants are 30-45 cm long. The inflorescence is 2-3 flowered. The flowers are fleshy, highly scented, fragrant and yellow barred with pale cinnamon brown. Flowers are available in the spring – summer season.
Arachnis flos-aeris: Distributed in Sumatra, Borneo, Malaya Peninsula and Java. A stout stemmed climbing plant, 5 m tall with leathery, bilobed lanceolate leaves and ascending or drooping simple or branched inflorescence. The flowers are fleshy, long lasting, and a pale yellow green colour with dark purple brown bars and spots. Flowers are available in the spring-summer season.
Arachnis Maingayi: This species is distributed in Singapore to Borneo. The stems are 10-15 cm tall with linear oblong leaves and 30-90 cm long inflorescence. The flowers are whitish or faintly pinkish with purple or pink blotches and produced during April to May.
Arachnis hookeriana (White Scorpion Orchid): Native to Borneo, Malayasia and Singapore. A hardy and frequently flowering species used as a parent for the development of hybrids for cut flowers.

11.3.3.2 Inter-generic Hybrids

These genera have the capacity to cross inter-specifically and multi-generically to evolve outstanding hybrids.
Arachnopsis = *Arachnis* x *Phalaenopsis*
Arachnostylis = *Arachnis* x *Rhyncostylis*
Aranda = *Arachnis x Vanda*

Aranthera = Arachnis x Renanthera
Sappanara = Arachnis x Phalaenopsis x Renanthera
Trevorara = Arachnis x Phalaenopsis x Vanda
Laycockara = Arachnis x Phalaenopsis x Vandopsis
Holttumara = Arachnis x Renanthera x Vanda

11.3.3.3 Common Commercial Hybrids

Arachnis 'Maggie Oei Red Ribbon', *Arachnis* 'Maggie Oei Yellow Ribbon', *Arachnis* 'Maroon Maggie', *Arachnis* 'Merry Maggie', *Aranthera* 'Bartha Braga', *Aranthera* 'Ishbel'.

11.3.3.4 Uses

Arachnis is commonly used in wedding decorations, flower arrangements etc. They are grown with stakes or beds with rods. Spider type early varieties are used as cut flowers.

11.3.3.5 Cultivation

They grow well in a temperature range of 18-20°C, light intensity of 2400-3600 foot candles and relative humidity range of 65 to 80%. A compost mixture consisting of tree fern, bark, charcoal, cowdung manure, bonemeal and sandy loam soil is good for their cultivation. The plants are grown in well drained raised beds and pots with compost and other components like leaf moulds, sphagnum moss and gritty white sand.

11.3.4 Ascocentrum

Ascocentrum consists of six species. These are dwarf monopodial epiphytes and characterized by large spur hangs from the tip. The plants are small, compact with small strap shaped leaves and short stalked erect and cylindrical covered with many closely spaced flowers. Flowers are long lasting and 1-2 cm in diameter.

11.3.4.1 Genetic Resources

Ascocentrum ampullaceum: Distributed in the Himalayas, Burma and China. Plants are up to 20-25 cm tall with many flowered short stalked inflorescence. The flowers are 2.5 cm in diameter and rose carmine in colour. Flowers are available during March to May.

Ascocentrum curvifolium: Distributed in the Himalayas and Java. Plants are stout, 12.5 cm tall with rigid and fleshy leaves and densely flowered inflorescence. The flowers are purple to orange scarlet in colour and produced in spring summer.

Ascocentrum miniatum: Distributed in Malayasia and the Himalayas, Borneo and the Philippines. Plants are dwarf (less than 15 cm tall) with linear, very fleshy leaves and cylindrical inflorescence crowded with densely flowers. Flowers are yellow to orange red and available in spring summer season.
Ascocentrum pumilum: An intermediate and cold loving dwarf species, 5 cm tall with needle like leaves and tiny pink flowers.

11.3.4.2 Hybrids:

It is cross compatible with *Aerides, Neofinetia, Vanda* and *Rhyncostylis.*

11.3.4.2.1 Bigeneric Hybrids

Aeridocentrum = Ascocentrum x Aerides
Ascofinetia = Ascocentrum x Neofinetia
Ascocenda = Ascocentrum x Vanda
Renancentrum = Ascocentrum x Renanthera
Asconopsis = Ascocentrum x Phalaenopsis

11.3.4.2.2 Trigeneric Hybrids

Kagawara = Ascocentrum x Renanthera x Vanda
Eastonara = Ascocentrum x Gastrochilus x Vanda
Christieara = Aerides x Ascocentrum x Vanda
Komkrisara = Ascocentrum x Renanthera x Rhyncostylis
Lowsonara = Aerides x Ascocentrum x Rhyncostylis
Mokara = Arachnis x Ascocentrum x Vanda
Nakamotoara = Ascocentrum x Neofinetina x Vanda
Vascostylis = Ascocentrum x Rhyncostylis x Vanda

11.3.4.2.3 Tetrageneric Hybrids

Alphonsoara = Arachnis x Ascocentrum x Vanda x Vandopsis
Bokchoonara = Ascocentrum x Arachnis x Phalaenopsis x Vanda
Darwinara = Ascocentrum x Neofinetia x Rhyncostylis x Vanda
Dominyara = Ascocentrum x Luisia x Neofinetia x Rhyncostylis
Knappara = Ascocentrum x Rhyncostylis x Vanda x Vandopsis
Moonara = Ascocentrum x Aerides x Neofinetia x Rhyncostylis
Okaara = Ascocentrum x Renanthera x Rhyncostylis x Vanda
Onoara = Ascocentrum x Renanthera x Rhyncostylis x Vanda
Pageara = Ascocentrum x Luisia x Rhyncostylis x Vanda
Robinara = Ascocentrum x Aerides x Renanthera x Vanda

Ronnyara = Aerides x Ascocentrum x Rhyncostylis x Vanda
Stamariaara = Ascocentrum x Phalaenopsis x Renanthera x Vanda
Yusofara = Arachnis x Ascocentrum x Renanthera x Vanda

11.3.4.2.4 Pentageneric Hybrids

Knudsonara = Ascocentrum x Neofinatia x Renanthera x Rhyncostylis x Vanda
Paulara = Ascocentrum x Doritis x Phalaenopsis x Renanthera x Vanda
Sutingara = Arachnis x Ascocentrum x Phalaenopsis x Vanda x Vandopsis

11.3.4.2.5 Common Commercial Hybrids

'Sagarik Gold'.

11.3.4.3 Uses

Mature plants producing many keikis are excellent for specimen culture. Species are ideal for bamboo baskets or coir basket culture and produce brilliant orange flowers in the early spring. They can be mounted on slabs of bark. Hybrids are long lasting and used as cut flowers.

11.3.4.4 Cultivation

The plants are grown under protected conditions in a temperature range of 24 to 29°C during daytime and 6-8°C during the night, with a light intensity of 2400-3600 foot candles and relative humidity of 70 to 90%. They thrive in strong bright light and can be acclimated to direct sunlight. Exposures of the plants below 15°C can cause delaying of flowering. Ascocentrum orchids are best grown in pots in a porous media consisting of osmunda tree fern fibre, tree bark, brick pieces and charcoal. Hanging plants require regular watering or misting. These orchids are heavy feeders and best fertilized weekly with ½ strength orchid fertilizers.

11.3.5 Bulbophyllum

Bulbophyllum consists of about 2500 species of orchids from tropical and subtropical Asia. The name indicates that the leaves are attached to the top of pseudobulbs. These orchids are sympodial epiphytes having rhizomatous stems with angled pseudobulbs. The pseudobulbs form a chain like growth on the rhizomes. Inflorescence is erect or pendent and arises from the base of the pseudobulb. The flowers are solitary or grow in an umbel or united on spurs. Flower size ranges from a few millimeters (*Bulbophyllum minutissimum, B. minututum*) to 40 cm in *Bulbophyllum echinolobium*.

11.3.5.1 Genetic Resources

Bulbophyllum barbigerum: Native to tropical West Africa. The pseudobulb is small, flat terminating in a single leaf. The 8 to 14 flowers are borne on inflorescence, brown in colour, and are available in July-August.
Bulbophyllum careyanum: Pseudobulbs are egg shaped and 4-angled. Leaves are thick, leathery and tongue shaped. Inflorescence is many flowered and densely arranged. Flowers are 1 cm in diameter, sour smelling, white-yellow and spotted with brown red and produced during October-December.
Bulbophyllum cornu-cervi: This species is native to Sikkim. Pseudobulbs are globular, minute and crowded. Flowers are small and borne on slender and erect 8-10 cm long inflorescence. Flowers are greenish and produced in July.
Bulbophyllum cylindraceum: Distributed in Sikkim and Meghalaya. The rhizome is stout with obtuse leaves. The inflorescence is erect, cylindric and catkin-like nodding raceme. Flowers are black purple and available during October-January.
Bulbophyllum eublephrum: Native to Sikkim Himalaya. The pseudobulbs are cylindric. The inflorescence is raceme and lax flowered. Flowers are green and produced in August.
Bulbophyllum falcatum: Native to tropical West Africa. The pseudobulbs produce one pair of leaves. Inflorescence is arranged with yellowish brown flowers marked with purple and produced in summer.
Bulbophyllum grandiflorum: Native to New Guinea. The single greenish brown flower is borne on a peduncle of 15-20 cm long and produced during July-August.
Bulbophyllum leopardinum: Distributed in Meghalaya. The rhizome is very short with pear shaped compressed bulbs. The inflorescence is short and 1 to 3 flowered. The flowers are fragrant, whitish, spotted with crimson and produced during OctoberNovember.
Bulbophyllum lobbii: Native to Burma and Malay Peninsula. The rhizome is stout with egg shaped pseudobulbs and leathery leaves. Flowers are solitary, fragrant, waxy and long lasting.
Bulbophyllum macranthum: Native to Java, Borneo, Sumatra and Malayasia. The flowers are whitish, densely spotted with purple and produced in spring season.
Bulbophyllum makoyanum: Native to Singapore. Pseudobulbs are ovoid with tongue shaped leaves. The flowers are scented, reddish with darker spots and fringed with yellow hairs.
Bulbophyllum medusae: Native to Malayasia. The plant is a creeper and arranged with 20 cm long flower stalk having a dense head of fragrant flowers. The flowers are white, cream white or pink or red spotted.
Bulbophyllum rigidum: Distributed in Sikkim Himalaya. The rhizomes are pseudobulbous and 25 cm long and 5 cm broad leaves arose from it. The inflorescence is sparsely flowered. The flowers are yellowish tinged with green and streaked with red and produced in May-June.
Bulbophyllum striatum: Distributed in Meghalaya. The pseudobulbs are ovoid. The inflorescence is 2 to 4 flowered. The flowers are yellow-green stripped with purple and produced in October.

Bulbophyllum thomsonii: Distributed in Coorg, Mysore, Nilgiris and the Annamalai Hills. The rhizome is slender with sub-globose pseudobulbs and round, ovate-lanceolate leaves. The inflorescence is erect, lax flowered and 6-10 flowered. The flowers are yellow tipped with purple.
Bulbophyllum appendiculatum: Distributed in North East states of India. Pseudobulbs are ellipsoid cylindric with fleshy, erect, sessile leaves. Inflorescence is one flowered and erect. Flowers are deep purple and produced in October-November.
Bulbophyllum bisetum: A dwarf species, distributed in all North East states of India. Pseudobulbs are small, oval shaped and unifoliate. Inflorescence is pendent racemose and 10-15 flowered. The flowers are dull purple with green yellow margins and produced in July-September.
Bulbophyllum cariniflorum: A dwarf species from Arunachal Pradesh and Sikkim. Pseudobulbs are ovoid to cylindric and ridged. Inflorescence is racemose and many flowered. Flowers are scented, greenish yellow with a bright yellow tip and produced in July-August.
Bulbophyllum crassipes: A robust species distributed in all states of North East India. Pseudobulbs are conical and many flowered with coriaceous oblong leaves. Flowers are yellow with purple spots and produced in September-October.
Bulbophyllum cauliflorum: Distributed in all North East states of India. Pseudobulbs are cylindric with linear-oblong leaves. Inflorescence is umbellate and 3-5 flowered. Flowers are pale yellow green with a yellow lip.
Bulbophyllum helenae: A native to all North East states of India. Pseudobulbs are ovoid with linear oblong leaves. Inflorescence is umbellate and 6-9 flowered. Flowers are orange red in colour and produced in May-July.
Bulbophyllum hirtum: Distributed in Arunachal Praadesh, Meghalaya, Nagaland and Sikkim. Pseudobulbs are crowded, ovoid cylindric with oblong-lanceolate leaves. Inflorescence is erect, racemose and many flowered. Flowers are scented, pale greenish white and produced in October-January.

11.3.5.2 Hybrids

11.3.5.2.1 Common Commercial Hybrids

'Ray Gabaldon', 'A dorbil Ring', 'Madeline Nelson', 'Tsiku Goldfinder', 'Fredensborg Delight', 'Swissthai Alamia', 'Jan Ragan', 'Wilmer Shear Magic', 'Jim Clarkson', 'Tsiku Taurus', 'Agathe', 'Short and Sweet', 'B.C.', 'Chua', 'Aquarius', 'Chanthaboon Glory', 'Meen Bulbul', 'Meen Candy Baby', 'Supat Sun Radiant', 'Supernova', 'Tee Dragon Fly', 'Tee Naga', 'Puch Golden Hydra', 'Tee Gigantic Condor', 'Grace Thoms'.

11.3.5.3 Medicinal Bulbophyllums

Bulbophyllum odoratissimum: This species is distributed in China, Nepal, Sikkim, Bhutan and Thailand. Plants are rich in phenanthrene, lignan, flavonoids, biben-

zyls, phenolic glycosides, aldehydes and acids. Plants are used to treat tuberculosis, chronic inflammations and fractures.
Bulbophyllum inconspicuum: In Japan, it is used as an expectorant and in the treatment of stomach cancers.
Bulbophyllum albidum: In Western Ghats, leaves and bulbs are used to strengthen a weak uterus for conception.
Bulbophyllum kwangtungense: In China and Japan, tubers are used for the treatment of pulmonary tuberculosis and fever. This species is rich in dihydrodibenzoxepins, densifloral A, cumulatin and plicatol B.

11.3.5.4 Uses

Cluster pseudobulb-bearing species grow well in shallow pans whereas long rhizome bearing species do better in baskets or slabs.

11.3.5.5 Cultivation

All tropical bulbophyllums grow in a flat pot, hanging or suspended on cork bark in a humid semi-shaded location where the temperature range is 15 to 27°C. These orchids like slightly drier conditions in between waterings and require a rest in winter when they are not growing. They also require good light but it is best if they are shaded from bright, direct summer sun. In summer, humidity level should be maintained at 70% and 50-60% in winter. Plants are easily propagated through division of clumps while repotting. A potting mixture of cocochips, bark pieces and brick pieces is ideal. During active vegetative growth, a half dose of balanced 20:20:20 fertilizer is applied twice a month.

11.3.6 Coelogyne

Coelogyne consists of 200 species of sympodial and pseudobulbous orchids. The pseudobulbs are topped by 2 to 4 leaves, slender in size and arranged along with creeping rhizomes. The leaves are coriaceous, thick and leathery with pronounced stalks. Inflorescences are erect or pendulous. The flowers are white or shades of brown, yellow and green and subtended by a papery bracts.

11.3.6.1 Genetic Resources

Coelogyne asparata: Distributed in Malay Peninsula, Borneo, Sumatra and New Guinea. Pseudobulbs are vigorous, ovoid, clustered and carrying each pair of huge leaves up to 60 cm long and 12.5 cm wide. Inflorescences are arching or pendulous and bear 12 to 18 flowers. The flowers are fragrant, heavily textured, creamy white with reddish brown lip and produced during spring-summer.

Coelogyne barbarata: Distributed in Nepal, Bhutan and India. The pseudobulbs are pale green in colour, clustered, ovoid with stalked leathery leaves. The inflorescence is densely flowered, erect or arching. The flowers are white, spotted deep brown and produced in autumn and winter.
Coelogyne cristata: Native to higher elevations of Himalayan region. The pseudobulbs are spherical or egg shaped and clustered with soft textured linear-lanceolate leaves. The inflorescences are semi-erect, many flowered. The flowers are large, pure white with yellow ridges on lip and are produced during January to May.
Coelogyne quadrangularis: Distributed in Thailand, Malaya Peninsula, Sumatra, Java and Borneo. The pseudobulbs are conical, strongly ribbed with short stalked leathery leaves. The flowers are fragrant, 5 cm in diameter, long lasting, dull yellow brown lined with whitish lips and produced in spring summer.
Coelogyne elata: Native to higher elevations of the Himalayan regions. The pseudobulbs are cylindrical to ovoid with stalked heavily textured leaves. Inflorescence is erect, 60 cm tall and arranged with 7 to 10 flowers. The flowers are fragrant, long lasting, milky white with an orange tip.
Coelogyne flaccida: Naturally distributed at higher elevations of Sikkim and Meghalaya. The pseudobulbs are oblong, bifoliate. The inflorescence is pendulous and loosely arranged with 5 to 12 medium sized creamy flowers. The flowers are highly scented and produced during March to May.
Coelogyne graminifolia: Native to India, Burma, Thailand and Malay Peninsula. The pseudobulbs are borne on creeping rhizomes with pairs of leaves. The inflorescence is 15 cm tall, arching and 2 to 4 flowered. The flowers are 5 cm across, scented, white and produced during January to April.
Coelogyne Huettneriana: Native to Burma and Thailand. The pseudobulbs are irregular dark green and bifoliated. The inflorescence is arching and arranged with 12 flowers. The flowers are fragrant, white, long lasting, 5 cm across with citron yellow lip and produced during summer and spring.
Coelogyne lawrenceana: Native to Vietnam. The pseudobulbs are bifoliate, ovoid, clustered and heavily ribbed. The inflorescence is terminal, arching and bearing 1 to 3 flowers. The flowers are 5 cm in diameter, pale yellow with orange lip and produced in July.
Coelogyne massangeana: Distributed in Sumatra, Java, Malaya, Thailand and Borneo. The pseudobulbs are ovoid, bifoliate, yellow green with elliptic leaves. The inflorescence is pendulous and 15 to 20 flowered. The flowers are fragrant, 6 cm in diameter, pale yellow with brown side lobed lip and produced in spring and summer.
Coelogyne odoratissima: Distributed in India and Sri Lanka. This species is bifoliate with sheathed pseudobulbs and linear leaves. The inflorescence is loosely arranged with 2 to5 flowers. The flowers are fragrant, white with yellow stripes and produced in spring and summer.
Coelogyne pandurata: Native to Malay Peninsula, Sumatra and Borneo. The pseudobulbs are compressed on both sides, smooth and leathery leaves. The inflorescence is arching, 20 flowered. The flowers are highly fragrant, emerald green in colour with white mottled lip and produced in autumn.

Coelogyne speciosa: Native to Sumatra, Java and the Lesser Sunda Islands. The pseudobulbs are ovoid, clustered and bifoliate with arrow elliptic leaves. The inflorescence is 2 to 3 flowered, slender and abbreviated. The flowers are long lasting, musk scented, 7.5 cm across, yellow brown with yellow tip and produced during summer and spring.

Coelogyne ocultata: A subtropical to temperate species of Arunachal Pradesh, Meghalaya, Nagaland and Sikkim. Pseudobulbs are green, tapering at both ends and bifoliate. The inflorescence bears 2-3 flowers. Flowers are 2.5 to 4.5 cm across and white with blotched yellow tips. Flowers are produced in June-July.

Coelogyne ovalis: A subtropical species from Arunachal Pradesh, Assam, Meghalaya, Nagaland and Sikkim. Pseudobulbs are oval to spindle shaped. The inflorescence bears 2-3 flowers of pale yellow green to pale brown with a brown marking lip. The flowers are produced in October-December.

Coelogyne corymbosa: A subtropical to temperate species from Bhutan, India and Nepal. Pseudobulbs are ovoid, clustered and bifoliate. Inflorescence is erect or pendent and 2-4 flowered. Flowers are fragrant and white or with four yellow markings bordered with orange red lip and produced in April-May.

Coelogyne fuscescens: A subtropical to temperate species from Bhutan, India and Nepal. Pseudobulbs are bifoliate. The inflorescence bears 2 to 10 flowers. The flowers are 5 cm across, pale yellow to pinkish brown with orange red lip and produced during October-December.

Coelogyne prolifera: A subtropical species from Sikkim, Meghalaya and Nepal. Pseudobulbs are round. The inflorescence is slender-peduncle and 3-8 flowered. The flowers are yellow, 1-1.5 cm in diameter and produced during May-August.

Coelogyne fimbriata: Native to Sikkim and Meghalaya. Inflorescence is erect raceme and bears 8-10 yellow flowers. Flowers are produced in October-December.

Coelogyne flavida: Native to Sikkim, Manipur and Meghalaya. Flowers are yellow and produced during March –May.

Coelogyne ochracea: Native to Assam and Sikkim. Inflorescence is erect and bears 8-10 flowers. Flowers are small, white and fragrant and produced in May-June,

Coelogyne nitida: Native to Bhutan, Assam, Meghalaya and Sikkim. Flowers are white with two yellow red margin blotches on lip. Flowers are produced during March-May.

Coelogyne rigida: Native to Nilgiri hills, Sikkim and Meghalaya. Inflorescence bears 5-7 flowers, pale yellow in colour and produced in June-July.

11.3.6.2 Hybrids

11.3.6.2.1 Inter-specific Hybrids

'Memoria W. Micholitz', 'Linda Buckley', 'Burfordiense', 'South Carolina', 'G. Beryl Read', 'John Leathers', 'Wood Fairy', 'Carolyn Sue', 'Golden Bug', 'Spring Showers', 'Professor Georges Morel', 'Joel Strandberg'.

11.3.6.2.2 Variety-species Hybrid
'Memoria Fukuba', 'Memoria Okami', Memoria Tokiko'.

11.3.6.3 Coelogyne for Hanging Baskets
Coelogyne asperata, Coelogyne cristata, Coelogyne foestermani, Coelogyne longifolia, Coelogyne pandurata.

11.3.6.4 Coelogyne for Herbal Medicines
- *Coelogyne cristata*: Juice of pseudobulb is applied to boils.
- *Coelogyne corymbosa*: A paste of pseudobulb is applied to forehead to relieve headache.
- *Coelogyne flaccida*: A paste of pseudobulb is applied to forehead to treat headache. Juice is also consumed for indigestion.

11.3.6.5 Cultivation
There are high elevation and low elevation species and hence they are divided into two major groups, and environment requirements vary between groups. Cool growing species require a summer day temperature of 21-24°C and night temperature of 14-15°C; winter day temperature of 11-13°C and night temperature of 2-5°C, light intensity of 2000-3000 foot candles and relative humidity of 60-70%. Strong air movement is essential. They are best grown in baskets containing a potting mixture of tree bark, charcoal and sphagnum moss. A balanced fertilizer mixed to ¼ to ½ of the recommended strength should be applied weekly during periods of active growth. Many growers recommend using a fertilizer lower in nitrogen and higher in phosphorus during late summer and autumn to promote better blooming the next season and to allow the new growths to harden before winter. Over watering should be avoided.

11.3.7 Doritis

Doritis is a monotypic epiphytic orchid from Burma to Sumatra. *Doritis pulcherrima* is a small, stemless and clump forming monopodial with oblong elliptic leathery leaves. The inflorescence is erect and arranged with 25 flowers. The flowers are 1.5 to 4.0 cm in diameter with pear shaped lips. The flowers are dark lavender, rose purple, or magenta rose in colour and produced during the autumn and winter seasons.

11.3.7.1 Primary Hybrids
Primary hybrids evolved through crosses between *Doritis pulcherrima* and *Phalaenopsis* spp. are listed below:

'Anna –Larati Soekardi', 'Annie Van Tweel', 'Asahi', 'Bandung Pink', 'Bonita', 'Celebrant', 'Charm', 'Imp', 'Jim', 'Jim Chan', 'Kelsey Blush', 'Keneth Schubert', 'Musick Surprise', 'Myriam Esther', 'Profusion', 'Purple Gem', 'Purple Passion', 'Purple Sum', 'Red Elf', 'San Shia Appendo' Sandrine Dream', 'Si Kancil', Siam Treasure', 'Suka Hati', 'Summer Red', 'Sweet Gem', 'Tan Swee Eng.', 'Tarina'.

11.3.7.2 Inter-generic Hybrids

Doriella = Doritis x Kingiella
Doritaenopsis = Doritis x Phalaenopsis
Rhyndoropsis = Doritis x Phalaenopsis x Rhyncostylis
Vandoritis = Doritis x Vanda

11.3.7.3 Commercial Uses

Ideal for pot plants or as cut flowers.

11.3.7.4 Cultivation

The plants require indirect sunlight and are shade loving plants that should be kept in a warm greenhouse with high humidity. For growing these plants, the optimum temperature range is 18 to 21°C, relative humidity 60 to 70% and light intensity ranges from 1500-1800 foot candles. The most suitable potting media is chunks of hardwood charcoal or a mixture of bark, tree fern fibre and charcoal. The spraying of liquid fertilizers once every fortnight is beneficial.

11.3.8 Epidendrum

Epidendrum consists of more than 1000 species of epiphytic orchids of America. The pseudobulbs are very elongated or short with coriaceous leaves. Some leaves are thick whereas in some cases they are leathery with 2 to 4 leaves at the apex of each pseudobulb. In the case of reed like types, the leaves are alternate. The inflorescence is terminal, simple or profusely branched, erect or pendulous. The number of flowers varies from one to many per inflorescence. The size of flower ranges from 2 cm to 7.5 cm in diameter and highly scented. The flowers are varied in colour.

11.3.8.1 Genetic Resources

Epidendrum atropurpureum: Distributed in Mexico to Panama. The pseudobulbs are egg shaped, glossy, purplish, and 10 cm tall with rigid, leathery leaves. Inflorescence is 60 cm tall, racemose, erect or arching and 2 to 10 flowered. The flowers are chocolate brown, marginated with apple grren, scented and produced in spring-summer season.

Epidendrum Brassovolae: Native to Central America. The pseudobulbs are clustered or borne on a stout rhizome and 25-30 cm long. The leaves are 2 to 3 leathery and yellowish green. Inflorescence is erect, long stalked, 45 cm long, 6 to 9 flowered. The flowers are fragrant, 10 cm across, deep yellowish brown to greenish and produced in March-April.
Epidendrum ciliare: Native to tropical America. The pseudobulbs are spindle shaped or cylindrical, 18 cm tall, borne on a creeping rhizome, terminated by two rigid elliptic oblong leaves. The inflorescence is racemose, 30 cm long with loosely arranged flowers. The flowers are 17 cm across, sweet scented, greenish white with trilobed white lip and produced during December-January.
Epidendrum cochleatum: Distributed in South Florida to Brazil. The pseudobulbs are 15 cm tall, pear shaped and clustered with paired light green leaves. The inflorescence is erect or arching, 60 cm tall, 8 to 12 flowered. The flowers are sweetly scented, greenish yellow with velvety red purple black lips.
Epidendrum fucatum: Native to Mexico, Guatemala and Cuba. The pseudobulbs are ovoid, 7.5 cm tall and clustered with 2 to 3 leathery, linear, 30 cm long leaves. The flowers are 4 cm across, long lasting, fragrant and yellowish or butter yellow in colour and produced in autumn-winter.
Epidendrum radicans: Native to Central and South America. The stems are elongated, 18 cm long, branched and offset forming rigid fleshy leaves. The inflorescence is erect or arching, terminal and 1.2 m tall. The flowers are 4 cm across, red, orange, scarlet, vermillion, orange red or yellow in colour and produced during summer.
Epidendrum vittellinum: Native to Mexico and Guatemala. The pseudobulbs are ovoid-conical, clustered with 2 to3 linear lanceolate to oblong, leathery leaves. Inflorescence is erect, branched, 45 cm tall and 10-12 flowered. The flowers are 4 cm in diameter, long lasting and vermillion with orange yellow lip.
Epidendrum xanthimum: A rare species. Flowers are bright yellow and orange and produced in masses.
Epidendrum parkinsonianum: It is pseudobulbless with pendent succulent leaves. Flowers are large, exotic, white and highly scented.

11.3.8.2 Inter-generic Hybrids

Brasso-epidendrum = Epidendrum x Brassovola
Vaughmara = Epidendrum x Brassovola x Cattleya
Yamadara = Epidendrum x Brassovola x Cattleya x Laelia
Epitonia = Epidendrum x Broughtonia
Epicattleya = Epidendrum x Cattleya
Arizara = Epidendrum x Cattleya x Domingoa
Epilaeliocattleya = Epidendrum x Cattleya x Laelia
Kirchara = Epidendrum Cattleya x Laelia x Sophronitis
Epidiacrinum = Epidendrum x Diacrinum

Epilaelia = Epidendrum x Laelia
Epiphronitis = Epidendrum x Sophronitis

11.3.8.3 Intervarietal Hybrids

'Happy Valley', 'Little Valley', 'Lovely Valley', 'Spring Valley', 'Lions Mane', 'Lions Pride', 'Peace Valley', 'Lucky Valley', 'Pacific Classic', 'Pacific Mandarin', 'Pacific Parade', 'Pacific Secret', 'Pacific Volcano', 'Apricot Valley', 'Rain Valley', 'Water Valley', 'Pacific Joy', 'Pacific Globe', 'Pacific Flame', 'Pacific Heat', 'Pacific Glory', 'Walnut Valley', Fireball', 'Autumn', 'Lunar Eclipse', 'Pacific Dream', 'Pacific Lights', 'Pacific Shore', 'Pacific Delight', 'Pacific Mulberry', 'Pacific Storm', 'Candy Valley', 'Holy Valley', 'Palm Valley', ' Milky Valley', Darling Jungle', 'Pacific Girl', 'Pacific View'.

11.3.8.4 Inter-specific Hybrids

'Cire Blanche', 'Thomas Ott', 'Little Miss Sunshine', 'Saranac Lake', 'Eddy Lick Run', 'Michael Riley', 'Ross Newman', 'Daves', 'Delectable Deliration', 'Orange Park', 'Cascade Blanche', 'Fuzzy Grapes', 'Doctor Ben', 'Suzuka Mars', 'Pink Lemonade', 'Golden Judy', 'Patricia Sander', 'Candy Dandy'.

11.3.8.5 Variety-species Hybrids

'Saint Nicholas', 'Wild Innovation', 'Chocolate Cherry Supreme', 'Costa Lense', 'Spirit of Volunteerism', 'Songbirds', 'Topaz Cherry', 'Topaz Prolific', 'Spirit of Giving', 'Topaz Gem', ' Hula Dancer', 'Pacific Sunshine', 'Topaz Butterball'.

11.3.8.6 Epilaelia Hybrids

'Royal Velvet', 'Topaz Sundaze', 'Irene', 'Sedona', 'Excaliber', 'Topaz Seduction', 'La Motte', 'Heart of Darkness'.

11.3.8.7 Epilaeliocattleya Hybrids

'Chocolate Kisses', 'Kopaa', 'Rainbow Sherbet', 'Pixford', 'Tiny Magic', 'Highland Canary', 'Pseudogold', 'Jackie Bright'.

11.3.8.8 Epicattleya Hybrids

'Cathy Meincer', 'Painted Hill Star', ' Calandria', 'Erin Routon', 'Burdekin Honey', 'Peles Flame'.

11.3.8.9 Medicinal Epidendrum

A Brazilian species namely *Epidendrum mosenii* is used to treat infective and dolorous processes.

11.3.8.10 Edible Epidendrum

The leaves of *Epidendrum ibaguense* are edible, resembling the taste of watermelon, and can be chewed like bubblegum.

11.3.8.11 Cultivation

The optimum temperature range is 12 to 21°C, relative humidity 50-60% and light 2400 to 3600 foot candles for successful growth and flowering. They grow in beds or pots with excellent drainage. Moderate shade during the summer and plenty of water during active growth are required. The pot mixture comprises of tree bark and charcoal mixture. Spraying with 0.3% NPK (10:10:10) every two weeks is beneficial.

11.3.9 Gastrochilus

Gastrochilus consists of 20 species of epiphytic orchids distributed in India, East Asia and Malaysia. These are short to long stemmed epiphytic orchids with 6 to 11 lance shaped, thick, leathery leaves. The inflorescences are solitary or spicate with one or more flower spikes arising from the axil of the leaves. The flowers are attractive, 1 to 4 cm in diameter and white in colour with reddish spots and blotched with yellow.

11.3.9.1 Genetic Resources

Gastrochilus bellinus: Native to Burma and Thailand. The stems are stout with leathery, tongue shaped leaves. The inflorescence is 4 to 7 flowered, erect and robust. The flowers are long lasting, 4 cm in diameter and fragrant. The flowers are greenish yellow with purple spots and blotches and are produced during winter and spring.
Gastrochilus calceolaris: Distributed in Meghalaya, Thailand, Sumatra, Java, Burma and Malayasia. The plants are short and stout stemed with tongue shaped leathery leaves. The inflorescence is stout, 3 to 6 flowered and purple spotted. The flowers are fragrant, waxy, 2 cm across, long lasting, greenish or yellowish with brown or purplish spots and produced during March-April.
Gastrochilus acutifolius: Native to Bhutan, India and Nepal. Stems are slender, pendent with oblanceolate leaves. Inflorescence is umbellate and many flowered. Flowers are fleshy, fragrant, spotted and flushed with brown and are produced during November-December.

Gastrochilus obliquus: Native to Bhutan, West Bengal, Sikkim and Arunachal Pradesh. A dwarf species, 20-15 cm tall with stout pendent stems bearing 3-5 oblong fleshy leaves. The inflorescence is pendent, corymbose to umbellate and many flowered. The flowers are bright yellow, purple spotted and produced during November-December.

11.3.9.2 Uses

The rhizomes of *Gastrochilus pandurata* are used by natives of Java as a medicine and as a spice.

11.3.9.3 Cultivation

These orchids grow well in small pots or baskets. They prefer a warm, humid atmosphere and well lighted and ventilated places. The optimum temperature is 15 to 18°C, light 1800-2400 foot candles and relative humidity is 60 to 80% for proper growth and flowering of these plants. They grow well on tree fern fibre or chunks of hard wood charcoal. Spraying of complete fertilizer mixture is recommended at fortnightly intervals. The plants should be liberally watered during their active period of growth.

11.3.10 Laelia

Laelia consists of 75 species of epiphytic orchids from Cuba and Mexico to Argentina. These sympodial orchids possess pseudobulbs of various sizes and shapes, e.g., *Laelia harpophylla* have needlelike pseudobulbs. Another group bearing flat and round pseudobulbs is *L. rubescens*, while other groups have ovoid or egg shaped pseudobulbs, such as *L anceps*. The number of leaves per pseudobulb ranges from 1 to 3. The leaves are leathery, entire, thick and 5 cm to 30 cm in length. The inflorescence bears more than 20 flowers in some species, while others bear solitary flowers close to the pseudobulb apex. The flowers are lavender, white, yellow and orange in colour.

11.3.10.1 Genetic Resources

Laelia anceps: Native to Mexico and Honduras. The pseudobulbs are stout, ovoid-oblong, glossy, purple flushed, quadrangular, clustered and borne on stout rhizomes. The leaves are solitary, leathery and glossy green and oblong lanceolate. The inflorescence is elongated, slender, erect or gracefully arching with 1 to 6 flowers, 60 cm long. The flowers are lilac-pink, scented, 10 cm across and produced during December-January.

Laelia autumnalis: Native to Mexico. The pseudobulbs are short, clustered, sub-conical and ribbed with 2 to 3 leathery, lanceolate leaves. Inflorescence is elongated, 90 cm long, loosely arranged, with 5 to 10 flowered raceme. The flowers are pinkish purple, waxy, long lasting, 10 cm across and produced in December-January.
Laelia cinnabarina: Native to Brazil. The pseudobulbs are elongated, 25 cm long and red purple with solitary, thick leathery rigid, 60 cm long purple leaves. The inflorescence is erect or arching and 5 to 20 flowered. The flowers are star shaped, orange red, long lasting, 7.5 cm across and are produced during March to May.
Laelia pumila: A dwarf species from Brazil. Pseudobulbs are terete and glossy dark green in colour. The leaves are fleshy, rigid, elliptic oblong and solitary. The inflorescence is 8 cm long and single flowered. The flowers are long lasting, fragrant, 10 cm across, pink purple in colour and produced in September to October.
Laelia purpurata: Indigenous to Brazil. The pseudobulbs are club shaped, robust, fusiform, compressed, glossy and yellowish green with leathery, solitary, dark green, thick leaves. The inflorescence is stout and 12 flowered. The flowers are fragrant, long lasting, 22.5 cm across, white, pale purple in colour and produced during May to July.
Laelia superbiens: A robust species from Mexico, Guatemala and Honduras. The pseudobulbs are elongated, fusiform, compressed and furrowed with one or two leathery, oblong to oblanceolate leaves. The inflorescence is 130 cm long, stout and 20-25 flowered, and borne in clusters in terminal raceme. The flowers are scented, long lasting, 12.5 cm across, rose coloured with purple yellow lips and are produced during January-February.
Laelia tibicinis: This is a robust species from Central America. The pseudobulbs are swollen –cylindrical, hollow, densely clustered, fusiform and yellow with 3 to 5, rigid yellow green, oblong elliptic, heavy textured leaves. The inflorescence is erect or arching and 150 cm tall bearing handsome flowers at the tip. The flowers are variable in colour, sweet scented, long lasting, 8.5 cm across and produced during the spring season.
Laelia milleri: A lithophytic orchid originating from Brazil. A robust plant bears small orange-red flowers measuring 5 cm across.
Laelia liliputiana: Native to Brazil. A very small plant with 1 cm long pseudobulb, 2 cm long leaves and flowers that are 3 cm across.

11.3.10.2 Bigeneric Hybrids

Brassolaelia = Laelia x Brassovola
Laelonia = Laelia x Broughtonia
Laeliocattleya = Laelia x Cattleya
Dialaelia = Laelia x Diacrinum
Sophrolaelia = Laelia x Sophronitis

11.3.10.3 Trigeneric Hybrids

Laeliocattkeria = Laelia x Barkeria x Cattleya
Brassolaeliocattleya = Laelia x Brassovola x Cattleya
Lawara = Laelia x Brassovola x Sophronitis
Hartara = Laelia x Broughtonia x Sophronitis
Dialaeliocattleya = Laelia x Cattleya x Diacrinum
Epilaeliocattleya = Laelia x Cattleya x Epidendrum
Sophrolaeliocattleya = Laelia x Cattleya x Sophronitis
Dillonara = Laelia x Epidendrum x Schomburgkia

11.3.10.4 Tetrageneric Hybrids

Iwanagara = Laelia x Brassovola x Cattleya x Epidendrum
Yamadara = Laelia x Brassovola x Cattleya x Epidendrum
Rechara = Laelia x Brassovola x Cattleya x Schombergkia
Potinara = Laelia x Brassovola x Cattleya x Sophronitis
Kirchara = Laelia x Epidendrum x Cattleya x Sophronitis

11.3.10.5 Natural Hybrids

Laeliocattleya elegans, a cross between *Laelia purpurata* and *Cattleya guttata*. The flowers are greenish white. Miniature types like *Laelia braderi, Laelia longipes* and *Laelia liliputiana* having butter yellow, and white and lavander flowers, respectively, are used as pot plants.

11.3.10.6 Common Commercial Hybrids

11.3.10.6.1 Laelia

'Tropical Mojo', 'Milpest', 'Rock Ruby', 'Akiras Spring', 'Bonnie Bonita', 'ViDoran', 'Newberry Glow', 'Rocket Sprite', 'Stirling Dickinson', 'Fran's Fuchsia', 'Flash', 'Gold Snood','William Curtis', ' Aphrodite', 'Venus'.

11.3.10.6.2 Laelio-cattleya

'Hidden Agenda', 'Magic Road', 'Samba Crown', 'Tropical Fashion', 'Higher Ground', 'Mini Case', 'Nobiles General', 'Spring Parade', 'Upstrart', 'Harmony Show', 'Julie Anne', 'Beautiful Park', 'Orange Sunset','Carbon Lines', 'Good Friend', 'Pat's Golden Dream', 'Sunlight Kiss', 'Sweet Cream', 'Penny Love', 'Big Lucy', 'Blue Boy', 'Gatton Park'.

11.3.10.7 Medicinal Laelia

Laelia autumnalis has antihypertensive and vasorelaxant activities. In Mexico, it is used for the treatment of diarrhea and abortion.

11.3.10.8 Cultivation

They are intermediate orchids requiring cool periods in winter and enjoy moderate shade and humidity during vegetative growth. Temperature requirements vary from 12 to 18°C depending upon species. They are best grown in earthen pots or in hanging baskets. The plants prefer bright sunlight with an intensity of 1800-2400 foot candles and a humidity of 50 to 65%. Plenty of water is required when the roots are in the active phase of growth. Watering is to be reduced during resting in winter. They grow well in tree fern fibre. Fertilizing with dilute, well rotten oil cake (1 kg in 10 gallons of water) at fortnightly intervals is beneficial.

11.3.11 Miltonia

Miltonia consists of 20 species of sympodial epiphytic orchids from Brazil. They are also called 'Pansy Orchids' at higher elevations. These orchids are characterized by sessile, flattened pseudobulbs and thin leaves arising from pseudobulbs. Inflorescence is single flowered or multiflorus raceme. The flowers are large, 10 cm across and yellow, reddish purple, white, pink or brown in colour.

11.3.11.1 Genetic Resources

Miltonia clowesii: Native to Brazil. The pseudobulbs are elongated, ovoid-oblong with linear-ligulate glossy leaves. Inflorescence is 7 to 10 flowered, dense, erect or arching and 60 cm tall. Flowers are yellow with whitish violet lip and produced in September to November.
Miltonia Roezlii: This species is native to Columbia and Panama. The pseudobulbs are ovoid-oblong with solitary narrow thin leaves. The inflorescence is 2 to 5 flowered, slender. The flowers are scented, whitish purple in colour and produced in October or spring.
Miltonia spectabilis: A robust rhizomatous orchid with flat pseudobulbs and paired linear ligulate leaves. Inflorescence is erect and 20 cm tall. The flowers are long lasting, heavily textured and whitish violet.
Miltonia vexillaria: Native to Coulmbia and Brazil. The pseudobulbs are clustered, tipped by 2 leaves and arranged basally with 4 to 6 leaves. Inflorescence is 4 to 12 flowered and 50 cm long. The flowers are fragrant, 8.5 cm long, rose- red or lilac- rose in colour and are produced in April-June.

11.3.11.2 Bigeneric Hybrids

Milpasia = Miltonia x Aspasia
Miltasia = Miltonia x Brassia
Odontonia = Miltonia x Odontoglossum
Miltodonium = Miltonia x Oncidium

11.3.11.3 Trigeneric Hybrids

Aliceara = Miltonia x Brassia x Oncidium
Colomanara = Miltonia x Odontoglossum x Oncidium

11.3.11.4 Tetrageneric Hybrids

Withnerara = Miltonia x Aspasia x Odontoglossum x Oncidium

11.3.11.5 Intervarietal Hybrids

'Puna Gold', 'Shaini's Star', 'Enzan Sunday', 'Mauri Star', 'Enzan Floss', 'Red Rock', 'Tear Drops', 'Yellow Sands', ' Black Waters', 'George W, Bush', 'Indian Hollow', 'Cast Sand', 'Emily's Delight', 'Ananda Apple', 'Whale Bay', 'Daydream', 'Solar Storm', 'Angel Eyes', 'Top Prize Capitula Redwood', 'Bingo', 'Kismet Sena', 'Saffron Butterfly', 'Dream River', 'Moon River', 'Coconut Cream', 'Spring Jewel', 'Blackout', 'Dear Lady'.

11.3.11.6 Interspecific Hybrids

'Marion Primmer', 'Wine Leopard FCA', 'Menina Taylor', 'Tropic Skies', 'Joyce Hill', 'Edwin Oka', 'Stellina'.

11.3.11.7 Variety-species Hybrids

'Invisible Man', 'Miami Spectacular', 'Walnut Valley Glade', 'Danny Denton', 'Princess Diana', 'Pink Sand', ' Enzan Lady', 'Spectacular Dawn', 'Candid Ruby', 'Komoda Marvel', 'Harry', 'David Manzur', 'Rick Hood', 'Pacific Wars'.

11.3.11.8 Natural Hybrids

Miltonia x bluntii, Miltonia x cogniauxiae, Miltonia x flava, Miltonia x leucoglossa, Miltonia x rosina, Miltonia x cuneata, Miltonia x binottii, Miltonia x lamarckiana, Miltonia x petersiana

11.3.11.9 Cultivation

The high elevation orchids require a specialized structure and must be grown in cool house. Bright diffused light is necessary for miltonias to bloom. Usually, a temperature range of 10-15°C, relative humidity of 50-70% and light intensity of 2400-3600 foot candles is ideal for these plants.

A perfectly drained compost consisting of chopped osmunda and chopped sphagnum moss is good. The plants are well grown in small pots. The tropical species require well lighted locations with warm temperature. It is also best to plant them in well drained, lightly packed compost mixture of equal parts of chopped tree fern fibre and dust tree bark preparation. They require abundant water and high humidity. Watering once a week during winter and twice a week during summer is sufficient. A complete fertilizer of NPK (30: 10: 10) in liquid form should be sprayed at monthly intervals.

11.3.12 Odontoglossum

Odontoglossum consists of 300 species of epiphytic and lithophytic orchids from Mexico to Bolivia and Brazil. The pseudobulbs are tipped by 1 or 2 leaves. Inflorescence arises from the base of the pseudobulbs, is erect or arching, and bears one too many large and beautiful flowers. The flower colour is variable from pure white spotted and blotched with other hues of yellow to chestnut brown.

11.3.12.1 Genetic Resources

Odontoglossum citrosmum: Native to Mexico. The pseudobulbs are round bearing two leathery leaves at the tip. The inflorescence is 10 to 15 flowered and drooping. The flowers are scented, white splashed with pink, with a pink lip and are produced in May.
Odontoglossum cordatum: Native to Mexico, Guatemala, Hondurus and Costa Rica. The pseudobulbs are clustered, ovoid-ellipsoid to ellipsoid with solitary, elliptic-lanceolate or oblong ligulate leaves. Inflorescence is erect with few too many flowered. The flowers are 7.5 cm in diameter, yellow in colour and produced in April-May.
Odontoglossum crispum: A beautiful species from Columbia. The pseudobulbs are egg shaped with compressed and furrowed edges. The leaves are 2 to 3, soft textured, narrow and linear ligulate. The inflorescene is erect or arching, and dense with many flowers. The flowers are white and tinted with pink, having pinkish white lip and produced during February to April.
Odontoglossum grande: This species is native to Mexico and Guatemala. The pseudobulbs are large, roundish and clustered and furrowed with 1 to 3 thick heavy textured leaves. The inflorescence is erect, stout, 6 to 8 flowered. The flowers are large, 15 cm across, waxy and long lasting, yellow streaked with brown or red purple and white lip. The flowers are produced during December to March.
Odontoglossum Krameri: The species is native to Mexico and Costa Rica. The pseudobulbs are small, compressed, clustered, roundish and tipped by single, leathery, elliptic lanceolate leaves. The inflorescence is short, erect and 3 to 5 flowered. The flowers are 4 cm across, lilac pink segment edged with white and produced during June to August.
Odontoglossum luteo-purpureum: A species native to Columbia. The pseudobulbs are robust, compressed, oval and tipped with two large ensiform leaves. The inflores-

cence is erect and 8 to 12 flowered. The flowers are 10 cm across, red chestnut brown in colour with white lip and produced during April-May.
Odontoglossum pulchellum: Native to Mexico and Guatemala. The pseudobulbs are clustered, compressed, and ellipsoid-oblong to ovoid with 2 to 3 very narrow leaves. The inflorescence is erect and 6 to10 flowered. The flowers are long lived, very scented, white and produced during March to April.
Odontoglossum rubescens: This species is native to Guatemala, Nicaragua and Mexico. The pseudobulbs are ovoid, clustered, compressed with solitary, ovate-lanceolate or oblong –elliptic leaves. The inflorescence is 2 to 4 flowered and bracted. The flowers are 7.5 cm across, narrow, white streaked with brown, have an undulated white lip, and are produced during January to March.
Odontoglossum stellatum: This species is found in Guatemala and Mexico. The pseudobulbs are clustered and tipped solitary leaves. The inflorescence is short, slender and 1 to 2 flowered. The flowers are 5 cm across, tangy brown with white or pink lip and are produced in the spring season.
Odontoglossum tripudians: This species is widely distributed in Columbia, Eucador and Peru. The inflorescence is many flowered and 7.5 cm long. The flowers are yellow and produced during the spring season.
Odontoglossum triumphans: This species is native to Colombia. The pseudobulbs are compressed with leathery dark green leaves. Inflorescence is many flowered and erect. The flowers are 12.5 cm across and golden yellow with reddish brown spots.
Odontoglossum wallisii: This species is native to Colombia and Venezuela. The pseudobulbs are compressed and round with arching leaves. Inflorescence is drooping and loosely 10 to 15 flowered. The flowers are pale yellow with reddish brown stripe, 6.5 cm across and whitish purple lip and produced during winter season.
Odontoglossum cirrhosum: A magnificent species from Eucador. Flowers are pure white with heavy cinnamon spots.
Odontoglossum harryanum: This species is native to South America. The inflorescence is erect, 50-90 cm tall and 12 flowered. Flowers are 7 cm across, reddish brown and segmented marked with yellow with a reddish brown and white lip.

11.3.12.2 Bigeneric Hybrids

Aspoglossum = Odontoglossum x Aspasia
Odontonia = Odontoglossum x Miltonia
Odontocidium = Odontoglossum x Oncidium
Odontobrassia= Odontoglossum x Brassia

11.3.12.3 Trigeneric Hybrids

Wilsonara = Odontoglossum x Cochlioda x Oncidium
Colmanara = Odontoglossum x Miltonia x Oncidium

11.3.12.4 Tetrageneric Hybrids

Withnerara = Odontoglossum x Aspasia x Miltonia x Oncidium
Watsonara = Odontoglossum x Brassia x Oncidiun x Tricocentrum
Bakerara = Odontoglossum x Miltonia x Brassia x Oncidium

11.3.12.5 Inter-specific Hybrids

'Cynthia Hill', 'Luis Salabazar', 'Mc Beans Elizabeth', 'Spotted Croc', 'Mayapan', 'Quito', 'Colocris', 'Mums Tiger', 'Ann', 'High Street', 'Uro-Day', Cormac', 'Gay Enchanment', 'Gay Starshooter'.

11.3.12.6 Inter-varietal Hybrids

'Italian Job', 'Joyce Stewart', 'Pepe Gerald', 'Purple Rain', 'Katherine Jenkins', 'Roy Wittwer', 'Laguna Blanca', 'Precocious', 'Snow Fall', 'Pesky', 'Queen of Mars', 'Stam Point', 'Holiday Yellow', 'Ronald Norman', 'Bridget Ring Lawless', 'John Hanry Hanson', 'Point Pesky'.

11.3.12.7 Variety –Species Hybrid

'Doctor Tom', 'Bob Hamilton', Winter Star', 'Age of Reality','Rocky Road', 'Black Diamond', 'San Damino Royal', 'La Platte', 'Oratia Bride', 'Golden Crisp', 'Offalman', 'Enzan Yankee', 'Ken Armour', 'Geneva Snow'.

11.3.12.8 Cultivation

The plants prefer a bright light with the intensity of 2000-2500 foot candles, a day temperature 25°C and night temperature of 15°C and 40-80% humidity with free movement of air. Bark mixes to be the standard potting media for odontoglossums. One mix consists of one part coarse sand, one part coarse shredded peat, one part perlite and four parts fine bark. In an 8-inch pot for these ingredients, a small handful each of bone meal and dolomitic lime can be added. Applications of NPK (30: 10: 10) twice a month is ideal for successful growth of plants.

11.3.13 Renanthera

Renanthera consists of 12 species of epiphytic or lithophytic orchids from China, Himalayas, the Philippines, Indonesia and New Guinea. Plants are monopodial, vine-like, pseudobulbless with ascending stems of intermediate growth. The leaves are leathery, distichous, thick and coriacious. A large number of aerial roots is produced from the stem of plants. The inflorescence arises from the axils of leaves (upper) and produce few to many flowers. The flowers are 9 cm across, scarlet, red orange, crimson to yellow in colour.

11.3.13.1 Genetic Resources

Renanthera coccinea: This species is native to Burma, South China and Thailand. The stems are ascending, erect and freely rooting with leathery, well spaced, yellowish green leaves. The inflorescence is produced from the stem opposite one of the upper leaves, 1.2 m tall with about 150 flowers. The flowers are large, 8.5 cm across, long lasting with red petals and vermillion sepals and yellow crimson lips and are produced during June –July.

Renanthera imschootiana: An endangered species, restricted to Manipur and other neighbouring states of North East India, which is commonly called 'Red Vanda'. The stems are solitary, 90 cm long arranged with a set of leaves closely packed on the stem. Inflorescence is horizontal, branched bearing more than 20 bright crimson flowers with red spots on a pale orange back ground of its dorsal sepal and petals. The flowers are long lasting, 6 cm across and are produced during April-May.

11.3.13.2 Bigeneric Hybrids

Renades = Renanthera x Aerides
Aranthera = Renanthera x Arachnis
Renancentrum = Renanthera x Ascocentrum
Renanthopsis = Renanthera x Phalaenopsis
Renantanda = Renanthera x Vanda
Renanopsis = Renanthera x Vandopsis

11.3.13.3 Trigeneric Hybrids

Sappanara = Renanthera x Arachnis x Phalaenopsis
Holttumara = Renanthera x Arachnis x Vanda
Moirara = Renanthera x Phalaenopsis x Vanda

11.3.13.4 Renanthera Hybrids

'SCGB Kylin', 'Datin Blanche', 'Olbery', 'Hooi Sew Yong', 'John Losgar', 'Red Leopard', 'Scarlet Belle', 'Chanachae', 'Serdang', 'Brady Crocker', '20th WOC Singapore-2011', 'Bart Motes'.

11.3.13.5 Renanstylis Hybrids

'Bangkok Beauty', 'Teo Choo Hong', 'Carl Niemann'

11.3.13.6 Renantanda Hybrids

'Forever Yvonne', 'Inspiration Ng Teng Fong', 'Ladda Glow'; 'Polyetheramine Singapore', 'Momon Shija', 'Paul Gripp', 'Science Arts', 'Memoria Charles Darwin', 'Prof. G.J. Sharma', 'Kebisana Shija', 'Mary Motes', 'Kofi Annan'.

11.3.13.7 Renanthopsis

'Bob Banister', 'John Mason SCBG Canary', 'Lieke Jjoa', 'Persian Carpet', 'Dancing Stars'

11.3.13.8 Renanapsis

'Santa Cruz'

11.3.13.9 Uses

Renanthera orchids are ideal for pot plants, hanging baskets and for mounting on trees.

11.3.13.10 Cultivation

The plants require a warm, moist atmosphere and airy and well lighted locations during the growing seasons. Full sun exposure is usually needed for proper flower production. The plants must be fastened on a block of fern stem or wood, to which they become firmly attached by their roots. These plants must be planted in a compost media consisting of tree fern fibre, mixture of fir bark and charcoal. The temperature of 15-30°C, light intensity of 4000 to 5000 foot candles and humidity of 60 to 70% are found to be optimum for their growth and flowering. The plants are to be watered daily during summer months and twice a week during winter months. A complete fertilizer of NPK (20:20:20) in liquid form should be sprayed at weekly intervals in summer and fortnightly intervals in winter.

11.3.14 Rhyncostylis

Rhyncostylis consists of 4 species of monopodial epiphytic orchids distributed in India, Burma, Thailand, Malayasia, Indonesia, Sri Lanka, the East Indies and the Philippines. These orchids are commonly known as 'Foxtail Orchids' and are considered the state flowers of Assam and Arunachal Pradesh. These are stout, short stemmed plants with thick and leathery linear-oblong leaves closely arranged on the stem. The inflorescence is erect or drooping, and densely covered with small, colourful showy flowers. The flowers are red, magenta, blue spotting, reddish lavender and white in colour.

11.3.14.1 Genetic Resources

Rhyncostylis gigantea: This species is native to Thailand and Burma. The stems are stout with tongue shaped, heavy, leathery leaves. The inflorescence is pendulous, 45 cm covered with many small flowers. The flowers are sweet scented, long lasting, waxy, 2.5 cm across with pure white sepals and petals spotted red-violet and magenta and produced during autumn and early winter.

Rhyncostylis retusa: This species is native to India, Sri Lanka and the Philippines. The stems are robust and woody with small stout aerial white roots. The leaves are leathery, strap shaped, linear, deeply channeled and arching. The inflorescence is pendulous, 60 cm long, many flowered and cylindrical. The flowers are waxy, short or long lived, sweet scented, 1 cm across with white sepals and petals spotted with bluish purple and are produced during April-May.

Rhyncostylis coelestis: Native to southeast Asia. The plants are dwarf with stiff folded leaves that curve downwards. The flowers are 2-tone blue, scented and long lasting.

Rhyncostylis violacea: This species bears many miniature lavender and white flowers, which are produced during winter and spring season.

11.3.14.2 Common Uses

In Assam, the flower spike of *Rhyncostylis retusa* known as 'Kopou Phul', is used by girls to adorn their hair during the summer festival. Hybrids are used as cut flowers or in bouquet making and flower arrangements.

11.3.14.3 Bigeneric Hybrids

Rhynchorides = Rhyncostylis x Aerides
Aranchostylis = Rhyncostylis x Arachnis
Neostylis = Rhyncostylis x Neofinetia
Rhynchonopsis = Rhyncostylis x Phalaenopsis
Renanstylis = Rhyncostylis x Renanthera
Rhyncovanda = Rhyncostylis x Vanda

11.3.14.4 Trigeneric Hybrids

Vascostylis = Rhyncostylis x Ascocentrum x Vanda
Rhyncostylis = Rhyncostylis x Doritis x Phalaenopsis
Yapara = Rhyncostylis x Phalaenopsis x Vanda

11.3.14.5 Rhynchorides Hybrids

'Thai Gem', 'Norma', 'Alice;s Wally', 'Bangkok Sunset'

11.3.14.6 Rhynchostylis Hybrids

'Charles Marden Fitch'

11.3.14.7 Rhyncholaeliocattleya Hybrids

'Gabgab Castro', 'Malvarosa Pearl', 'Robinson's Delight', 'Chief Journey', 'Aguti Gold', 'Little Circle', 'Butterfly Wings', 'Magic Morning', 'Chief Birde', 'Chief Pink', 'Crimson Circle', 'Bold As Love', 'Julian's Shaw', 'Jims Beauty', 'Omi Melody', 'Proda Flora', 'Scarlet Bay', 'Salmon Sensation', 'Himalayan Aspiration', 'Innocent Mastiff', 'Lhonak Sunset', 'Pradhan's Delight', 'Rumtek Jewel', 'Scented Serendipity', 'Sikkim Sentinal', 'Sikkim Majestic', 'Sikkim Accolade', 'Sukhim Song', 'Big Shot','Deception Girl', 'Fire Rose', 'Fire Lake', 'Agatha Fire', 'Agatha Lim', 'Volcano Glory', 'Wild Promise', 'June Simpson', 'Tinker Bell'.

11.3.14.8 Rhynchosophrocattleya Hybrids

'Chief Blue', 'Chief Heart', 'Comet Pink', 'Long Run', 'Golf Pink', 'Gorgeous Lady', 'Alpha Plus Peacock', 'Exotic Ruby', 'Red Bay', 'Alphs plus Pink', 'Skip Wilson', 'Tropical Punch', 'Giant Beauty', Fruit Island', 'Lovely Air', 'Sakura Grand', 'Ashahi Delight'

11.3.14.9 Rhynchovola Hybrid

'Green Pixie'

11.3.14.10 Rhyncattleanthe Hybrids

'Buddy Bay', 'Burana Fire', 'Samba Script', 'Typical Worth', 'Shigfong White', 'Charm', Alpha Plus Buddha', 'Purple Rainbow', 'Golden Bell', 'Peter Lin', 'Nobile's Teens', 'Nobile's Coralina', 'Nobile's Toffee', 'Algestor Gold', 'Sikkim Treasure', 'Tendong Glow', 'Life's Charm', 'Laughing Magic', 'Silent Moon', 'Atomic Jungle', 'Hot Blooded', 'Sea of Love', 'Uptown Girl', 'Alpha Plus Candy', 'Brown Tone', 'Clinton Lewis', 'Little Treat', 'California Love', 'Chocolate Rose', 'Heart Warmer', 'Sweet Sound'.

11.3.14.11 Rhynchovanda Hybrids

'Wilton Hill', 'Jammie Harper', 'Apichart', 'Noo Noi', 'Peter Draper', ' Brighton's Albino', 'Prairie Lady'.

11.3.14.12 Rhyncattleya Hybrids

'Olor de Otoño', 'Cinco de Mayo', 'Rios De Oro', 'Tadong Delight', 'Hsinying Moves', 'Irvine Sunset', 'Deception Sweetheart', 'RIO's Treasure', 'Petit Prisme', 'RIO's Spectacle', 'Chuchemen'.

11.3.14.13 Rhyncobrassoleya Hybrids

'Chief Cindy', 'Starry', 'Taurus', 'Tetraspotts', 'Bosque Antiguo', 'Ramar's Happiness', 'Roberts Dodson', 'Sedona's Surprise', 'Tranquility', 'Golden Angel Lemon Tang', 'Pink Blush'.

11.3.14.14 Medicinal Rhyncostylis

Roots of *Rhyncostylis retusa* are effective against rheumatisms. Plants are also used against asthma, tuberculosis, cramps, epilepsy, vertigo, palpitation, kidney stones and menstrual disorders.

11.3.14.15 Cultivation

These plants require high humidity, moderate shade and plenty of water during active phase of growth. They require bright light with an intensity of 3000-4000 foot candles and strong air movements all the times. These orchids are ideal for hanging baskets. They grow well in large chunks of tree fern fibre or in a medium of chunks of hardwood charcoal.

A day temperature of 30-32°C and night temperature of 24-25°C and relative humidity of 80% in summer and 60-70% in spring and winter are ideal for their active vegetative growth and flowering. Plants should be watered heavily while actively growing, but aeration around the roots must be excellent, allowing the roots to dry rapidly after watering. For plants grown in pots or baskets, the medium must never become water logged or soggy. These plants are heavy feeders and respond very well with the dilute solution of 0.3% NPK (20: 20: 20) sprayed twice a month.

11.3.15 Zygopetalum

Zygopetalum consists of 25 species of terrestrial, lithophytic or epiphytic orchids from Brazil, Bolivia, Peru, Paraguay, Venezuela and New Guinea. The pseudobulbs are ovoid with a distinct sheath, 5 cm to 7.5 cm tall and strong with lanceolate, disticous leaves. The inflorescence develops from the base of pseudobulbs, and is arching or erect in nature. The flowers are fragrant and long lasting in shades of brilliant green, blue or purple. The flowers are 5 cm to 7.5 cm across with wavy margins. They are excellent for cut flowers and corsages.

11.3.15.1 Genetic Resources

Zygopetalum cerinum: This species is native to Columbia with degenerative pseudobulbs and oblong leaves. Flowers are borne singly, creamy white with yellow lip streaked with purple.

Zygopetalum crinitum: This species is native to Brazil with ovoid to conical pseudobulbs and fleshy. Coriaceous, lanceolate and glossy leaves. The inflorescence is 45 cm long and 3 to 10 flowered. The flowers are 8 cm across, scented with green segments, maculated brown and have a white lip streaked with violet and are produced in December-January.
Zygopetalum mackayi: A rare and beautiful species from Brazil with long, erect and curving leaves. The inflorescence is 90 cm tall and 5 to 10 flowered. The flowers are large, 8.5 cm across, scented, long lived with yellowish green petals and sepals maculated with violet purple and white lip intricately veined with red and blue, and produced during December.
Zygopetalum intermedium: A popular species, native to Brazil with ovoid conical pseudobulbs and glossy bright green leaves. The inflorescence is 60 cm long, 10-12 flowered and attractive. The flowers are 7.5 cm across, long lasting, fragrant, yellowish green with purple brown blotches and white lips with dots of blue and produced in December.
Zygopetalum wendlandii: This species is native to Costa Rica and has degenerative pseudobulbs and lanceolate leaves. The flowers are 10 cm across, pale green in colour and produced at the end of summer to the beginning of autumn.

11.3.15.2 Intergeneric Hybrids

Zygonisia = Zygopetalum x Aganisia
Chondropetalum = Zygopetalum x Chondrorhynca
Zygocolax = Zygopetalum x Colax
Zygocaste = Zygopetalum x Lycaste

11.3.15.3 Common commercial Hybrids

'Blue Blood', 'Blue Bear', 'Imagination', 'Tanzanite', 'Millie', 'Black Plague', 'Kiwi Black', 'Kiwi Choice', 'Pioneer', 'Tasman', 'New Era', 'Dark Star', 'Impulse', 'Intuition', 'Great Eisen', 'Blue River', 'Indigo Skies', 'Blue Banks', 'Bon Voyage', 'Centenary', 'Hot Springs', 'Blackjack', 'Night Hawk', 'Violet Moon', 'Leopard Prince', 'Blue Blood', 'Big Country', 'Hawker'

11.3.15.4 Cultivation

Zygopetalums thrive well in a temperature range of 20-26°C during day time and 10-15°C during night. They require bright light in the 3000-4000 foot candles range. During summer season, it needs 40% shade cloth. They love water during their active growth period and are watered at 5-7 days intervals. A potting mixture consisting of cocopeat, cocochips and tree barks is ideal. They are easily propagated through division of pseudobulbs.

12 Medicinal and Aromatic Orchids

12.1 Medicinal Orchids

Orchids are the most diverse group among the angiosperms and are cultivated for their attractive flowers. There is no doubt that the Chinese were the first to cultivate and describe orchids, and they were almost certainly the first to describe orchids for medicinal use. Reinikka in 1995 reports a Chinese legend that Shên-nung described *Bletilla striata* and a *Dendrobium* species in his Materia Medica of the 28th century BC.

Some species like *Dendrobium nobile*, *Eulophia campestris*, *Orchis latifolia*, *Vanda roxburghii* and *Vanda tessellata* have been documented for their medicinal value. Phytochemically, orchids have been reported to contain alkaloids, triterpenoids, flavonoids and stilbenoids. Ashtavarga, a group of eight medicinal plants, is a vital part of Ayurvedic formulations like Chyvanprasha and four of these plants viz, Riddhi, Vriddhi, Jivaka and Rishbhaka belong to the family Orchidaceae (Tab. 12.1).

Tab. 12.1: Medicinal plants used in Ashtavarga, composite Ayurvedic formulation (Singh and Duggal, 2009)

S.No.	Ayurvedic name	Botanical name	Family	Part used
1.	Jivaka	*Malaxis muscifera*	Orchidaceae	Bulb
2.	Rishbhaka	*Malaxis acuminata*	Orchidaceae	Pseudo-bulb
3.	Meda	*Polygonum verticillatum*	Polygonaceae	Rhizome
4.	Mahameda	*Polygonum cirrhifolium*	Polygonaceae	Rhizome
5.	Kakoli	*Roscoea procera*	Zingiberaceae	Root
6.	Kshira Kakoli	*Fritillaria royeli*	Liliaceae	Root
7.	Riddhi	*Habenaria intermedia*	Orchidaceae	Root
8.	Vriddhi	*Habenaria edgeworthii*	Orchidaceae	Root

Orchids are widely used in traditional Chinese medicines. In India, chemical analyses have been conducted on some medicinally important orchids like *Eulophia campestris*, *Orchis latifolia*, *Vanda roxburghii*. *Dendrobium macraei* is another important orchid used in Ayurvedic medicine as it is reported to be the source of Jivanti. *Cypripedium parviflora* is widely used as an aphrodisiac and nervine tonic in Western herbal medicines.

Many medicinal orchids are reported to contain alkaloids and have antimicrobial activities. Recently, studies have focused on the isolation of anthocyanins, stilbenoids and triterpenoids from orchids. Orchinol, hircinol, cypripedin, jibantine, nidemin and loroglossin are some important phytochemicals extracted from orchids. Some of the medicinal orchids along with distribution, parts used, and medicinal properties have been tabulated below (Gutierrez, 2010; Singh and Duggal, 2009; Rao, 2004) (Table 12.2).

Tab. 12.2: Medicinal orchids and their medicinal properties

Sl. No.	Botanical Name	Distribution	Parts used	Medicinal properties
1.	*Acampe papillosa*	North Eastern India	Roots	Root is used for rheumatism, sciatica, neuralgia, syphyllis and uterine diseases.
2.	*Acampe praemorsa*	Western Ghat of India	Roots	Anti-rheumatism
3.	*Aerides crispum*	Western Ghat of India	Whole plant	Its plants are powdered, boiled in neem oil, filtered, 2-3 drops of oil are put into the ear once at night as a cure for earache.
4.	*Aerides multiflorum* Roxb	Himalaya (Garhwal to Sikkim), Assam, India and Burma	Tubers	Antibacterial
5.	*Anoectochilus formosanus* Hayata	Taiwan	Tubers	Chest and abdominal paints, diabetes, fever, nephritis, hypertension, impotence, liver spleen disorders, and pleurodynia, anti-inflammatory agent
6.	*Arundina graminifolia* (D. Don) Hochr.	Himalayas of Nepal, Sri Lanka, Thailand, Laos, Cambodia, Vietnam, southern China, Japan, Taiwan and south to Malaya and Java	Rhizome	Antibacterial
7.	*Bletilla striata* (Thunb.) Rchb.f.	Taiwan, Nepal, Tibet, China	Tuber	Treatment of sores, ulcers and chapped skin, heal wounds, reduce swelling, and promote regeneration of tissue
8.	*Calanthe triplicata*	North East India	Roots, flowers & pseudobulbs	Roots are ingredient of local medicine to treat swollen hands; with other ingredients roots chewed for diarrhea, Flowers as a painkiller in caries, Pseudobulbs as a masticatory, gastrointestinal disorders.
9.	*Coelogyne ovalis*	Western Ghat of India	Whole plant	The whole plant is used in Western and Southern parts of India for cough, urinary infections and eye disorders.
10.	*Cypripedium calceolus pubescens* (Willd.) Correll	N. America to E. Asia - Japan	Roots	Antispasmodic, diaphoretic, hypnotic, nervine, sedative, tonic
11.	*Dendrobium chrysanthum*	China	Leaves	Antipyretic, eyes-benefiting, immuno-regulatory purposes, skin diseases

continued Tab. 12.2: Medicinal orchids and their medicinal properties

Sl. No.	Botanical Name	Distribution	Parts used	Medicinal properties
12.	*Dendrobium jenkisii*	North East India	Stems	Fresh and dried stems used in preparation of Chinese drug Shih-hu
13.	*Dendrobium macraei* Auct	Himalayas	Tubers	Tonic for general debility
14.	*Dendrobium nobile* Lindl.	Himalayas and China	Stems	Antiphlogistic, pectoral, sialogogue, stomachic and tonic
15.	*Dendrobium ovatum*	Western Ghat of India	Stems	Juice obtained by hand crushing the stems is used on patients suffering from constipation and stomachache
16.	*Epidendrum Mosenii*	China & Korea	Stems	Analgesic
17.	*Eulophia nuda* Lindl.	Himalayas	Tubers	Demulcent and anthelmintic
18.	*Gastrodia elata*	Asia	Whole plant	Treatment of epilepsy
19.	*Goodyera schlechtendaliana*	India	Whole plant	Tonic for internal injuries and to improve circulation
20.	*Habenaria edgeworthii* Hook.f. ex Collett.	E. Asia - Himalayas	Leaves & roots	Cooling and spermopiotic
21.	*Habenaria pectinata* D.Don	Himalayas	Leaves & tubers	The leaves are crushed and applied in snake bites. Tubers mixed with condiments are used in arthritis
22.	*Malaxis acuminta* D.Don	Himalayas 1800 m to 3500 m eastwards to Sikkim	Pseudobulb	Cooling, febrifuge and spermopiotic
23.	*Malaxis muscifera* (Lindl.) Kuntze	Himalayas 1850 m to 2300 m Himachal Pradesh to Arunachal Pradesh	Bulb	Cooling, febrifuge and spermopiotic
24.	*Maxillaria densa*	Mexico	Whole plant	Treatment of painful complaints. Relaxant agent
25.	*Orchis latifolia* L.	Western Himalayas, Afghanistan and Iran	Roots	Treatment of diabetes, diarrhea, dysentery, paralysis, convales-cence, impotence and malnutrition

continued Tab. 12.2: Medicinal orchids and their medicinal properties

Sl. No.	Botanical Name	Distribution	Parts used	Medicinal properties
26.	*Orchis laxiflora* Lam.	South Europe, North Africa and West Asia.	Bulb	Treatment of diarrhea, bronchitis and convalescence
27.	*Satyrium nepalense*	North East India	Tubers	Tubers eaten by Monpa tribe for Malaria, dysentery, also aphrodisiac
28.	*Spathoglotis plicata*	North East India	Whole plant	Decoction of the boiled plant used for rheumatism and used in hot as a foment.
29.	*Spiranthes sinensis* var *amoena*	Nepal, China & Taiwan	Roots	Aphrodisiac, treatment of hemoptysis, epistaxis, headache, chronic dysentery and meningitis
30.	Vanda roxburghii	India	Leaves & roots	The paste applied to the body to bring down fever. The juice is dropped in the ear for the treatment of otitis. The roots are used in dyspepsia, bronchitis, rheumatim and sciatica
31.	*Vanda tessellata* (Roxb.) Hook. Ex Don	India, Sri Lanka and Burma	Whole plant	Paste of leaves is used as application in fevers. It is ingredient of *Rasna Panchaka Quatha*, Ayurvedic formulation used in the treatment of arthritis and rheumatism. Expressed juice of the leaves is sued in the treatment of otitis media. The root is used as antidote against scorpion sting and remedy for bronchitis
32.	*Vanilla planifolia*	Mexico	Sheath	Used as for the treatment of hysteria, fever, impotence, rheumatism, and to increase the energy, of muscular system

12.1.1 Pharmacological Profile of Orchids

Throughout the ages, several health-promoting benefits, including diuretic, anti-rheumatic, anti-inflammatory, anti-carcinogenic, hypoglycemic activities, anti-microbial, anticonvulsive, relaxation, neuroprotective, and antivirus, activities have been reported with the use of orchids extracts. Orchid species attributed to medicinal properties of various ailments are given below (Gutierrez, 2010):

Anti cancer/Anti-tumor: *Anoectochilus formosanus, Bletilla striata, Bulbophyllum kwangtungense, Dendrobium chrysanthum, Dendrobium fimbriatum, Dendrobium*

nobile, Ephemerantha ionchophylla, Gastrodia elata, Spiranthes australis, Bulbophyllum odoratissimum
Convulsive diseases: *Gastrodia elata, Goodyera schlechtendaliana, Anoectochilus formosanus*
Anti-microbial: *Vanilla planifolia, Galeola foliata, Cypripedium macranthos* var. rebunense, *Spiranthes mauritianum, Gastrodia elata*
Anti-inflammatory: *Anoectochilus formosanus, Gastrodia elata, Dendrobium moniliforme, Pholidota chinensis*
Antioxidant: *Anoectochilus formosanus, Anoectochilus roxburghii, Dendrobium amoenum, Dendrobium moniliforme, Gastrodia elata, Pholidota yunnanensis*
Antidiabetic: *Anoectochilus formosanus, Dendrobium candidum*
Diuretic: *Cymbidium goeringii*
Antihepatotoxic: *Anoectochilus formosanus, Goodyera schlechtendaliana, Goodyera matsumurana, Goodyera discolor*
Neuroprotective: *Coeloglossum viride, Gastrodia elata*
Pain treatment: *Maxillaria densa, Scaphyglottis livida, Epidendrum Mosenii*
Anti-viral: *Epipactis helleborine, Listera ovata, Gastrodia elata, Cymbidium spp.*
Relaxation: *Scaphyglottis livida, Gastrodia elata, Maxillaria densa*
Antiplatelet aggregation: *Dendrobium loddigesii, Den. densifiorum, Ephemerantha lonchophylla, Gastrodia elata*
Anti-allergic: *Gymnadenia conopsea*
Antipyretic: *Dendrobium moniliforme*
Antimutagenic activity: *Dendrobium nobile*
Endurance capacity: *Anoectochilus formosanus*
Ameliorative: *Anoectochilus formosanus*
Anthelmintic: *Bletilla striata*
Anti-aging: *Coeloglossum viride* var. bracteatum
Gastric: *Dendrobium nobile, Gastrodia elata*
Herbicidal agent: *Epidendrum rigidum*
Maturation: *Anoectochilus formosanus*
Phytoalexin: *Coelogyne cristata*
Skin blood flow: *Calanthe discolor*
Wound healing: *Vanda roxburghii*

12.2 Aromatic Orchids (De, 2014)

Orchid fragrance is a relatively volatile substance found in plants. It is stored as essential oils in special cells (osmopheres) at the periphery of flowers, leaves or roots. Only small amounts are present as the substance can be toxic to the plant. These fragrant oils can consist of volatile compounds (Tab. 12.3). Being volatile, they readily change into vapour at ordinary temperature, which allows us to smell them.

Tab. 12.3: Aromatic chemicals and fragrant orchids

Aromatic chemicals	Aromatic species
cineole medicinal (citronellol rose-like)	*Brassavola nodosa, Brassavola digbiana*
benzyl acetate (jasmine)	*Stanhopea tricornis, S. Grandiflora, S. Reichenbachiana, Cycnoches ventricosum, C. Warscewizii, C. Loddigesii, C. Chlorochilon*
d-carvone (rye bread)	*Catasetum discolor*
methyl salicilate(wintergreen)	*Catasetum collare, Catasetum gnomus, Catasetum candida*
Methyl cinnimate	*Catasetum roseum, Stanhopea saccata, Gongora quinquenervis*
Eugenol	*Gongora quinquenervis*
1,8-cineole	*Stanhopea cirrhata*
Linalool	*Brassovola digbiana, Gongora quinquenervis*

12.2.1 Scent Production

It has been estimated that as many as 75% of all orchids are 'fragrant'. They emit detectable chemical compounds; some are extremely fragrant while in some instances they are extremely repulsive smells. Only some of the odoriferous compounds released by a flower are detectable by the human sense of smell, since these are complex substances closely related to the body chemistry of the pollinator they are 'supposed' to attract. Fragrances are produced in specialized glands (osmopheres) which can be located anywhere on a flower or bud, depending on function. These are glands of intense physiological activity and are a large drain on the plant's energy. When non-fragrant flowers become isolated geographically, fragrance may evolve as a pollinator attractant. There is, for example, a fragrant form of *Phalaenopsis amabilis* from New Guinea, although all other known forms of the species from other locations are without scent.

All flower parts can produce odours, from sepals and petals to calluses and basal spurs. Osmopheres in orchids may be diffuse and function only in very general attraction, or they may be confined to certain regions of the flower so that pollinators are attracted to these specific areas and collect or deposit pollinia in the process.

Scent glands are most often situated on the lip - e.g. *Stanhopea, Herschelia* and *Catasetum.*

Members of the *Catasetinae* and *Gongorinae* subtribes produce the most voluminous quantities of scent known amongst orchids.

The fragrance of *Catasetum* flowers is interrupted within a few hours of pollination to conserve energy by limiting osmopheric activity.

The intricate flowers of the scented *Gongoras* last only for two or three days but compensate for this by several of them opening in succession. It is found that if the lip (where the scent is produced) is removed, the flower lasts for two to three weeks. A urine-like smell is produced at the tips of the long tepals in *Phragmipedium caudatum* and this could be to attract ants who aid in pollination. The long tails of the sepals of *Cirrhopetalum ornatissimum* give rise to an odour of whale oil, while the lip smells of fresh herring.

Orchid floral fragrances are produced in a daily cycle with the time of maximum fragrance production generally being during the time when the pollinator of that species would be active. Fragrance production requires energy, therefore the timing of scent production often coincides with the time of visitation of pollinators to use the least energy to achieve the maximum effect. Lady of the Night orchid (*Brassavola nodosa*) will perfume a warm summer evening with its heavy fragrance. The medicinal sweet odour is released shortly after sunset, reaching maximum strength around midnight, and fading quickly after sunrise. The scent release is strictly a light-controlled phenomenon and is regulated by a photochrome trigger. Fragrances may change throughout the day both quantitatively and qualitatively as well as from day to day:

Clowesia rosea smells of Vicks Vapo rub in the morning and cinnamon in the afternoon. *Catasetum expansum* smells of turpentine in the morning and rye bread in the afternoon.

Bee-pollinated flowers are fragrant early in the day. *Cattleya luteola*, for example, is very fragrant between 4:00 and 8:00 am. Some orchids such as *Epidendrum difforme* are moderately fragrant throughout the day with a peak fragrance production at night. Others such as *Epidendrum falcatum*, change fragrance quality and intensity during the day, from the delicate, haunting scent of jasmine in the morning to a stronger note resembling that of Easter lilies or narcissi during the afternoon.

Fragrant compounds can be manufactured synthetically and used to attract pollinators in the field. This helps to identify pollinators where field observations may be lacking. *Rhyncolaelia (Brassavola) digbiana* is a wonderfully fragrant and handsome parent producing a strong lemon-like perfume. *Rhyncolaelia glauca* emits a rosy-floral scent. *Neofinetia falcata*, which is fragrant during the day and night, awards most of its progeny with fragrance.

12.2.2 Other Examples of Aromatic Orchids

Maxillaria tenuifolia, Lycaste aromatica, Lycaste cruenta, Lycaste locusta, Thunia marshalliana, Eria hyacinthoides, Masdevallia triangularis, Masdevallia glandulosa, Angraecum distichum, Zygopetalum intermedium, Calauthron bicornutum, Cycnoches chlorochilon, Dendrobium anosmum, Diaphanante fragrantissima, Cyrtorchis arcuata, Pterygodium caffrum, alatum and *catholicum, Disa cooperi, Satyrium neglectum, Satyrium odorum, stenopetalum lupulinum, bracteatum, muticum* and *erectum Mystacidium capense* and *Mystacidium venosum, Aerides multiflorum, Aerides odoratum, Aeranthes, Bulbophyllum odoratissimum, Cattleya maxima, Coelogyne cristata, Coelogyne ochracea, Cymbidium ensifolium, Dendrobium nobile, Epidendrum cristatum, Epidendrum floribundum, Epidendrum nocturnum, Lycaste, Oncidium spaceolatum, Phaius tankervilleae, Rhyncostylis retusa, Vanda cristata, Vanda tessellata.*

13 Post–harvest Management of Cut Flowers of Commercial Orchids

13.1 Introduction

The valuable orchid genera that are highly priced in the international flower trade due to their incredible range of diversity in size, colour, shape, forms, appearance and long post-harvest life of flowers include *Cymbidium, Cattleya, Dendrobium, Phalaenopsis, Vanda* and *Paphiopedilum*.

Post-harvest life of orchid cut flowers is influenced by pre-harvest factors like varietal differences, light intensity, sugar level of flowers, temperature and water loss. It is also affected by harvest factors such as time and stage of harvest, and post-harvest factors viz. ethylene production, pre-cooling, pulsing, use of preservatives, packaging and storage. Vase life of cut flowers ranges from 18 to 56 days depending upon types of species and hybrids. The optimum harvesting stage of commercial orchids occurs when the flowers are fully open and mature. Ethylene is the main factor responsible for early senescence; de-capped and emasculated flowers produce more ethylene. There is an obvious increase in flower sensitivity to ethylene following pollination.

Indian orchid species having high ornamental values used directly as cut flowers are *Cymbidium eburneum, C. hookerianumm, C. lancifolium, Paphiopedilum venustum, P. spicerianum, P. hirsutissimum, P. insigne, Phaius wallichii, Renanthera imschootiana, Thunia alba, Vanda cristata, Vanda coerulea* and *Vanda coerulescens* (Singh, 1990).

In India, several native genera like *Cymbidium, Paphiopedilum, Vanda, Arachnis* and *Dendrobium* are cultivated on a large scale for cut flower production. The *Cymbidium* is mainly grown in the NEH Region, Sikkim, Darjeeling hills and Arunachal Pradesh. Tropical orchids are cultivated in Kerala and some parts of Tamil Nadu. Those species which flower during winter and spring months are preferred because they allow the export flowers to temperate regions from December to May. The orchid has taken a significant position in the cut flower industry due to its attractiveness, long shelf life, high productivity, right season of bloom, and ease of packing and transportation.

13.2 Hybrids and Varieties for Cut flower

Among the cut flower crops, *Cymbidium, Dendrobium, Phalaenopsis, Odontoglossum, Oncidium, Cattleya, Paphiopedilum, Vanda, Aeridovanda, Aranda, Mokara, Arachnis, Vascostylis, Renanthera, Rhyncicentrum, Rhyncovanda* etc. are important.

A good quality cut flower of an orchid should have the following characteristics (Sarkar *et al.*, 2009).

- A minimum of eight standard blooms per stem
- Flowers must be cleaned, evenly coloured and free from physiological disorders

- Stem must have flowers evenly arranged around the stem.
- Two thirds of the stem should be covered with flowers.
- Flowers must have a firm texture and a luminescent sheen
- Stems must be firm when held up
- The minimum base diameter of the stem should be 10 mm

Important varieties and hybrids under different genera of orchids for cut flower purposes are listed in Tab. 13.1 & Plate 13.1 (Bhattacharjee and De, 2005).

Tab. 13.1: Important varieties and hybrids under different genera of orchids

Sl No.	Genera	Hybrids/Varieties
1	*Aeridovanda*	'Doctor Poyck', 'Vieng Ping', 'Bensiri', 'Noreen', 'Early Bird', 'Shiv Sidhu', 'New Dawn', 'Harrison Luke Somsri Sunlight'
2	*Aranda*	'City of Singapore', 'Hilda Galistan', 'Urmila Nandey', 'Christine', 'Thailand Sunspot', 'Millenium Dawn', 'Broga Giant', 'Salaya Red', 'Propine White', 'Propin Spot', 'Lueng Cholburi', 'Ishbel Manisaki', 'Baytown', 'Chao Praya Blue', Chao Praya Dot Com', 'Chao Praya Beauty', 'Ethan Pride', 'Taksari Chandrabir', 'Thailand Sunspot'
3	*Arachnis*	'Ishabel', 'Maggie Oei', 'Maroon Maggie', 'Bartha Braga'
4.	*Ascocenda*	'Apinantat Red Berry', 'Pralor Tuyen', 'Pak-Kred', 'Bangkok', 'Surin', 'Karnada', 'Crownfox', 'Sundancer', Laksi 'Red Ruby', 'Guo Chia Long' 'Spotty', 'Fuchs Angel Frost', 'Carol Belk', 'Renuka Angle', 'Joyce Bevins', 'Adisak Blue', 'Renu Gold', 'Tipi Blue Boy', 'Bobs Fortune', 'Rubychai', 'Shah Rukh Khan', 'Yang Sophia Firuz', 'Abdul Ghani Othman', 'Chunika', 'Fuch's Star'
5.	*Cattleya and allied*	'Lovely Bangkok', 'Admiration', 'Bob Belts', 'General Patton', 'Joyce Hannington', 'Little Angel', 'Margaret Stewart', 'Nillie Roberts', 'Pearl Harbour', 'Primma Donna', 'Queen Sirkhit' 'Diamond Crown', 'Secret Love', Ladda Belle 'Pink Pearl', 'Maikai', 'Pastoral', 'Robert', Prism Palette 'Tricolour Magic', 'Ahmad Seikhi', 'Hsinging Catherine', 'Chia Lin New City', 'Chinese Beauty Orchid Queen'
6.	*Cymbidium*	'Levis Duke Bella Vista', 'Madrid Forest King', 'Sparkle Late Green', 'Angelica December Gold', 'Sleeping Nymph', 'Pine Clash Moon Venus', 'Soul Hunt', 'Dr. H. C. Aurora', 'Susan Highes', 'Tia Gaig Suther Land', 'Miss Sanders', 'Amesbury', 'Kenny Wine', 'Red Star', 'Red Princess', 'Show Girl', Jungfrau 'Snow Queen', Jungfrau 'Dos Pueblos', Lilian Stewart 'Coronation', Lilian Stewart 'Party Dress', Orkney 'Pink Heather', Ensikhan 'Alpha Orient', 'Fire Storm Ruby', 'Fire Storm Blaze', Bob Marlin 'Lucky'
7.	*Dendrobium*	'Emma White', 'Thongchai Gold', Julie', 'Erika, 'Sonia-17', 'Sonia-28', 'Kasem White', 'Madam Pompadour', 'Bangkok Blue', 'Ann', 'Gold Twist', 'Candy Stripe Pink', 'Genting Blue', 'Bengal Beauty', 'Sakura Pink', 'Burana Charming', 'Blue Fairy', 'Channel', 'Nette White', 'Dang Saard', 'Big White 4N', 'Ear Sakul', 'Erika', 'Lervia'

continued **Tab. 13.1:** Important varieties and hybrids under different genera of orchids

Sl No.	Genera	Hybrids/Varieties
8.	*Mokara*	Walter Oumae 'Seksan', Thailand, 'Sayan', Walter Oumae 'Royal Sapphire', Susan 'Orange', Walter Oumae 'Calypso', 'Eng Ling', 'Madame Panne', 'Mak Chin On', 'Bangkok Gold', 'Bibi', 'Chao Praya Gold', 'Chark Kuan Orange', 'Chark Kuan Pink', 'Chark Kuan Rose', 'Chark Kuan Super', 'Dinah Shore', 'Kelvin Red', 'Kelvin Orange', 'Luenberger Gold', 'Margaret Thatcher', 'Pink Star', 'Sayan Pink', 'WTO', 'Jiti', 'Happy Beauty', 'Salaya Gold'
9.	*Odontoglossum*	Carroll, Ismene, Palnina, 'Italian Job', 'Joyce Stewart', 'Pepe Gerald', 'Purple Rain', 'Katherine Jenkins', 'Roy Wittwer', 'Laguna Blanca', 'Precocious', 'Snow Fall', 'Pesky', 'Queen of Mars', 'Stam Point', 'Holiday Yellow', 'Ronald Norman', 'Bridget Ring Lawless', 'John Hanry Hanson', 'Point Pesky'
10.	*Oncidium*	'Aloha Iwanga Dogasima, Goldiana', 'Gower Ramsey', 'Golden Shower', 'Sum Lai Who Jungle Queen', 'Taka H & R', 'Sharry Baby Sweet Fragrance AM/AOS', 'Golden Glow', 'Popki Red', 'Irine Gleason Red', 'Vision Brownish Red', 'Catherine Wilson x New Calidonia Brownish Red, 'Robson Orchid Glad'
11.	*Paphiopedilum*	Niveum, Concolor, P. *rothschildianum* (3 to 5 flowers), P. *sanderianum* (3 to 5 flowers), 'Prince Edward of York', 'Michel Koopwitz', 'Saint Swithin', 'Mount Toro', 'Sorcerers Apprentice', 'Grande', 'Don Wimber', 'Elizabeth March', 'Hanne Popow', 'Jason Fischer', 'Living Fire'
12.	*Phalaenopsis*	'Taisuco Crane', 'Taisuco Kochdian', 'Cygnus', 'Yukimai', 'Sogo Musadian', 'White Dream', 'Florida Snow', 'Nobby's Pink Lady', 'Minho Valentine', 'Minho King Beauty', 'New Cinderella', 'Taisuco Firebird', 'Sogo Smith', 'Carol Campbell', 'Emil Giles', 'Brother Lawrence', 'Taipei Gold', 'Golden Bells', 'Sogo Managers', 'Brother Passat', 'Be Glad', 'Cassandra', 'Vilind', 'Carmelas Pixie', 'Zuma's Pixie', 'Timothy Christopher', 'Be Tris', 'Quevedo', 'Strawberry', 'Detroit', 'Maki Watnabe', 'Kaleiodoscope'
13.	*Renanthera*	Brookie Chandler, Manila T-Orchids, Kilauea, Mok Yark-Seng, Poipu, Tom Thumb, 'Red Leopard', 'Scarlet Belle', 'Chanachae', 'Serdang', 'Brady Crocker', '20th WOC Singapore-2011', 'Bart Motes'
14.	*Renantanda*	'Forever Yvonne', 'Inspiration Ng Teng Fong', 'Ladda Glow'; 'Polyetheramine Singapore', 'Momon Shija', 'Paul Gripp', 'Science Arts', 'Memoria Charles Darwin', 'Prof. G.J. Sharma', 'Kebisana Shija', 'Mary Motes', 'Kofi Annan'
15.	*Rhynchovanda*	'Wilton Hill', 'Jammie Harper', 'Apichart', 'Noo Noi', 'Peter Draper', ' Brighton's Albino', 'Prairie Lady'
16.	*Vanda*	'Annette Jones', 'Antonio Real', 'Golamcos Blue Magic', 'Fuch's Charmer', 'Jimmy Millers RF Orchids', 'Keree Delight', 'Memoria Lyle Swanson', 'Motes Indigo x Merrillii, Motes' 'Honeybun', 'Motes Primerose', 'Miss Joaquim, V. Rothschidiana', 'VTMA –Red', 'Pink', 'White', 'Vasco, Johnny Miller', 'Veerawan', 'Roberts Delight', 'Rasriprai', 'Pat Delight', 'Pakchong Blue', 'Mimi Plammer', 'Manuvade', 'Lumpini Red', 'Kultana Gold x Thongchai Gold', 'Fuchs Delight', 'Charles Goodfellow', 'Pine River', 'Adisak', 'Doctor Anek', 'John Club', 'Bill Sutton', 'Ellen Noa', 'Emily Notley', 'Evening Glow', 'Honomu', 'Honolulu', 'Hilo Blue'
17.	*Vascostylis*	Paragon Joy x Kasems Delight, Precious, Veeraphool, Crown Fox 'Red Yen', Aroon Fairy, Viboon Velvet, Chao Praya Lime', 'Lanna Rosy', 'Jeans Delight', 'Bay Sapphire', 'Spring Hill'

13.3 Physiology of Cut Flowers

Vase life or longevity of a cut flower is determined on the basis of attributes like diameter and length of florets, opening of flowers, changes in fresh weight, diameter or length of stem or pedicel, senescence pattern, colour of petals, total longevity and foliage burning. In general, cut flowers complete their life cycle in two distinct phases: (i) bud swelling to bud opening and (ii) maturation, senescence and wilting. Flower bud development to swelling involves growth or a change in the orientation of petals or subtending tissues and may also require abscission of protective structure (Evans and Reid, 1986). When an inflorescence is cut from the plant, a number of physiological processes are affected which include the supply of water, depletion of stored substrates and production of ethylene. The most common symptom of flower senescence is wilting i.e. loss of turgor pressure of the cells due to failure of water uptake (Eze *et al.*, 1986). The failure of water uptake as a result of stem blockage may be due to air blockage, microbial growth or physiological plugging. In addition, senescence is accompanied by a dramatic increase in the leakage of several molecules such as amino acids, sugars, inorganic ions and anthocyanins, and activity of petal ACC synthetase and the disintegration of tonoplast and mitochondria (Bieleski and Reid, 1992). Two major metabolic and biochemical changes occurring in senescing petals are increase in respiration and hydrolysis of cell components. Vase life of cut flowers depends upon the rate of transpiration through open stomata of leaves and solutes present in vase water (Van Doorn, 1997).

Experimental evidences have shown that cuticular transpiration plays an important role in the water loss of orchid flowers. The transpiration rate of tropical orchid flowers ranges from 0.15 to 0.17 mg water $cm^{-2}h^{-1}$ or 0.4 to 1.9 g of water per inflorescence per day depending upon the total floral surface area. Halevy (1986) classified flowers into climacteric or non-climacteric based on the presence or absence of an increased rate of ethylene production associated with petal senescence. In most vegetative tissues, the overall synthesis of ethylene is the conversion of SAM (S-adenosylmethionine) to ACC (1-aminocyclopropane-1-carboxylic acid). Water loss in orchid flowers is considerably lower than that reported for roses and carnations due to the absence of supporting leaves in orchid sprays. It has been found that carbohydrate levels in mature flowers are lower than the levels in the tight buds. Moreover, the level of carbohydrates in the flower decreases markedly with time after harvest as reflected in the decreasing rate of respiration. This problem can be overcome by the exogenous supply of sucrose.

In orchid flowers, ethylene production is an autocatalytic process and ethylene level of 1 ppm caused premature fading of flowers (Lindner, 1946, Fischer, 1950). The premature fading of petals may be induced by pollination and by removing pollinia (Akamine and Goo, 1981). In *Cymbidium*, fading of flowers due to pollination is characterized by the formation of anthocyanins in both column and lip, swelling of column, stigmatic closure and wilting of sepals and petals. The anthocyanin level begins to decrease with age (Arditti *et al.*, 1971, 1973).

Developmental processes associated with post-pollination events include senescence of perianth, pigmentation changes, ovary maturation, ovule differentiation and female gametophyte development (O'Neil, 1997). Flowers of *Dendrobium* 'Pompadour' develop premature petal and sepal senescence following pollination. Pollination induces an ethylene climacteric accompanied by a small respiratory climacteric, epinasty and increased flower and inflorescence fresh weight and water uptake (Ketsa and Rugkong, 1999). The orchid flowers harvested in the tight bud stage had a lower rate of respiration than open flowers. The respiration rates continuously declined during the post-harvest period until the flower faded (Sheehan, 1954). Da Silva (2003) reviewed the some aspects of changes, and found that programmed cell death occurred in petal senescence.

13.3.1 Cellular Structural Changes

- Membrane rupturing and increase in cytoplasm debris and loss of permeability and fluidity due to oxidation.
- Invagination of tonoplast and endocytosis of cytoplasmic contents and disappearance of cortical microtubules.
- Reduction in cytoplasm volume, cessation of cytoplasmic streaming and change in proton flux across plasma membrane.
- Degeneration and collapse of organelles and increase in number of peroxisomes.

13.3.2 Biochemical and Structural Molecular Changes

- Increase in proteinases and nucleases; upregulation of phospholipases, acyl hydrolases and lipoxygenase neutral lipids; sterol/phospholipid ratio; lipid peroxidation; reactive oxygen species; water leakage and cell wall cross linking.
- Decrease in phospholipids, chlorophylls, proteins, thiol groups, nucleic acids and RNA.

Post-harvest life of orchid cut flowers is influenced by pre-harvest factor like varietal differences, light intensity, sugar level of flowers, temperature, nutrition and water loss; harvest factors like time, method and stage of harvest; and post-harvest factors including ethylene production, pre-cooling, water quality, pulsing, use of preservatives, bud opening, packaging and storage.

13.4 Pre-harvest Factors

It is estimated that a third of the post-harvest life of flowers is programmed by pre-harvest handling.

13.4.1 Varietal Differences

Varietal differences in cut flowers have been reported due to variations in water uptake, fresh weight, flower diameter, stem lignification, vase life and senescence behaviour. Among different species, the vase life ranges: *Lycaste* spp. (9 days), *Phaius tankervillae* (24 days), *Zygopetalum intermedium* (18 days), *Aerides multiflorum, A. odoratum, Cymbidium irridioides, Dendrobium nobile* and *Renanthera imschootiana* (28-56 days), *Paphiopedilum hirtussimum, P. wardianum* (56 days) and *Vanda coerulea, Vanda teres* (28-42 days) (Tab. 13.2).
Out of nine hybrids of *Cymbidium* ('PCMV', 'Red Princess', 'White Beauty', 'H.C. Aurora', 'Sun Gold', 'Ensikhan, Florance', 'Valley Legend' and 'Platinum Gold') evaluated at NRCO, Pakyong, Sikkim during 2008-2009, 'Florance' had the highest vase life of 54 days followed by 'White Beauty' (53 days) and the lowest occurred in 'Platinum Gold' (22 days) (Fig.13. 1). The present day orchid hybrids of *Dendrobium, Vanda* and *Mokara* remain perfect for 7 to 30 days. The flowers of *Cattleya* and *Phalaenopsis* remain fresh for 1 to 4 weeks whereas *Aranda* lasts for 18 to 28 days.

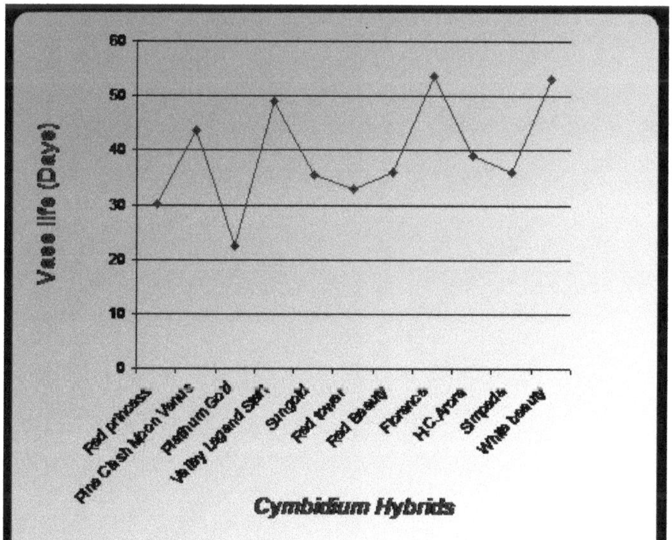

Fig. 13. 1: Evaluation of *Cymbidium* hybrids for their vase life

Cymbidium, ‘Fire Storm Ruby’ for cut flower

Dendrobium, ‘Dang Saard’ for cut flower

Vanda, ‘Roberts Delight Blue’ for cut flower

Oncidium, ‘Sharry Baby Sweet Fragrance’ for cut flower

Blc, ‘Hsinging Catherine’ for cut flower

Phalaenopsis, ‘Strawberry’ for cut flower

Plate 13.1: Important orchid hybrids for cut flowers

Tab. 13.2: Vase life of orchids due to species and varietal differences

Name of Species/hybrids	Vase life (days)
Aerides odoratum, Aerides multiflorum, Cymbidium iridioides, Dendrobium nobile, Renanthera imschootiana	28-56 days
Paphiopedilum hirtussimum, P. wardianum	56 days
Phaius tankervillae	28-42 days
Vanda coerulea, Vanda teres, Zygopetalum intermedium	14-21 days
Cymbidium hybrids	20-55 days
Dendrobium hybrids	14-21 days
Vanda, Mokara hybrids	14-30 days
Cattleya hybrids	10-20 days
Phalaenopsis hybrids	25-30 days
Aranda	18-28 days

13.4.2 Light Intensity

Light determines the carbohydrate levels before harvest which in turn influence the keeping quality. Flowers containing relatively higher amounts of carbohydrates, especially mobile sugars, last longer in the vase. Plants having few leaves, or leather-like leaves (like most cattleyas and oncidiums), require a high-light environment. If the leaves are soft and limp (like some phalaenopsis and most paphiopedilum), the plants are probably very light-sensitive, and should not be placed in a sunny south-facing window. Most orchids prefer indirect or filtered light and 50% shading.

- **Low light orchids (1200-2000 foot candle):** *Phalaenopsis, Calanthe*
- **Medium light orchids (2000-3000 foot candle):** *Cattleya, Laelia, Brassovola*
- **High light orchids (3000 foot candle or more):** *Cymbidium, Vandaceous groups.*

13.4.3 Temperature

Generally, higher temperatures result in a higher level of respiration. Cooling is essential to reduce other metabolic changes, such as enzymatic activity, and to slow the maturation of flowers. Cooling prior to packaging and transport reduces ethylene production and improves longevity. Based on temperature requirements, orchids are classified into three groups:

- **Warm orchids** (*Aerides, Vanda, Rhyncostylis* & *Dendrobium*): 32.2°C day temperature and 15.5°C night temperature
- **Intermediate orchids** (*Cattleya, Laelia, Oncidium, Miltonia*): 26.6°C day temperature and 12.8°C night temperature

- **Cool orchids** (*Cymbidium, Odontoglossum, Cypripedium*): 24°C day temperature and 10°C night temperature.

13.4.4 Humidity

As a rule of thumb, orchids require 80-85% humidity for satisfactory growth. Monopodial orchids require higher humidity than sympodial ones. Many sympodial orchids like *Cattleya, Oncidium* and *Dendrobium* form pseudobulbs, which are swollen shoots that store water and nutrients to help the plant survive periods during prolonged drought. Insufficient humidity during summer may lead to shriveling of the pseudobulb. Excessive humidity during winter may lead to spotting of flowers usually caused by *Botrytis.* Most orchids prefer water of pH 5.0-6.5; watering with lower or higher pH or with high levels of dissolved minerals can hamper nutrient uptake. Rain water is best for watering. Regular watering is essential under high sunlight and high temperature conditions. Sprinkling or misting may be practiced during hot summer. Watering should be reduced in late summer and the plants should be kept barely moist during winter. Watering orchids with thick leaves and CAM activities (such as *Aranda* and *Dendrobium*) in the late afternoon prior to harvesting season improved the keeping quality.

13.4.5 Nutrition

Orchids are light feeders and require nitrogen for the first two-thirds of their life cycle. During rest periods they do not need any fertilizers, while during flower initiation and inflorescence development plants are fed with less nitrogen, more phosphorus and potassium. During the blooming time, a small level of nitrogen and phosphorus and high levels of potassium are maintained. In orchids, foliar feeding is found to be ideal. Frequent application of fertilizers in low concentrations is the best way to feed orchids. A concentration of 0.2 to 0.3% of 30:10:10 (N:P:K) at the vegetative stage and 10:20:20 (N:P:K) at the blooming stage are applied for quality flower production. Sometimes fresh coconut water, diluted cow urine and fish meal emulsions are also useful as foliar spray.

13.5 Harvest Factors

13.5.1 Time of Harvest

Flowers should be harvested in mild temperatures because high temperature causes rapid respiration rates and excessive water loss. Flowers should be harvested in the early morning or in the evening. In the early morning, flowers remain turgid due to transpiration at night and higher sugar levels. Similarly, flowering stems retain a higher amount of stored carbohydrates if cut in the afternoon and retain more vase life.

13.5.2 Method of Harvest

Sharp tools or secateurs should always be used to detach the stems of flowers from the mother plant. The angle of the cut should be slanting and the stem should not be crushed during harvesting, especially hard wood stems. The spikes should be dipped in a bucket containing water immediately after harvest.

13.5.3 Stage of Harvest

The optimum harvesting stage of the commercial orchids is mostly fully open and mature flowers. The stage of harvest, spike length and number of flowers of some commercial orchids are given in Tab. 13.3.

Tab. 13.3: Stage of harvest, spike length and no of flowers of some commercial orchids

Orchid hybrid	Commercial stage of harvest	Spike length (cm)	No of Flowers
Aranda	50% bloom	45-60 cm	8-10
Cattleya	2-4 days before bud open	25-40 cm	1 or more
Cymbidium	75% bloom or two buds open stage	60-90 cm	10-15
Dendrobium	All flowers except top bud	40-60 cm	8-12
Oncidium	80% bloom	60 cm	Many
Paphiopedilum	3 to 4 days after opening of flowers	25-40 cm	1-5
Phalaenopsis	Fully open flowers	40-60 cm	8-10
Vanda	Fully open flowers	50-75 cm	8-15

Out of three *Cymbidium* hybrids (namely 'Pine Clash Moon Venus', 'Valley Legend Stefi', 'Pure Inca Gold') flower spikes harvested at four stages: fully open, 75% open, 50% open and 25% open to standardize the stage of harvesting. Vase life was noted to be highest in the 75% open stage. A maximum vase life of 59 days was recorded in 'Pine Clash Moon Venus', followed by 48.83 days in 'Valley Legend Steff' and 53 days in 'Pure Inca Gold'.

13.6 Post-Harvest Factors

13.6.1 Temperature

Opening of flower buds and rate of senescence accelerate at higher temperatures. At lower temperatures, the respiration rate decreases and the flowers produce less ethylene. Temperature plays an important role in the expansion of buds for flowers harvested at the immature stage, therefore flower buds are kept at temperatures as low as 0.5 to 4.0°C in *Cymbidium* and *Paphiopedilum*, 5-7°C in *Dendrobium* and 7-10°C in *Cattleya*.

13.6.2 Light

Light is essential for long distance transport or prolonged storage of cut flowers. Similarly, high light intensity is essential for opening of tight bud cut flowers. Flowers like carnations can be stored in darkness for a longer period without affecting quality. Florists should maintain a light intensity of 2000-3000 lux for 12-24 hours in their shops for illuminations.

13.6.3 Humidity

Cut flowers should be kept at 90-95% relative humidity for maintaining turgidity; flowers begin to show wilting symptoms when they have lost 10-15% of their fresh weight. The rate of transpiration from leaves is reduced with the increase of high relative humidity.

13.6.4 Water Quality

Water quality is defined by pH and EC value, hardness contents of phytotoxic elements and the presence of microorganisms causing vascular occlusions that affect the longevity of cut flowers. Saline water decreases the vase life of cut flowers. In case of cut gladiolus, the longevity of flowers decreases when the concentration of salts in the water reaches 700 ppm, whereas for cut roses, chrysanthemum and carnations, 200 ppm is harmful. At salt levels over 200 ppm, each 100 g per litre increase in salinity shortens vase life by half a day. . The presence of basic ions like $Ca^{++,}$ Mg^{++} in hard water is less harmful to flowers than soft water that contains sodium ions. Fluorine is very toxic to most of the cut flowers and causes injury to freesias, gladiolus and gerberas at 1 ppm, and chrysanthemums, roses, poinsettias and snapdragons at 5 ppm. Flowers like lilacs, cymbidiums and daffodils, however, are resistant to fluorine ions. Vase life increases in tap or well water that has been passed through a de-ionizer. The importance of low pH of the holding solutions is well known for improving vase life. A holding solution of pH 3.0-5.0 is optimum for increasing vase life of cut flowers.

13.6.5 Ethylene

Ethylene plays an important role in the regulation and co-ordination of senescence in climacteric flowers. Less of this hormone is produced but is more stable in floral buds and young flowers. A sharp increase in ethylene evolution is found during flower maturation, opening and senescence. Afterwards, ethylene production decreases and remains static. Basically, ethylene is first produced in the pistil and the evolved ethylene acts on the petals and induces expression of genes for ACC synthase, ACC oxidase and cysteine proteinases, resulting in the auto-catalytic ethylene production from the petals, in-rolling of petals and wilting of flowers (ten Have and Woltering, 1997). The gynoecium has been shown to produce a significant amount of ethylene before its production in the petals, possibly induced by factors such as ABA or IAA. This suggests its importance in controlling ethylene production in the flower during natural and pollination induced senescence, with emasculation hastening the release of ethylene (Shibuya *et al.*, 2001). A wide range of flowers is affected by ethylene. Some typical symptoms are: sleepiness of carnation and kalanchoe petals, fading and in-rolling of the corolla of *Ipomoea*, fading and wilting of sepal tips in orchids, and induction of anthocyanin formation in female reproductive parts and abscission of flowers and petals. Orchid flowers are highly sensitive to ethylene. High levels of ethylene production are due to herbivore damage, mechanical injuries and pollination. Decapped and emasculated flowers produce more ethylene than untreated ones. Sometimes, forced unfolding of flower buds in orchids reduces vase life.

13.6.6 Diseases and Insect- pests

Fungi, bacteria and insects affect the quality of cut flowers by causing the production of higher amounts of ethylene. Microbes accelerate flower senescence by the plugging of xylem vessels with pectin degraded products, and by producing ethylene and toxic compounds. Among bacteria genera, *Alcaligenes, Pseudomonas, Enterobacter, Erwinia, Bacillus, Corynaebacteria, Aeromonas, Acetinobacter* and *Flavobacterium* are commonly found in vase water. Some fungal species *Botryris cinerea, Fusarium oxysporum, Mucor, Penicillium* spp, *Rhizopus, Aspergillus* spp, *Alternaria alternata* and *Acremonium strictum* are responsible for early senescence of flowers and wilting and decaying of potted plants (De and Bhattacharjee, 2000, 2002).

13.7 Treatments for Improving Longevity of Cut Orchid Flowers

13.7.1 Physical Treatments

13.7.1.1 Pre-cooling

It is the fast removal of field heat and is an important operation in post-harvest handling and transport of cut flowers, wherever flowers are held dry pack. All flowers should be pre-cooled immediately after harvest by placing them in cold storage without packing or in open boxes until they reach the desired temperature (Bhattacharjee, 1997). The temperature varies with the species and cultivar: *Cattleya* (7-10°C), *Cymbidium* and *Paphiopedilum* (0.5 to 4°C), *Dendrobium* (5-7°C). Pre-cooling lowers respiration rate; decreases the break-down of nutritional and other stored materials in the stems, leaves and petals; and delays bud opening and flower senescence. It also prevents rapid water loss and decreases flower sensitivity to ethylene. Several pre-cooling techniques such as room cooling, forced air cooling, hyder-cooling, vacuum cooling and ice bar cooling etc. are available.

13.7.1.2 Vase Life of Cut Flowers as Affected by Stem Cut Ends

The resistance to water flow through stem segments increases predominantly in the lower one to two centimeters of the cut stems. Re-cutting stems under water improves longevity due to the elimination of air from the conducting vessels.

13.7.2 Chemical Treatments

13.7.2.1 Conditioning

Conditioning or hardening is a simple process where flowers are kept standing loosely in a big container so that air can circulate around the stems. The purpose of the treatment is to restore the turgidity of cut flowers with water stress during storage and transport. Conditioning is achieved by treating the flowers with de-mineralized water supplemented with germicides and acidified with citric acid to pH 4.5 to 5.0, with or without sugar. Hydration is improved when water is de-aerated or acidified, or when a wetting agent like Tween 20 at the rate of 0.01 to 0.1% is added. Flower stems should be placed in warm water or in a preservative solution in plastic jars at a depth of 2-4 cm and held at room temperature or in cold storage for several hours.

13.7.2.2 Impregnation

Sometimes, the cut ends of flower stems are impregnated for a short time with chemicals. This treatment prevents the blockage of water vessels in the stem by microbial growth and stem decay. Impregnation of cut bases of flowers with high concentrations (1000-1500 ppm) of silver nitrate, nickel chloride or cobalt chloride for 10-15 minutes improves the longevity of several flowers such as aster, gerbera, gladiolus, carnation, chrysanthemum, phalaenopsis and snapdragon. In *Cymbidium* 'Baltic Glaciers Mint Ice', the highest longevity compared to the control (39 days) was recorded when $CoCl_2$ (1000 ppm) was used for 15 minutes (46 days) followed by $CoCl_2$ (1500ppm) for 15 minutes (44days) (De *et al*, 2013).

13.7.2.3 Pulsing

The absorption of chemical solutions containing sugars and germicides through the lower cut bases of flower stems is known as pulsing. Pulsing may be used by growers, wholesalers or retail florists in order to enhance the subsequent vase life of cut flowers in water. Pulsing is employed with higher concentrations of sugar, mainly sucrose, the percentage of which varies with species and cultivars. Other chemicals used in pulsing treatments are STS, $AgNO_3$, HQ, MH, AOA, $CaCl_2$, $CoCl_2$, nickel sulphate, aluminium sulphate and benzyladenine. Pulsing is found to be of great value in prolonging life, promoting opening and improving the colour and size of petals through osmo-regulation. In the *Cymbidium* hybrid 'Red Princess' pulsing with 5% sucrose increases vase life (56 days) followed by sucrose at 8% (54.78 days). In *Cymbidium* 'Baltic Glaciers Mint Ice', pulsing of flowers with 5% sucrose followed by 150 ppm 8-HQS increased the vase life of flowers both with pollina (49.33 & 46.33 days) and without pollinia (44.00 & 41.67 days). The vase life of cut flowers is increased by pulsing with 4 mM STS for 10 minutes in the case of *Aranda* (Hew *et al.*, 1987), and in *Dendrobium* hybrid 'Pompadour' by pulsing with 25 ppm $AgNO_3$ + 135 $Na_2S_2O_3$, $5H_2O$ for 30 minutes (Hew and Yong, 2004). In *Oncidium* 'Goldiana', pulsing cut sprays with $AgNO_3$ for 30 minutes improves vase life (Ong and Lim, 1983). In *Phalaenopsis*, pulsing with 0.5 mM STS for 24 hours blocks the deleterious effect of ethylene. In *Dendrobium* cv. 'Sonia', when pulsing was conducted with 4% sucrose + 400 ppm HQ the highest recorded vase life was 21.33 days. When inflorescences were pulsed with 6% sucrose + 400 ppm HQ, the highest recorded sugar content in the flowers was 27.64% (Jomy and Sabina, 2002).

13.7.2.4 Bud Opening

In this procedure, flowers are harvested earlier than the normal cutting stage and then the buds are opened off the plant. Such types of post-harvest handling may be applied by growers or wholesalers. Bud opening of flowers increases longevity of cut flowers by reducing the sensitivity of flowers to extreme temperatures, low humidity and ethylene, by saving space during shipment and by extending the useful storage life. The sugar concentration used is lower than the concentration of pulsing and the optimum temperature is kept lower. In *Dendrobium* hybrid, 'Thongchai Gold', opened flowers had 29%, half opened flowers had 28.25% and buds had 16.17% reducing sugars. In *Dendrobium* hybrids, HQS or $AgNO_3$ (50 ppm) is effective for opening tight bud cut flowers. Ketsa *et al.* (2001) reported that a preservative solution containing 225 ppm HQS, 30 ppm $AgNO_3$ and 4% glucose increased bud opening and the time to wilting of the open florets of *Dendrobium* Cv. 'Ceasar'. In *Cymbidium* Cv. 'Ensikhan' 4% sucrose + 100 ppm acetyl salicylic acid or 4% sucrose + 100 ppm $Al_2(SO_4)_3$ improved bud opening. In *Cymbidium* hyb. 'PCMV', sugar (4%) + salicylic acid (200 ppm) showed maximum per cent of flower opening (75%) and vase life (45 days), followed by sugar (4%) + $Al_2(SO_4)_3$(100 ppm) (57% & 44 days) and sugar (4%) + 8-HQS (200 ppm) (53.8% & 44 days for flowering and vase life, respectively) over control (bud drop and senescence on 27th days) (Tab. 13.4). The highest carbohydrate content (140 mg/g) was estimated in fresh condition at bud stage (140 mg/g) followed by at bud stage at senescence in control (131 mg/g). A minimum carbohydrate content of 60 mg/g was observed with sugar 4% + 8-HQS

(200 ppm), followed 64 mg/g carbohydrate with sugar 4% + salicylic acid (200 ppm). In *Dendrobium* hybrid 'Thongchai Gold', the maximum percentage of fully opened buds (66%) was recorded with sucrose (4%) + $Ca(NO_3)_2$ (1%), followed by 60% when sucrose (4%) + acetyl acetic acid (100 ppm) was used. Longest vase life (36 days) was found with sucrose (4%) + $Al_2(SO_4)_3$ (100 ppm) followed by 33 days with sucrose (4%) + acetyl acetic acid (100 ppm).

Tab. 13.4: Effect of chemicals on bud opening in Cymbidium hyb. 'Pine Clash Moon Venus'

Treatment	Days to first floret opening	Diameter of first floret (cm)	% of half opened buds	% of fully opened buds	Vase life (Days)
Distilled water	---	----	0	0	27
Sugar 4%	20	5.1	5.8	44	37.8
Sugar 4% + $Al_2(SO_4)_3$(100 ppm)	21	5.4	0	57	44
Sugar 4% + 8-HQS (200 ppm)	18	5.5	30.7	53.8	44
Sugar 4% + Salicylic acid (200 ppm)	21	6.6	0	75	45
Sugar 4% + Ca $(NO_3)_2$ (1%)	25	4.85	2	22.8	37.8
Sugar 4% + Boric acid 200 ppm + K_2SO_4 (2 mM)	20.5	5.5	6.25	25	36.2

13.7.2.5 Preservatives

Preservatives are used in holding solutions in the form of tablets containing a mixture of chemicals such as sugars, germicides, salts, growth regulators etc. In addition, the chemicals are employed during conditioning, pulsing and for making bud opening solutions to improve flower shape, size, colour and opening.

Sugar, biocide, anti-ethylene compounds and hydrated compounds are used for conditioning. The sugar and biocide solutions are effective for opening of bud cut flowers.

The vase solution should contain sugars, acidifying agent and a biocide. Citric acids are mainly used as acidifying agents and hydroxy quinoline as biocide. Metallic salts like silver nitrate, cobalt chloride, aluminium sulphate, zinc sulphate, calcium nitrate and nickel chloride have been found to prolong post-harvest life of various cut flowers. Among several growth regulators used to increase vase life of cut flowers, BA, IAA, NAA, 2,4,5-T, GA3, B9, CCC are common.

New chemicals that have been found promising as floral preservatives are ethylene inhibitors like amino- oxyacetic acid, 1-amino cyclopropane, aminotriazole, aminoethoxy vinyl glycine, alpha aminoisobutyric acid, diazocyclopentadiene and phenidone.

Different chemicals used in holding solution for improving vase life of orchids are listed in Tab. 13.5.

Tab. 13.5: Holding solutions for different types of orchids

Name of orchid	Holding solution
Oncidium	8-HQC (100-200 ppm) + 4% sucrose
	kinetin (50ppm) + 4% sucrose
Cymbidium	8-HQC 200 ppm + sucrose 2%,
	1-MCP (500ppb)
	1% Sucrose + STS (1 mM)
Arachnis, Aranda, Aranthera, Cattleya	STS (1 mM) + 1% sucrose
Paphiopedilum	8-HQC (200 ppm) +2% sucrose
Vanda	AgNO3 (30 ppm) + 1.5% sucrose
Dendrobium	8-HQC (200 ppm) + sucrose (2%),
	0.5 mM AOA + 4% sucrose,
	AgNO3 (30 ppm) + 4% sucrose
	400ppm HQ + 30ppm AgNO3 + 2% sucrose
	200 ppm 8-HQS + 50ppm AgNO3 + 8% sucrose

Tab. 13.6: Effect of chemicals on post-harvest life of Cym. 'PCMV'

Treatments	Loss in wt (g)	Longevity of first floret (days)	Vase life (days)	Solution uptake (ml)
Distilled water)	12.6	37.2	44.4	21
$Al_2(SO_4)_3$ (100 ppm)	18.4	45.2	53.4	33
$Al_2(SO_4)_3$ (500 ppm)	21.4	28.4	36.8	21
Ca $(NO_3)_2$ (100 ppm)	16.4	42.0	54.8	27
Ca $(NO_3)_2$ (500 ppm)	29.6	32	35.6	24
8-HQS (100 ppm)	27.4	33.2	45.6	29
8-HQS (200 ppm)	26.8	42.2	49.8	33
BA (25 ppm)	24.0	49.2	48.0	27

In *Cymbidium* 'PCMV', the use of 2% cane sugar resulted in the highest depletion of stored carbohydrates, maximum longevity of first floret (54 days), zero % of flower dropping, maximum solution uptake (24 ml) and the highest vase life (61.2 days). This was followed by 4% cane sugar with medium longevity of first floret (40.8 days) and a vase life of 52.8 days. The least longevity of first floret (27.2 days) was found when 8% cane sugar which also resulted in a vase life of 36.2 days. Longevity of first floret was recorded maximum with BA (25 ppm) (49.2 days) followed 45.2 days when Al_2 $(SO_4)_3$ (100 ppm) was used (Tab. 13.6). Maximum vase life (54.8 days) was found with $Ca(NO_3)_2$ (100 ppm) followed by Al_2 $(SO_4)_3$ (100 ppm) (53.4 days). The maximum solution uptake (33 ml) was observed with Al_2 $(SO_4)_3$ (100 ppm) and 8-HQS(200 ppm). In *Cymbidium*, 'Baltic Glaciers Mint Ice', 8HQS at 150 ppm increased the vase life of flowers with pollinia (48.33 days) followed by 8 HQS 200 ppm (46.80 days). 8 HQS 200 ppm increased the vase life of flowers without pollinia to 36.67 days.

In another experiment, in the *Cymbidium* hybrid, 'PCMV', out of six treatment combinations, the maximum vase life (77.6 days) was obtained with 2% sucrose + 200 ppm 8-HQS, followed by 2% sucrose + 100 ppm $Al_2(SO_4)_3$ (77.4 days) over control in tap water (65 days).

13.8 Grading and Packing

Grading is done based on parameters like appearance, stage of maturity, blemishes or injuries due to diseases, infestations caused by insect pests, colour and size of the bud, and straightness, strength and length of stem. Flowers are generally grouped into bunches of 5, 10, 12 or 20 stems and loosely tied with rubber bands. Before placing in the package, individual flower bunches are wrapped with suitable packing materials such as cellophane paper, kraft paper, newspaper, tissue paper or corrugated cardboard sheets. For local markets, bunches are held in buckets containing water or preservative solution. It is advisable that for long distance transport and storage, flower bunches are held in dry cardboard boxes. The minimum length of boxes should be about twice the width, and the width should be about twice the height. The use of telescope-style boxes made of CFB is ideal. Grading of *Cymbidium* and *Dendrobium* cut flowers are given in Tab. 13.7 and Tab. 13.8, respectively.

Tab. 13.7: Grading of Cymbidium Cut Flowers

Cymbidium	Grades	Flower count	Spike length	Other consideration
Cymbidium Standard	AAA	12-15	1.25 m	– Strong straight stems, uniform length, no marks on flowers.
	AA	8	90 cm	
Cymbidium Miniature	XL	>15	65+	– Bent crooked spikes but with perfect flowers.
	L	12-14	55-64	
	M	8-11	40-54	
	S	<5	30-39	

Tab. 13.8: Grading in Dendrobium Cut Flowers

Grade	Spike length	No. of opened flowers
Small (S)	30 cm	4-5
Medium (M)	40 cm	6-8
Large (L)	45 cm	8-10
Extra Large	50 cm	>10

13.9 Packaging in Orchids

Cut flowers are inserted in a tube containing water with or without preservatives or simply wrapped in wet cotton swabs covered with a piece of plastic and tied with rubber band to keep in it place (Plate 13.2). Flower spikes are grouped into bunches of 5 or 10 or so. Bunches or individual spikes are placed inside the box in in alternate fashion (Plate 13.3). Ethylene scrubbers with $KMnO_4$ or Purafil may also be kept in the box. For export purpose, packing of flowers in two piece boxes is the best option. In *Cymbidium*, single flowers backed by a fern leaf are inserted in small flasks containing preservative solution. The flasks are then packed in a 3-sided box with a display window. In the *Dendrobium* hybrid 'Sonia-17' a low gauge polyfilm of 100 gauge thickness along with the cotton dipped in 8-HQS (25ppm) covering the base of the spike had maximum vase life and flower quality (Jawaharlal *et al.*, 2006). A glass flute containing a flowering mini *Cymbidium* and stylish setting is called a Cilindra, and is commonly used as festive packaging for special occasions like Birthdays.

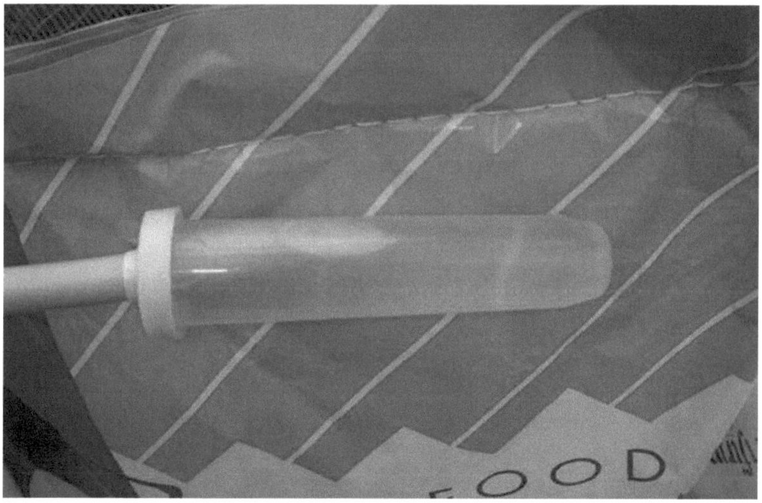

Plate 13.2: Cut flowers are inserted in tube containing water or water with preservatives

Plate 13.3: Individual spikes are placed inside the box in alternate fashion

13.10 Storage of Cut Flowers

Low temperature treatment during the storage or shipment period reduces the entire metabolism in the tissues, slows down the respiration, transpiration and ethylene action and retards the multiplication of bacteria and fungi.

In general, temperate orchids are stored at temperatures as low as 5°C in cold chambers whereas tropical orchids are stored at 7-10°C. A 90-95% relative humidity is necessary during storage to minimize moisture loss and to prevent wilting.

There are two types of cold storage methods, namely ‘wet storage’ and ‘dry storage’. In wet storage, flowers are stored with their bases dipped in water or preservative solution for a short time, whereas dry storage methods are used for long term storage. In this method, fresh flowers are harvested in the morning, graded and sealed in plastic bags or boxes to prevent the loss of moisture. In Controlled Atmosphere (CA) storage, cut flowers are kept in gas tight cool chambers equipped with cooling systems at a higher level of CO_2 and lower level of O_2 to reduce the respiration rate and the production and action of ethylene. Generally, the concentration of CO_2 should be maintained at levels higher than 4% and not below 0.4% in CA storage (Tab. 13.9).

Tab. 13.9: Storage of Orchid Cut Flowers

Name of Orchid	Storage Temperature	Storage period
Oncidium, Phalaenopsis, Odontoglossum, Cattleya	7-10°C	2 weeks
Dendrobium	5-7°C	10-14 days
Cymbidium	1-4°C	14 days
Paphiopedilum	-0.5-3.0°C	20 days
Arachnis, Aranda, Aranthera, Ascocenda, Epidendrum	8-13°C	10-14 days

13.11 Transport

Flowers are short lived and perishable in nature and should be delivered to their destination as early as possible immediately after harvest. For long distance markets, cut flowers are transported by cargo planes, merchant ships and trucks. Other modes of transportation are head loads, bicycles, two-three wheelers, cars, vans etc. Hence, for long distance transportation, advanced methods of post-harvest handling like cooling, conditioning, impregnation, pulsing, bud opening and packaging are followed.

Short time pulsing of flowers with optimal concentration of sucrose, $AgNO_3$, STS and growth regulators is important for long term truck and sea shipments. Flowers such as standard and spray carnations, chrysanthemum with non-hardy stems of full flowers, gerbera, coloured bud lily, miniature and floribunda roses are suited for dry transport over a period of several days; while chrysanthemum with hardy stems and single flowered freesias, iris, narcissus and H.T. roses are unsuited for dry transportation over a long period. Other than tropical flowers, the best method of transport of most cut flowers is under refrigeration from the grower to final consumers.

14 Value Addition in Orchids

14.1 Introduction

Value addition is a way to increase the value of a raw product anytime between harvesting and sales of the final product. A typical value addition includes processing in some ways like cleaning, cutting, packaging, smoking, drying, freezing, extracting or preserving. Value added products give a higher return, open new markets, create brand recognition and add variety to a farm operation. Value addition does not offer any guarantee on profitability. Careful planning and management are required to promote profitability. These key factors for the success of value added enterprises are quality products, good marketing and sufficient capital.

Other factors required for value added enterprises are:

- A unique product
- An enthusiastic promoter of the product
- The right kind of labeling and packaging
- Aggressive marketing
- A full time presence on the farm
- Strong agricultural or livestock knowledge
- Ability to cater to customers
- Assistance from agencies and universities
- A strong relationship with the local community
- Safe food handling and food safety regulations
- Product liability insurance

Value addition in floriculture increases the economic value and consumer appeal of any floral commodity. In floriculture, value addition is accomplished through genetic changes, processing or diversification. The profitability of a commodity is increased when a raw material is converted into a unique product. Although it requires more time, labour and skill it can significantly increase the net cash return of a small scale floriculture enterprise. Value addition gives high premium to the grower and provides quality products for the domestic and export markets. Recently, the consumption pattern is becoming diversified and leaning towards value added products such as essences, perfumes and other by-products from flowers. There is an urgent need for value addition in floricultural products through processing, packaging and supply chain management to increase farm income and generate employment (De, 2011).

The value added products from non-conventional floricultural crops like essential oils of rose, tuberose, jasmine, marigold and plant extracts used in medicines and pharmaceutical industry are unique and export-import opportunities.

Orchids comprise the largest family of flowering plants with 25,000 to 35,000 species belonging to 600-800 genera and covers 6.83% of the flowering plants. They are prized for their incredible diversity in the size, shape, colour and attractiveness of their flowers and their high keeping qualities even up to 10 weeks.

14.2 Orchids - A Component Adapted to the Diversified Climate

Orchids are found in nearly every environment in the world. Epiphytic orchids like *Thunia, Coelogyne, Cattleya, Laelia, Dendrobium, Calanthe, Bulbophyllum, Aerides, Phalaenopsis, Aranda* and *Aranthera* with thick leaves and succulent stems have CAM and are drought tolerant with higher water use efficiency. Rhizomatous orchids like *Habenaria, Eulophia* etc require a terrestrial climate.

14.3 Orchids-organically Viable

Each orchid genus has different requirements for potting media collected from locally available organic sources. It is very important to have the correct medium for each type of orchid, depending on whether it is terrestrial or epiphytic. Growing media commonly include fir bark, coconut husk, sphagnum moss, tree fern fibres, coco peat, saw dust and perlite, and frequently a mixture of two or three of these materials. All orchids potted in a typical bark medium need to be repotted every 18 to 24 months, depending on the needs of the individual plant.

14.4 Landscaping with Orchids

Orchidscaping is the use of orchids permanently planted into specially prepared beds or attached to trees, shrubs or rocks in the appropriate spot in the garden. Combined with other traditional ornamentals such as palms, ferns, flowering perennials, shrubs, trees and herbs etc. it is easy to create some of the most interesting and beautiful gardens imaginable, depending upon the cost involvement and microclimatic factors. Many orchids can be grown on rocks and logs for placing in the landscape. They are attached to either cut wooden logs, coconut logs or living trees and shrubs. Once the orchids are established they will attach to the trees and logs (Teoh, 2005). In order to create visual impact in landscaping, the orchids should be planted in a single bed of one type and one colour. If someone has only one or two plants of a type, it is advisable to growing them in pots. Almost all spider orchids (*Arachnis* and their inter-generic hybrids, terete and semi-terete vanda, *Phaius tankervillaea*, *Calanthe* spp, and Ladies Slippers) perform well, if they are grown on the ground in full sun with liberal watering and fertilization. Sloping or flat ground with good drainage provides the ideal location for orchid beds.

14.5 Colour Scheme with Orchids

To develop an orchidscape, a gardener should aware of the flowering period of each orchid. Some gardeners enjoy seasonal bursts of colour. In this case, cymbidiums and dendrobiums which flower from winter to spring should be the first choice (Friend, 2004).

14.5.1 Winter Flowering Orchids

Bulbophyllum hirtum, B. putidum, Cymbidium lowianum, C. mastersii, Eria bambusifolia, Paphiopedilum farrieanum, P. insigne, P. spicearum, Pleione maculata, P. praecox.

14.5.2 Spring Flowering Orchids

Ascocentrum ampullaceum, Calanthe plantaginea, Coelogyne cristata, Cymbidium devonianum, C. eburneum, Paphiopedilum hirsutissimum, P. villosum, Phalaenopsis lobii, Pleione humilis.

14.5.3 Summer Flowering orchids

Coelogyne corymbosa, C. cristata, C. nitida, C. ochracea, Cymbidium aloifolium, Dendrobium fimbriatum, D. heterocarpum, D. nobile, P. mannii, Pleione hookeriana, Phaius flavus, P. tankervillae, Renanthera imschootiana, Rhyncostylis retusa, Spathoglotis plicata, Vanda coerulea, Vanda cristata, Vanda stangeana, Vanda tessellata.

14.6 Orchids in Balcony Gardens

In Balcony gardens, lithophytic orchids can be grown by attaching them in free standing rocks or to the balcony's masonry walls. Genera suitable for shady location may include *Bulbophyllum, Coelogyne, Eria, Maxillaria*, some *Oncidiums, Sarchochilus* hybrids, *Phalaenopsis* and *Cattleya* hybrids.

14.7 Orchid Tree

An orchid tree is a variation on mounting orchids that involves the placement of many orchids on a branch or branches to give a completely natural look (Taylor, 2009). It is used in those areas of the country where orchids are grown outdoors most of the year. Usually, the larger plants are attached to the bottom and the smallest on the upper portions for aesthetic reasons and to provide extra weight at the bottom to balance the weight of the structure. It is better to select those plants which require similar light, temperature and humidity conditions. Another factor that has to be considered is flowering times, to get different colours on the tree throughout the year. The chosen plants are mounted on the tree with sphagnum moss and fishing wire. Thorough misting and maintenance of humidity are essential for a month to establish the plants on the structure.

14.8 Orchids – Genetic Materials for Breeding and Species Trade

Several local species of *Vanda, Cymbidium, Ascocentrum, Paphiopedilum, Calanthe, Dendrobium, Coelogyne, Renanthera, etc* are in great demand in international market for breeding materials (Tab. 14.1) (Bose and Bhattacharjee, 1980; Kumar and Sheela, 2007).

Tab. 14.1: Orchid species suitable for breeding

Arachnis cathcartii	*Ascocentrum ampullaceum*	*Bulbophyllum leopardinum*	*Bulbophyllum putidum*
Calanthe chloroleuca	*Calanthe herbacea*	*Calanthe masuca*	*Calanthe plantaginea*
Calanthe triplicata	*Coelogyne barbata*	*Coelogyne corymbosa*	*Coelogyne cristata*
Coelogyne fuscescens	*Coelogyne nitida*	*Coelogyne ochracea*	*Cymbidium devonianum*
Cymbidium eburneum	*Cymbidium hookerianum*	*Cymbidium iridioides*	*Cymbidium lancifolium*
Cymbidium longifolium	*Cymbidium lowianum*	*Cymbidium munronianun*	*Cymbidium tigrinum*
Cymbidium tracyanum	*Cymbidium whiteae*	*Dendrobium bensoniae*	*Dendrobium candidum*
Dendrobium densiflorum	*Dendrobium farmeri*	*Dendrobium formosum*	*Dendrobium gibsonii*
Dendrobium infundibulum	*Dendrobium nobile*	*Dendrobium parishii*	*Dendrobium pendulum*
Dendrobium primulinum	*Dendrobium wardianum*	*Dendrobium williamsonii*	*Paphiopedilum fairrieanum*
Paphiopedilum hirsutissimum	*Paphiopedilum insigne*	*Paphiopedilum spicearum*	*Paphiopedilum venustum*
Paphiopedilum villosum	*Papilionanthe teres*	*Pecteilis gigantea*	*Phaius flavus*
Phaius tankervillea	*Phalaenopsis decumbens*	*Phalaenopsis lobii*	*Phalaenopsis mannii*
Pleione hookeriana	*Pleione humilis*	*Pleione maculata*	*Pleione praecox*
Renanthera imschootiana	*Spathoglottis plicata*	*Thunia alba*	*Thunia marshalliana*
Thunia venosa	*Vanda coerulea*	*Vanda corulescens*	*Vanda cristata*
Vanda pumila	*Vanda stangeana*	*Vanda tessellata*	*Vandopsis undulata*

14.9 Orchid Species and Hybrids for Cut flower

Orchid species like *Cymbidium eburneum, Cymbidioum iridioides, Paphiopedilum insigne, Paphiopedilum villosum, Paphiopedilum venustum, Paphiopedilum hirsutissimum, Paphiopedilum spicerianum, Paphiopedilum fairrieanum, Renanthera imschootiana, Vanda coerulea, vanda tessellata, Zygopetalum intermedium* are used as cut flowers. Orchid hybrids of *Cymbidium, Dendrobium, Vanda, Phalaenopsis, Oncidium, Cattleya, Paphiopedilum, Mokara, Aranda, Renantanda* etc. with different colour and forms are used as cut flowers, floral displays and exhibits (Bhattacharjee and De, 2005; De, 2011; De, *et al*, 2013).

14.10 Orchids – as Potted plants/Hanging Baskets/ Trays

Potted orchids last for longer than cut flowers, their shelf life being three weeks to four months depending upon species and hybrids (Nagrare and Ram Pal, 2008). Tall growing monopodial orchids are best grown in large clay pots up to 30 cm in diameter. Terrestrial and semi-terrestrial plants like *Paphiopedilum* and *Cymbidium* perform better in deep pots.

Orchid plants, as a rule grow, to be near one another to aid in creating a microclimate higher in humidity. Basket culture is useful for those orchids like *Vanda, Rhyncostylis, Arachnis* with pendent flower spikes and long dangling roots. Clay pots are best suitable for terrestrial orchids. Plastic pots are used for epiphytes. Slabs or logs of tree fern are effective for cool growing orchids.

Important orchid genera used as potted plants in the international market are *Phalaenopsis, Oncidium, Miltonia, Cymbidium, Paphiopedilum, Dendrobium, Cattleya, Ascocenda, Vanda, Brassia, Miltonia* and *Epidendrum* (Lopez and Runkle, 2005).

14.11 Orchids – as Herbal Medicine and Aromatic Products

Tribal people of the North eastern hill region use wild orchids for a variety of folk medicine, as orchids are rich in alkaloids, flavonoids, glycosides, carbohydrates and other phytochemicals (Tab. 14.2).

Tab. 14.2: Common medicinal orchids (Rao,2004)

Name of species	Parts used	Uses
Acampe papillosa	Root	Rheumatism, Sciatica and uterine diseases
Aerides odorata	fruit, leaves	The ground fruit used for healing wounds. Juice of leaves is used to heal boils in ear and nose
Cymbidium aloifolium	whole plant	Ground plant to cure chronic illness, weakness of eyes, vertigo and paralysis

continued **Tab. 14.2:** Common medicinal orchids (Rao,2004)

Name of species	Parts used	Uses
Dendrobium nobile	Stems	Fresh and dried stems used in preparation of chinese drugs for longevity and as aphrodisiac, stomachic and analgesic
Habenaria acuminata	Roots	Roots are used as tonic
Phaius tankervilleae	Pseudobulbs	Pseudobulb is used to heal swellings of hands and legs, poultice to soothe pain of abscess
Pholidota imbricata	Pseudobulbs	Pseudobulbs are mixed with mustard oil and applied on joints for rheumatic pain
Rhyncostylis retusa	Roots	Roots are effective against rheumatism, asthma, tuberculosis, cramps, epilepsy, vertigo, kidney stone, menstrual disorder
Arundina graminifolia	Stems	Bulbous stems are used to heal cracks
Dendrobium densiflorum	Leaves	Leaves crushed to paste with salt and applied on fractured area to set bones
Vanda corulea	Leaves	Leaf juice is used against diarrhea, dysentery and external application for skin diseases
Vanda teres	Leaves	Leaf paste to reduce temperature in fever
Vanda cristata	Leaves	Leaves are used as tonic and expectorant
Dendrobium moschatum	Leaves	Leaf juice is used as ear drop
Aerides multiflorum	Tubers	Anti-bacterial
Anoectochilus formosanus	Tubers	Hepatitis, hypertension, cancer
Bletilla striata	Pseudobulbs	Anti-bacterial, anti-inflammatory, demulcent, skin styptic
Cypripedium pubescens	Roots	Anti-spasmodic, diaphoretic, hypnotic, sedat-ive, tonic
Orchis latifolia	Roots	Aphrodisiac, expectorant, nervine tonic, dia-betes, diarrhea, dysentery
Eulophia nuda	Tubers	Worm infestation, Scrofula
Habenaria edgeworthii	Leaves and roots	Blood diseases
Habenaria intermedia	Leaves and roots	Blood diseases
Habenaria pectinata	Leaves and tubers	Arthritis
Malaxis acuminata	Pseudo-bulb	Bleeding diathesis, burning sensation, fever
Orchis laxiflora	Bulb	Diarrhea, bronchitis, convalescence
Vanda spathulata	Flowers	Consumption, asthma, mania
Vanda tessellata	Whole Plant	Fever, arthritis, rheumatism, bronchitis
Calanthe dlscolor	Whole plant	Halr restorlng
Dendrobium chrysanthum	Leaves	Antipyretic, Immunoregulatory, skin diseases
Dendrobium loddigessii	Leaves	Stomach tonic
Habenaria repens	Tubers	Aphrodisiac
Pholidota chinensis	Pseudobulbs	Scrofula, toothache and stomachache
Vanilla planifolia	Sheath	Hysteria, fever, impotence, rheumatism
Cymbidium aloifolium	Rhizomes	Salep; used as nutrient and demulcent; as emetic and purgative

continued Tab. 14.2: Common medicinal orchids (Rao,2004)

Name of species	Parts used	Uses
C. ensifolium	Rhizomes & Flowers	Eye sores
C. longifolium	Pseudobulb	As emetic and demulcent
C. giganteum	Leaf juice	Blood clotting

14.12 Fragrant Orchids

Fragrant orchids are delightful in outdoor living areas. *Brassovola* species are perfumed at night and the Australian native Dendrobiums perfume the air on warm spring mornings. Other aromatic orchids are *Aerides multiflorum, Aerides odoratum, Aeranthes, Bulbophyllum odoratissimum, Cattleya maxima, Coelogyne cristata, Coelogyne ochracea, Cymbidium ensifolium, Dendrobium nobile, Epidendrum cristatum, Epidendrum floribundum, Epidendrum nocturnum, Lycaste, Oncidium spaceolatum, Phaius tankervilleae, Rhyncostylis retusa, Vanda cristata, Vanda tesselata, Zygopetalum intermedium.*

14.13 Orchids – as Source of Phytochemicals

Many medicinal orchids are rich in alkaloids. Experimental studies have been conducted on the isolation of a number of alkaloids like anthocyanins, stilebnoids and triterpenoids from orchids. Orchinol, hircinol, cypripedin, jibantine, nidemin and loroglossin have been isolated from orchids.

Several phytochemicals isolated from orchids along with active ingredient are listed in Tab. 14.13.

Tab. 14.3: Phytochemicals from orchids

Sl. No.	Name of orchid	Phyto-Chemical class	Name of phytochemical
1	*Aerides crispum*	Phenanthropyran	Aeridin
2	*Agrostophyllum brevipes*	Triterpenoid	Agrostophyllinol
3	*Agrostophyllum callosum*	Triterpenoid	Isoagrostophyllol
4	*Agrostophyllum callosum*	Stilbenoids	Orchinol, 6-methoxycoelonin, imbricatin, flaccidin, oxoflaccidin, oxoflaccidin, isooxoflaccidin, flaccidinin, agrostophyllin, callosin, callosinin, callosumin, callosuminin, callosumidin

continued Tab. 14.3: Phytochemicals from orchids

Sl. No.	Name of orchid	Phyto-Chemical class	Name of phytochemical
5	*Arundina graminifolia*	Stilbenoids	Arundinan
6	*Cypripedium calceolus pubescens*	1-4 phenan-threnequinone	Cypripedin
7	*Orchis latifolia*	Glucoside	Loroglossin
8	*Dendrobium macraei*	Alkaloid	Jebantine
9	*Dendrobium nobile*	Bibenzyl	Gigantol
10	*Dendrobium nobile*	Bibenzyl	Moscatilin
11	*Dendrobium nobile*	Alkaloid	Dendrobine
12	*Dracula chimaera*	Anthocyanins	
13	*Eulophia nuda*	Phenanthrene	Nudol
14	*Vanda roxburghii*	Glycoside	Melianin
15	*Nidema boothi*	Triterpenoid	Nidemin
16	*Anoectochilus formosanus*	Glycoside	Kinsenoside
17	*Dendrobium moschatum*	Phenanthrene	Rotundatin and moscatin
18	*Bulbophyllum gymopus*	Phenanthrene	Gymopsin

14.14 Orchids – Used as Foods

Leaves, tubers and pseudobulbs of different species are used for edible purposes. Vanilla, a major spice crop and source of vanillin, comes from *Vanilla planifolia. Anoectochilus* leaves are used as vegetables in Indonesia and Malayasia. Pseudobulbs of *Cymbidium maladimum* and *Dendrobium speciosum,* and tubers of *Microtis uniflora* and *Caladenia carnea* are eaten. The popular beverage called as 'Faham' or 'Madagascar Tea' on the islands of Mauritius and Madagascar is prepared from the orchid *Jumellea fragrans*. The tubers from orchid genera such as *Acianthus, Dipodium, Glossodia, Lyperanthus, Prasophyllum* and *Thelymitra* have been used as food by the inhabitants of Australia. In Africa, the tubers of *Cynorchis, Eulophia, Disa, Habenaria* and *Satyrium* are used as food or juice is extracted from them. Roots, tubers or rhizomes of *Eulophia, Gastrodia, Habenaria, Orchis, Pholidota, Platanthera* and *Spiranthes* are used as food in Asia. Tubers of *Disa engleriana, D. robusta* and *D. zambica, Habenaria clavata, Satyrium ambylosacco, S. buchananii* and *S. carsonii* are used as foods in Malaysia. In Bhutan, the inflorescence or the flowers and pseudobulbs of *Cymbidium* spp. are eaten (Bhattacharjee and Das, 2008).

14.15 Orchids – a Special Item for Value Addition

Cilindra - a gift of a glass flute containing a flowering mini Cymbidium.
Stylish setting - Festive packaging for special occasions like Birthday

14.16 Orchids – for Festivals and Special Uses

People of Assam and Arunachal Pradesh use *Rhyncostylis retusa, Papilionanthe teres, Vanda roxburghii, Aerides odoratum* and many *Dendrobium* species in their religious and cultural festivals. In Assam, the flowering spike of *Rhyncostylis retusa* known as 'Kopou Phul' is used by girls to adorn their hair during the spring festival. The flowers of some other orchids like *Vanda roxburghii* and *Coelogyne nitida* are also used to adorn the hair of girls of Assam and Arunachal Pradesh in local festivals. The flowers of *Papilionanthe teres* are offered to Lord Buddha and spirits by the Khamtis and other Tai ethnics of Assam and Arunachal Pradesh. In the Kameng district of Arunachal Pradesh, *Dendrobium hookerianum, Dendrobium nobile* and *Dendrobium gibsonii* are considered as the symbol of purity and sanctity by the local people. Monpas consider the flowers of *Cymbidium grandiflorum* important for holy worship. The young naga women of Manipur wore the orange flowers of *Dendrobium densiflorum* behind their ears. Similarly, the flowers of *Vanda coerulea* are used by the women of Manipur in their hair during the autumn puja festival.

In several countries, orchid species and hybrids are used as National Flowers. For example, *Vanda* 'Miss Joaquim' in Singapore, *Peristeria elata* in Panama and *Lycaste skinneri* var. *alba* in Guatemala. Orchids are depicted on the stamps of several countries like Venezuela, USA, New Zealand, Australia, Indonesia, India, Singapore, Japan, Russia, Thailand, Malaysia and many others.

14.17 Orchids for Dry Flowers

Orchids are highly attractive, delicate, available in a variety of colours, and can be preserved by drying for use in flower arrangement and dried flower craft. The best method for drying orchids is using silica gel for microwave drying or by freeze drying. Drying orchids is a challenging task as these flowers are considered difficult to preserve. Dried orchids are used for different purposes, such as for use in vases and baskets and sometimes in shadow boxes. Bright flowers of orchid genera like *Dendrobium, Cymbidium, Paphiopedilum Cattleya, Pholidota* etc. can be used for drying.

14.18 Orchid Flower Arrangements

Orchids symbolize wealth, beauty and social status. Orchids flower arrangements are good table decorations and weddings. Among orchids, *Cymbidium, Dendrobium* and *Phalaenopsis* are excellent for wedding centerpieces.

An arch decorated with chic white silk combined with white orchids can be considered an admirable orchid flower arrangement. In the home, they can be displayed in three ways viz. single flower vases, plants in pots and traditional mixed flower arrangements.

14.19 Other Uses

In the Philippines and New Guinea, the stems of some Dendrobium species are used to make baskets and bracelets. In some tribes, the sap of *Cattleya labiata* var. *autumnalis* is used as glue for musical instruments. In Central America, the empty pseudobulbs of Schomburgkias are used to make horns.

15 Marketing

15.1 Introduction

Marketing is one of the most important aspects of nursery crop production. Marketing of any business is broken down by using the 4P's: Product, Price, Promotion and Placement.

Product: A nurseryman must let everyone know about the business and how your product is superior to the already available product in the trade. First, every business should have a plan to market the product. A marketing plan should cover the following questions:

What or Who is your market?

How will you identify your product to the market?

Price: In the first instance, the price should be fixed high enough to generate a profit for the business. In fixing price, the following factors should be taken into considerations:

- What are customers willing to pay?
- Are all costs covered; where do you break even?
- What does the competition charge?
- How much profit do you want to generate?

Promotion: This is the strategy that a person uses with the customer to encourage them to purchase the product. Promotion includes advertisement, public relations and personal contact.

Placement: When the product is ready for sale, has been priced and consumers know that it exists, how do they get? This is called placement of the product or getting the goods to the customers i.e., distribution. A nurseryman must rely upon distribution channels like wholesalers, retailers, distributors, brokers and cooperatives.

15.2 Marketing Channels

There are various markets for nursery crops including landscape contractors, garden centres, mass merchandisers, other growers, brokers, agricultural producers and home owners. Most nursery crops are sold at the wholesale level. The major wholesale market is the garden center. Brokering is a small but useful market channel for new growers with a limited product line, insufficient market contacts or poor marketing skills. Direct selling of product to the consumer is attractive because it maximizes the price per plant. Those engaging in direct farm marketing should have a building permit to ensure compliance with building codes, a business license or registration, liability insurance and signs that meet standards with respect to placement, size, location and type of sign used.

Mail order selling is another form of retailing that does not require face to face business with the client. Internet websites can bring the world market place to the door of even the smallest nursery business at a reasonable cost. Nursery management has attained a status of a commercial venture where retail nurseries sell planting materials to the growers and general public. It includes the following aspects:

- Capital investment in nursery development
- Process of distributed nursery production
- Knowing the government loan and subsidies available for nursery development
- Analysis of income, expenditure and profit in nursery management
- Understanding the role of horticultural nurseries in entrepreneurship development

15.3 Capital Investments in the Nursery Development

A large volume nursery requires a full time manager and considerable capital investment, while a small volume nursery requires a part time job and less capital investment. A plant nursery needs various capital components of expenditure like land, building, road, fencing, polyhouse, mother plant block, shade net, equipment, machinery, office stores, well/ tube well, motors, irrigation systems etc. The cost and expenditure components for non-recurring heads that are included under capital components are seed beds, nursery beds, potting mixtures and potting yards, polyhouse, mist chambers, polytunnels.

15.4 Marketing of Plants from Nurseries

Based on demand survey, the cost of various nursery plants are fixed depending upon the age of plants, height of plants, variety, propagation techniques used, production costs, maintenance costs etc. There should be adequate publicity and advertisement of availability of various grafts and seedlings in the nursery. This includes advertisements in the sales area, advance booking, information through gram sabhas, exhibition through stalls, advertising through newspaper and magazine, publicity in weekly bazaars, posters, appointing sale agents, information brochure, and printing catalogues.

15.5 Government Regulations and Support Nursery

The availability of quality planting materials is the main criteria for the success of horticulture industry. Nursery Registration Act is at present in force in the states of Punjab, Maharashtra, Himachal Pradesh, Uttar Pradesh, Uttarakhand, J&K, Orissa and Tamil Nadu. A system of monitoring exists for horticulture nurseries in the states of Andhra Pradesh, Assam, Bihar, Goa, Haryana, Karnataka and Kerala while there is no horticulture nursery Act in the states of Arunachal Pradesh, Chhatisgarh, Jharkhand, M.P., Manipur, Meghalaya, Mizoram, Nagaland, Rajasthan, Sikkim, Tripura and West Bengal. It is essential to adopt quality standards for Horticulture Nurseries to facilitate, promote and monitor production and trade of quality planting materials of horticultural crops which are propagated vegetatively.

A nursery must set certain standards and these standards are categorized into three main groups, cost efficiency, quality and size. Cost efficiency includes the cost of production, profit and sales price. Quality standards include general plant health, hardness or softness of plant tissue, uniformity, labeling etc. Size standards include pot size for trees, shrubs, herbs and indoor plants; and plastic punnets for seedling vegetables and flowers etc.

15.6 Support for Nurseries under National Horticulture Mission (NHM)

The National Horticulture Mission has given important focus on production and distribution of good quality seeds and planting materials. Most of the States have a network of Central or State assisted nurseries for producing planting materials. Assistance for infrastructure in nurseries includes the following:

- Proper fencing.
- Mother stocks block maintenance under polycover to protect from adverse weather conditions.
- Raising rootstock seedlings under shade net house conditions.
- Propagation house, tropical polyhouse with ventilation having insect proof netting on sides and fogging and sprinkler irrigation systems.
- Hardening / maintenance in insect proof net house with light screening properties and sprinkler irrigation systems.
- Pump house to provide sufficient irrigation to the plants and water storage tank to meet at least two days requirement

Under the revised guidelines of NHM, assistance can be provided for setting up a nursery of 1 ha costs Rs. 6.25 lakhs and maximum area of 4ha with a total cost of Rs. 25 lakhs. The nurseries in the public sector are given 100% assistance and for those under the private sector 50% assistance is provided as credit linked back and subsidies. Nurseries that are 1 ha are classified as small nurseries and beyond that are classified as large nurseries. Each nursery has to produce a minimum 50,000 plants per ha per year of the mandated perennial fruit plant /tree spices/ plantation crop through vegetative propagation. The nurseries have to supply quality planting materials. In order to ensure the supply of quality planting material, the planting material must be purchased from the nurseries which are accredited by the National Horticulture Board, SAU's and ICAR institutes.

15.7 Nursery Expenditure, Income and Profit Analysis

The nursery enterprise is an important source of employment and income generation and depends upon proper planning of work, man power, raw material and proper financial appraisal. Entrepreneurs may use business analysis as a tool for important decisions such as business expansions, financing, marketing strategies, operations planning and product selection. With proper implementations, this information may increase nursery business profitability, control costs, reduce the risk of business failure, boost employee productivity and job satisfaction, enhance physical efficiency and increase management professionalism. Financial analysis assists in identifying some common problems in wholesale plant nurseries such as low output, slow crop growth or poor pricing, excessive costs, waste or overuse, poor cash flow, over-capitalization or under-capitalization and imbalanced debt structure.

15.8 Expenditure Estimate in Nursery

The estimate of expenditure in a nursery consists of two major components fixed costs and variable costs. Fixed cost components include one-time capital investment include fencing, workshed, mother plant block, irrigation with pipe line, office-cum-store, shade net house, mist chamber, polytunnel, seedbeds, nursery beds, internal roads, pathways, polling yard, water storage. Variable cost is the amount used by a business unit to fulfill its daily requirements. Variable cost components include labourer, raw materials, like media, polybags, fertilizers, manures, pesticides and consumable items like labels, stationary, electricity and telephone.

15.9 Nursery Income

A nursery enterprise can have various sources of income such as the sale of saplings, grafts, vegetable seedlings, potted plants, bagged plants and ball seedlings, planting materials like seeds, bulbs and corms. They can sell other raw material and equipment like pots, bags, pruning and grafting implements etc. Other sources of income can be obtained from realization of commission, consultancy fee, subsidy, donation and minor sale of weed grass, fuel wood etc, and from the nursery.

15.10 Maintenance of Records for Sale and Income

The nursery should keep a record of mother plants illustrating both rootstock and scion trees, production of grafts and seedlings and sale of plants. Book keeping is another important function of financial accounting and it maintains proper records with complete details of business transactions. Business transactions are classified into several major activities or groups, e.g., sales, purchases, assets etc. Separate books are maintained for recording transactions. Details of transactions are recorded into respective headings.

Purchase books are maintained to record all transactions pertaining to purchases on credit or cash. *Sales books* are maintained to record all transactions related to credit or cash sales. *Ledgers* are maintained to record all accounts involved in the transactions.

A Cash book is maintained in a subsidiary book of the ledger for recording the account of cash. A *Bank book* is a subsidiary book of the ledger where the account of the bank is maintained. The *Stock register* is a register where the movement of stock is maintained.

15.11 Profit Analysis and Financial Ratios (De, 2013)

To make a nursery business economically viable and profitable, various financial ratios are taken into consideration and these ratios help the lenders to analyse the viability and profitability of the nursery enterprise. These financial ratios include return on investment covering both borrowed capital and owned capital, Debt Service Coverage Ratio (DSCR), Internal Ratio of Return (IRR), Break Even Point and Debt Equity Ratio.

$$\text{DSCR} = \frac{\text{Net profit} + \text{Interest} + \text{Tax}}{\text{Interest} + \text{Principal loan}}$$

A project remains viable if the minimum value of DSCR is 1.7.
IRR (Internal Return Rate) is the interest rate at which the net present value of costs of the investment equal the net present value of the benefits of the investments. IRR are used to evaluate the desirability of investments or projects.
Break Even Point (BEP): This is the level of activity where the total contribution is equal to the total fixed cost: Contribution = Sales - Variable Cost
The BEP indicates the risk involved in a project.
Debt-Equity Ratio: This indicates the extent to which the funds of the promoter are leveraged to procure loans.

$$\text{DER} = \frac{\text{Total long term debt}}{\text{Total promoters fund}}$$

A higher debt enquiry ratio indicates more risk due to a higher fixed cost of interest.

15.12 Entrepreneurship Development through Nursery

Nursery enterprise creates sizeable employment opportunities for the rural masses and the areas are production of seeds, seedlings, grafts, potted plants, ball plants, plants for aesthetic purposes in landscaping or management program on rental basis; sale of plants to retail nurseries; starting plant libraries; providing vegetable seedlings, ornamental plants and planting materials, medicinal and aromatic plants and tissue culture plants; landscape gardening and consultancy. A nursery entrepreneur undertakes a nursery business or an enterprise.

The qualities required for a nursery entrepreneur are as follows:

- Calculated risk takers
- Perceive opportunities very quick, synthesize the available information and analyze emerging pattern that escape the attention of others
- Desire to hit new goals
- Convert adversities into opportunities
- Seek out experts for assistance
- Take immediate feedback on performance and record prompt and accurate data
- Achievement driven person
- A successful entrepreneur plans while keeping an open mind to achieve goals in pre-scheduled time frame.
- A successful entrepreneur influences others and motivates them to think and act in his way.

15.12.1 Role of Nursery Entrepreneurship

A nursery enterprise aids in providing plants and planting material and other processes that indirectly help in the national economy. It also assists in gaining a more balanced financial development as the business flourishes both in urban and rural sector. It aids in empowering people by providing increased community participation. Another aspect of nursery enterprise is to contribute national economy by providing employment opportunities for the skilled and unskilled workers.

15.12.2 Risk Analysis in Nursery Enterprise

A nursery enterprise can be carried out successfully provided the raw material, labour, capital requirement, planning of expenditure and income and knowledge of market are carefully taken into consideration.

A number of factors that can lead to losses in a nursery enterprise and the factors are underutilization of capacity, untimely availability and sale of plants, production not as per demand, increasing competition poor recovery of credit sales, increasing cost of labour and raw material and poor technical knowledge.

15.12.3 Profitability in a Nursery Enterprise

The profitability of a nursery enterprise depends upon assumptions of demand of plants and availability of various production inputs like raw material and labour.
Points to be considered to bringing nursery enterprise into a profitable venture

- Production of plants in a nursery in a commercial venture.
- Balancing the technical aspects and practical feasibility of production and sale of plants in a nursery
- Proper planning and execution of operations
- Maintenance and critical analysis of records of expenditure and income at regular intervals
- Monitoring of daily activities in nursery
- A proper coordination among production of plants, rearing of plants and sale of plants
- Increasing sales through aggressive marketing, publicity and advertisement in newspaper, magazines and other mass media.

15.13 Export Market Views of Orchids

- Southeast Asian countries such as Thailand, Singapore, and Malaysia are the major suppliers of tropical orchids to the world. Japan is considered one of leading orchid-importing countries, consuming 40% of the world supply. The European countries also consume 40% of the world supply while Australia, Hongkong, and the United States take up the remaining 20%. Recognizing the immense potential of floriculture export, governments in the Asia-Pacific countries continue to support research and development by providing sophisticated laboratory testing and control apparatus to help growers improve quality and varieties. Distribution and transportation facilities are constantly upgraded and promotion activities conducted to nurture appreciation for plants and flowers.
- *Dendrobium and Vanda* are the most preferred orchids while the rose had the highest demand among ornamental plants. The price of flowers fluctuates widely depending on the occasion or season. Various factors affect pricing include supply and demand situation, cut-flower quality, and grading of cut orchids. In grading the cut orchids, size, color, freshness, and texture are considered.
- Proximity to export markets such as Japan and Taiwan is considered an advantage to orchid exporters. The bulk of the products are sold to commercial buyers and flower retailers.
- The creation of the Cut-flower Development Plan, and the enactment of the High Value Crop Law can enhance growth of the cut-flower industry.
- The serious problems that have been identified for growers, breeders, retailers, and traders are the lack of quality planting materials, improved production tech-

nologies, inadequate capital investment, proper diagnosis and control of plant pests and diseases.

- To minimize problems that hinder the growth of the cut-flower industry, researchers recommended collaborative efforts among growers, breeders, traders, and retailers with assistance from the different government agencies, private sector, academia, financial institutions, and the media. There is also an urgent need to form cut-flower cooperatives, associations, and federations. The creation of allied services could be beneficial to the orchid industry in terms of having greater access to opportunities and addressing requirements of the cutflower growers. The cut-flower association's approach would be to provide technical and marketing assistance to grower members and would-be growers.

15.14 Development of International Orchid Market

The production of pot orchids and cut orchid flowers has increased rapidly since the early-1990s. More first-time orchid growers are becoming involved in producing orchids than ever before. Orchid production is turning into a true international business.

Since the signing of the CITES in the late 1970s, orchid trade across international boundaries has been regulated by this international treaty. Although CITES was designed to protect the threatened and endangered species of plants and animals, it plays a vital role on growers and traders who export and/ or import orchids. Through the efforts of the American Orchid Society and countries having large orchid-exporting businesses, CITES has slowly relaxed the control of trading on a limited number of commercially-important orchid genera.

During the 8th Asia-Pacific Orchid Conference, held in March 2004 in Tianan, Taiwan, Mr. Andy Matsui first proposed the idea of forming a worldwide orchid producers' organization. With the help of the organizers of the Taiwan International Orchid Show (TIOS) and through the coordination of Dr. Yin-Tung Wang, Professor of Horticulture at Texas A&M University, an organizational meeting was held in March at the 2006 TIOS. All of the attendees of this preparatory meeting expressed high interest in forming an international organization to represent the orchid growers. A second meeting was held during March: the 2007 TIOS. Mr. Andy Matsui, Dr. David Beck (a cut-flower cymbidium grower in New Zealand) and Dr. Yin-Tung Wang presided at this meeting. Over 70 representatives from several countries around the world in attendance elected Mr. Matsui the first President of the International Commercial Orchid Growers Organization (ICOGO). Dr. Yin-Tung Wang was elected the Executive Secretary, and Mr. Matsui donated $50,000 to get the organization off the ground and to start functioning. Two orchid growers, Yupin Biotechnology and Chian Huei Flowers in Taiwan, pledged to donate to help with ICOGO.

The Objectives of ICOGO

- Develop business models from the experience of successful growers and regions.
- Increase orchid trade around the world through more extensive marketing efforts.
- Enhance the availability of orchid varieties to benefit all parties in the industry.
- Disseminate information related to orchid production and orchid trade.
- Cooperate and support applied research and conservation that help the advancement of commercial orchid production with local, regional, national and international research agencies and institutions.
- Provide the necessary trade information among the members to enhance orchid trade and cooperation.

The Functions of ICOGO

- To develop a system to help and protect the exclusive orchid varieties and the exchange of breeding materials among orchid breeders, propagators, and producers worldwide.
- To assemble up to date orchid information and assist in developing new techniques for producing orchids more profitably.
- To promote the eradication and control of insect pests and diseases and mutations for producing quality plants.
- To participate in regulatory activities, i.e., CITES, government regulations, etc. to ease on international orchid trade.
- To provide the necessary trade information among members to enhance orchid trade and cooperation.
- To conduct trade shows and publish newsletters to enhance communication among the members.

Geographical Regions

- Region I: Europe
- Region II: The Americas
- Region III: North Asia
- Region IV: South Asia, Oceania, Australia, New Zealand
- Region: V: Africa, Middle Asia

Membership Classes

- Full Members
  - Commercial orchid growers, production area ≥ 1,000 m^2.
  - Have voting rights.

- Associate Members
  - Allied suppliers, equipment & greenhouse manufacturers, orchid-related businesses, employees of Full Members, educators, researchers, members of government and regulatory agencies, orchid organizations, etc.
  - Have no voting rights.

ICOGO's headquarters is currently located in the United States and it has been registered with the State of California. The orchid trade has just passed its infancy stage and there is a tremendous avenue for growth in the years to come.

15.15 Domestic Orchid Market

In India, it took a century for Sikkim and the northeastern region to realize the potential of *Cymbidium*, together with five other orchids - *Dendrobium*, *Vanda*, *Cattleya*, *Phalaenopsis* and *Oncidium* - as revenue earners. Flowers became a commercial prospect for us 10 years ago. It is a late start but is turning out to be a promising sector.

Sikkim annually produces Rs 20 million worth of orchids on an average. A cut stalk (spike) costs between Rs.20 and Rs.150 depending on the quality. The state government has introduced a special package in 18 clusters under which 50 flori-farmers have been given 500 orchid plantings (saplings) each with necessary support for cultivation, training and marketing. The state has to overcome several challenges to send the produce out to the national market. The first hurdle is transportation to the nearest marketplace. An effective cold chain is another area of concern.

The state requires refrigerated vans to the cold storages, and then the flowers are flown out. Orchids and other horticultural products are marketed by SIMFED - the government's national sales network. Veteran flower trader Andy Warren, managing director of the New Zealand-based company Bloomz, proposed that Sikkim and the northeastern states – (the country's traditional orchid hothouses) should concentrate on the domestic market and explore its "optimal capacity" for high returns instead of eyeing bigger export shares globally. This region lacks post-harvest facilities such as good packaging houses and cold storages. During hot summer logistics to move the harvested products are inadequate. Flowers like orchids wilt. The region needs to look at expanding markets in New Delhi, Mumbai, Bangalore and Chennai to fetch good prices. The critical thing is to ensure top quality in the market every day. This observation was made by Warren, who has been trading flowers with India for the last 20 years, to IANS at the Sikkim International Flower Show 2013. According to his concern, going to the international market is a "whole new ball game" with its own stringent set of standards.

Producers

- Open to certification
- Require support and extension service

Government strategies

- Clear Good Agricultural Practices (GAP) framework at national level
- Strengthen Standard knowledge & dissemination
- Promote production of high quality orchids: GAP farm

Which information is still needed?

- Do producers benefit from certification? (i.e. welfare, access to export markets)
- Information on environmental and social impacts
- Views of other stakeholders in importing countries i.e. buyers in EU

References

Abraham, A. & Vatsala, P. (1981). Introduction to orchids with illustrations and descriptions of 150 South Indian Orchids. TBGRI, Trivandrum, Kerala, pp.533.

Akamine, E.K. & Goo, T. (1981). Controlling premature fading in Vanda 'Miss Joaquim' flowers with potassium permanganate. Univ. of Howaii, College of Tropical Agric. and Human Resources. Res. Series, No. 007, 8.

Akurugoda (2013). Bio-piracy and its impact on biodiversity: A critical analysis with special reference to Sri Lanka. *International Journal of Business, Economics and Law*, 2 (3): 48–52.

Ando, T. & Ogawa, M. (1987). Photosynthesis of leaf blades of *Laelia anceps* Lindl. is influenced by irradiation of pseudobulb. *Photosynthetica*, 21: 588–590.

Arditti, J. (1992). Fundamentals of Orchid Biology. Available at Mcquerry Orchid Books, 5700, W. Salerno Road, Jackson Ville, FL 3 2244–2354, USA.

Arditti, J., Flick, B.H. & Jeffrey, D.(1971). Post-pollination phenomena in orchid flowers-II. Induction of symptoms by abscissic acid and its interaction with auxin, gibberellic acid and kinetin, *New Phytol*, 70: 333–341.

Arditti, J., Hogan, N.M. & Chaelwick, A.V. (1973). Post-pollination flowers in orchid flowers-III. Effects of ethylene. *Amer. J. Bot.*, 60: 883–888.

Avadhani, P.N., Nair, H., Arditti, J. & Hew, C.S. (1994). Physiology of orchid flowers. In: Arditti, J. (Ed.), Orchid Biology – Reviews and Perspectives, vol. 5. Wiley, New York, pp. 189–362.

Baker, M.L. & Baker, C. O. (1991). *Orchid Species Culture*. Timber Press, Portland.

Basiran, M.N., Bangi, Mys and Ariffin, S. (2002). Molecular techniques as complementary tools in orchid mutagenesis. Nippon Genshiryoki Kenkyujo, JAERI Conference, 91–102.

Beattie, A.J. (1985). The Evolutionary Ecology of Ant-Plant Mutualisms. Cambridge University Press, Cambridge.

Belarmino, M.M. & Mii, M. (2000). Agrobacterium mediated transformation of a Phalaenopsis orchid. *Plant Cell Reports*, 19: 435–442.

Bhattacharjee, S. K. & De, L. C. (2010). *'Advanced Commercial Floriculture'(2nd and Enlarged Edition,)*: In 2 volumes. pp. 798. Published by Aavishkar Publishers & Distributors, Jaipur, Rajasthan.

Bhattacharjee, S.K. & Das, S.P. (2008). In: Orchids- Botany, Breeding, Cultivation, Uses and Post-harvest Management. Aavishkar Publishers, Distributors, Jaipur, India, pp. 396.

Bhattacharjee, S.K. & De, L. C. (2005). *Post-harvest Technology of Flowers and Ornamental Plants*. Pointer Publishers, Jaipur, Rajasthan (India), pp. 440.

Bhattacharjee, S.K. & De, L.C. (2003). In: Advanced Commercial Floriculture, Part I, aavishkar Publishers, Distributors, Jaipur, India, pp. 330.

Bhattacharjee, S.K. (1997). Packaging fresh cut flowers. *Indian Horticulture*, 41 (4): 23–27.

Bielaski, R.K. & Reid, M.S. (1992). Physiological changes accompanying senescence in the ephemeral daylily flower. *Plant Physiol.*, 98: 1041–1049.

Bobisud , C.A. & Kamemoto, H. (1982). Selection and inbreeding in apmhidiploid Dendrobium (Orchidaceae). *J. Amer. Soc. Hort. Sci.*, 107: 1024–1027.

Bose, T.K. & Bhattacharjee, S.K. (1980). In: *Orchids of India*, Naya Prakash, Kolkata, India, pp. 538.

Bose, T.K., Bhattacharjee, S.K., Das, P. & Basak, U.C. (1999). In: *Orchids of India* (Second Edition), Naya Prakash, Kolkata, India, pp. 487.

Chai, M.L. & Kim, D.H. (2004). Establishment of Agrobacterium mediated transformation in Phalaenopsis. *Acta Horticultura Sinica*, 31 (4): 537–539.

Chan, Y.L., Chen, W.H. & Chan, M.T. (2003). Phalaenopsis orchid gene transformation (I) – optimization of transient gene expression. *Journal of the Chinese Society for Horticultural Science*, 49 (10): 33–44.

Cheamuangphan, A., Panmanee, C. & Tanusuchat, R. (2013). Value Chain Analysis for Orchid Cut Flower Business in Chiang Mai. *Business and Information*, July 7–9.
Chen, L., Hatano, T. & Niimi, Y. (2002). High efficiency of *Agrobacterium* mediated rhizome transformation in *Cymbidium*. *Lindleyana*, 17 (3): 130–134.
Chia, T.F., Chan, Y.S. & Chua, N.H. (1994). The firely luciferage gene as a non invasive reporter for Dendrobium transformation. *The Plant Journal*, 6: 441–446.
Chowdhery, H.J. (1998). Orchid Flora of Arunachal Pradesh. Bishen Singh Mahendra Pal Singh, Dehradun, India.
Chowdhery, H.J. (2001). Orchid diversity in north-east India. *J. Orchid Soc. India*, 15: 1–17.
Chowdhery, H.J. (2009). Orchid diversity in northeastern states of India. *J. Orchid Soc. India*, 23 (1–2): 19–42.
Chung, SY, Choi SH & Kim, MJ (2006). Genetic relationship and differentiation of *Paphiopedilum* and *Phragmepidium* based on RAPD analysis. *Scientia Horticulturae*, 109: 153–159.
Da Silva, J.A.T. (2003). The Cut Flower: Post-harvest considerations. *J. Biol. Sci.*, 3(4): 406–442.
Davidson, D.W.& Epstein, W.W. (1989). Epiphytic associations with ants. In: Luttge, U. (Ed.), Ecological Studies: Evolution and Ecophysiology of Vascular Plants as Epiphytes, Springer, Berlin, pp. 200–233.
Davidson, O.W. (1960). Principles of orchid nutrition. In: *Proceedings of the Third World Orchid Conference*. Staples Printers Ltd, Rochester, pp. 224–233.
De, L. C. & Bhattacharjee, S.K. (2011). *'Ornamental Crop Breeding'*, Pp. 438, Published by Aavishkar Publishers & Distributors, Jaipur, Rajasthan.
De, L. C. (2011). *'Value Addition in Flowers and Orchids'*, pp. 294, Published by New India Publishing Agency, Pitampura, New Delhi-110088.
De, L. C. (2013). In: *'Nursery and Landscaping'*, pp. 248, published by Pointer Publisher, Jaipur, Rajasthan.
De, L. C. (2014). *'Production Technology of Commercial Flowers'*: In 2 volumes Pp. 599. Published by Pointer Publisher, Jaipur, Rajasthan.
De, L.C. & Medhi, R.P. (2012). Cymbidium-A Diversified High Value Orchid for North Eastern India. *Agrobios Newsletter*, 10: 30–32.
De, L.C. Barman, D. Medhi, R.P. Chhetri, Geetamani & Pokhrel,H. (2013). Production Technology of *Phalaenopsis*. Technical Bulletin No. 15. NRC for Orchids, Pakyong, Sikkim, pp. 26.
De, L.C. D. Barman, D., Medhi, R.P. Chhetri, Geetamani & Pokhrel, H. (2013). *Production Technology of Dendrobium*. Technical Bulletin No.13, NRC for Orchids, Pakyong, Sikkim. pp. 30.
De, L.C.& Bhattacharjee, S.K. (2002). Vase life of cut rose 'Queen Elizabeth' as affected by aquatic fungi and bacteria. *Indian Rose Annual*, 18: 86–90.
De, L.C. & Bhattacharjee, S.K.(2000). Methods for prolonging vase life of cut flowers-a review. *Orissa Journal of Horticulture*, 28(1): 73–78.
De, L.C., Chhetri, Geetamani & Medhi, R.P. (2011). Orchid-A Wonderful Crop for Diversification. In: Proceedings of National Conference on 'Orchids in India: Diversity, Charaterization and Resource Development for community Livelihood & Orchid Show' held at National Academy of Sciences, Allahabad, December 21–23, 2011, pp. 17–25.
De, L.C., Deb, P. Chhetri, Geetamani, Medhi, R.P. & Pokhrel, H. (2013). *Post-harvest management in orchids*. Technical Bulletin No.16, NRC for Orchids, Pakyong, Sikkim, pp. 37.
Devi, J. and Deka, P.C. (2001). Hybrids in orchids: an overview: In Souvenir and Abstract, *Orchid Diversity in India, Science and Commerce and Orchid Show*, 6th National Seminar, TOSI, Palampur, India.
Devi, J. & Deka, P.C. (2000). Induction of polyploidy in *Phaius tankervilliae*. *J. Orchid Soc. India*, 14 (1–2): 1–5.
Devi, J. & Deka, P.C. (1992). Pollen viability, stigma receptivity and cross compatibility of some Indian orchids. *J. Orchid Soc. India*, 6(1–2): 79–84.

Drake, B.G., Gonzalez, Miller, M.A. & Long, S.P. (1997). More efficient plants: a consequence of rising atmospheric CO_2? *Annual Review Plant Physiology and Plant Molecular Biology*, 48: 609–639.
Duncan, E, Robert (1959). Orchids and Cytology, In: *The Orchids-A Scientific Survey* (Withner, C.L. ed.). The Ronald Press Company, New York, pp. 529–587.
Duncan, E. Robert (1947). The hybrid lady slipper. Orchid Digest, Sept-Oct, 199–207.
Evans, R.Y. & Reid, M.S. (1986). Control of petal expansion during diurnal opening of roses. *J. American Soc. Hort. Sci*, 111: 55–63.
Eze, J.M.O., Mayak, S., Thompson J.E. & Dumbroff E.B. (1986). Senescence in cut carnation flowers, temporal and physiological relationship among water status, ethylene, abscissic acid and membrane permeability. *Physiol. Plant.*, 68: 323–328.
Fenton, K.J. (1951). Colour inheritance in orchids. *Am. Orchid Soc. Bull.* 20: 519–523, 587–591.
Fischer (1950). Presence of ethylene gas poses problems in cut-flower storage. *Florists Rev.,* 106: (3814): 31–32.
Friend, G.M.R. (2004). *Gowing orchids in your garden*. Published by Timber Press, USA, pp. 218.
Fuchs, R.F. (1997). Fabulous Vandaceous intergenerics. *Orchids,* 66: 350–357.
Gautam, P.L., Singh, Ajay Kumar, Srivastava, Manoj & Singh, P. K. (2012). Protection of Plant Varieties and Farmers Rights-A Review. *Indian J. Plant Genet. Resour.*, 25 (1): 9–30.
Ginsberg, H.S. (2000). Breeding trends in red Phalaenopsis- the incomparable Phalaenopsis, 'Brother Purple'. *Orchids*, 69 (11): 1050–1060.
Goh, C.J. & Arditti, J. (1985). Orchidaceae, pp. 309–336. In: A.H. Halevy (Ed.). *Handbook of Flowering*, vol. I, CRC Press, Florida.
Goh, C.J., Strauss, M.S. & Arditti, J. (1982). Flower induction and physiology in orchids. In: Arditti, J. (Ed.), Orchid Biology – Reviews and Perspectives, vol. 2, Cornell University Press, Ithaca, pp. 213–241.
Granier, E. (2002). Blue Cattleya breeding.*Orchid Digest*, 66(2): 59–61.
Griesbach, R.J. (1985). An orchid in every pot. *Florists' Rev.*, 176(4548): 26–30.
Griffiths, H., Smith, J.A.C., Luttge, U., Popp, M., Cram, W.J., Daiz, M., Lee, H.S.J., Medina, E., Schafer, C. & Stimmel, K.H. (1989). Ecophysiology of xerophytic and halophytic vegetation of a coastal alluvial plain in Northern Venezuela IV. *Tillandsia flexuosa* Sw. and *Schomburgkia humboldtiana* Reichb., epiphytic CAM plants. *New Phytologist*, 111: 273–282.
Guignard, L. (1982). Recherches sur le development de l'anthere et dn pollen des orchidees. *Ann. Sci. Nat. Bot.*, (Ser VI) 14 : 26–45.
Gupta, R. (1993). Conservation and utilization of Indian medicinal plants. *Indian J. Plant Genetic Resources*, 6: 131–138.
Gutierrez, R.M.P. (2010). Orchids: A review of uses in traditional medicine, its phytochemistry and pharmacology. *Journal of Medicinal Plant Research*, 4 (8): 592–638.
Hajra, P.K. & De, Aparajita. (2010). Phyto-geographic analysis of orchid flora in India, *J.Orchid Soc. India*, 24 (1/2): 43–46.
Halevy, A. H. (1986). Flower senescence. In: *Process and Control of Plant senescence*, (Eds).Lesham, Y.Y., Halevy, A.H. and Frenkel, C. Amsterdam, The Netherland: Elsevier.
Harper, T. (1993): Multiflora Phalaenopsis, *Am. Orchid Soc. Bull.*, 62 (2): 126–133.
Hedren, M., Fay, M.F. & Chase, M.W. (2001). Amplified fragment length polymorphisms (AFLP) reveal details of polyploid evolution in Dactylorhiza (Orchidaceae). *Am. J. Bot.*, 88 (10): 1868–1880.
Hedren, M., Klein, E. & Teppner, H. (2000). Evolution of polyploids in the European orchid genus Nigritella: evidence from allozyme data. *Phyton Horn*, 40(2): 239–275.
Hegde, S.N. (2012). *Ex-situ* and *in situ* conservation of orchids in India. *J.Orchid Soc. India*, 26(1–2): 1–4.
Henke de Greef. (2008). Details about D.U.S. Testing for Plant Breeders Rights in Orchids in Europe. Abstracted in Taiwan International Orchid Symposium.
Herman, D.E. (1997). The species behind standard cattleyas. *Orchids,* 66(3): 234–243.

Hew, C.S. & Ng, C.K.Y. (1996).Changes in mineral and carbohydrate content in pseudobulbs of the C_3 epiphytic orchid hybrid *Oncidium* 'Goldiana' at different growth stages. *Lindleyana*, 11: 125–134.
Hew, C.S. & Yong, J.W. H. (2004). *The Physiology of Tropical Orchids in relation to industry*. World Scientific A1 Books. Co.in, pp. 370.
Hew, C.S. & Yong, J.W.H. (1994). Growth and photosynthesis of *Oncidium* Goldiana. *J. Horticultural Sci.*, 69: 809–819.
Hew, C.S. & Yong, J.W.H. (1997). The physiology of tropical orchids in relation to the industry. World Scientific Press, Singapore, New Jersey, London, pp. 341.
Hew, C.S. (1987). Respiration in orchids. In: Arditti, J. (Ed.), Orchid Biology – Reviews and Perspectives, vol. 4. Cornell University Press, Ithaca, pp. 227–259.
Hew, C.S., Ng, C.K.Y., Gouk, S.S., Yong, J.W.H. & Wong, S.C. (1996). Variation in $\delta^{13}C$ values for different plant parts of an *Oncidium* orchid. *Photosynthetica*, 32: 135–139.
Hew, C.S., Soh, W.P. & Ng, C.K.Y. (1997). Variation in photosynthetic characteristics along the leaf blade of *Oncidium* Goldiana, a C_3 tropical epiphytic orchid hybrid. *Int. J. Plant Sci.*, 159: 116–120.
Hew, C.S., Wee, K.H.& Lee, F.Y.(1987). Factors affecting the longevity of cut Aranda flowers. *Acta Horticulturae*, 205: 195–202.
Hew, C.S., Ye, Q.S. & Pan, R.C. (1989). Pathways of carbon dioxide fixation in some thin-leaved orchids. *Lindleyana*, 4: 54–157.
Hsieh, R.M., Chen, W.H., Hsu, H.M., Lin, Y.S., Tsai, W.T., Fu, Y.M., Chan, M.T. & Yu, S.M. (1997). *Agrobacterium tumefaciens* mediated transformation of a Phalaenopsis orchid. *Taiwan Sugar*, 44: 12–18.
Hurst, C.C. (1925). In: *Experiment in Genetics*. Cambridge.
Jalal, J.S., Kumar, P. Rawat, G.S. & Pangtey, Y.P.S. (2009b). *Habenaria pubescens* Lindley: an interesting orchid from Western Himalaya. Richardiana 9 (2): 76–84.
Jalal, J.S., Kumar, P. Tiwari, L.M. & Pangtey, Y.P.S. (2009a). Conservation status of the endemic orchid *Peristylus kumaonensis* Renz. (Orchidaceae) of Western Himalaya. Nature and Science 7 (5): 86–89.
Jalal, J.S., Rai, I.D., Kumar, P. Rawat, G.S. & Pangtey, Y.P.S. (2010a). *Platanthera leptocaulon* (Hooker f.) Soo: an addition to the orchid flora of western Himalayas, India. *Richardiana* 10 (2): 85–93.
Jalal, J. S., Kumar, P. Rawat, G. S. & Pangtey, Y. P. S. (2010b). *Bulbophyllum hirtum* (Sm.) Lindl. (Orchidaceae) – New record from western Himalaya, India. *Indian Journal of Forestry* 33 (3): 447–448.
Jawaharlal, M., Dinesh Babu, M. & Indhumathi, K. (2006). Postharvest packaging techniques for Dendrobium Hybrid Sonia -17. *J. Orn. Hort.* 9 (1): 16–19.
Johnson, D.C. (1999). Floriculture and environmental horticulture situation and outlook report, October, 1999. Economic Research Service, United States Dept. of Agriculture, Horticulture Year Book Stock # 99003.
Jomy, T.G.& Sabina,G.T. (2002). Effect of conditioning and pulsing on vase life of Dendrobium Sonia inflorescences. *J. Orn. Hort.*, 5(1): 80–81.
Kaiser, R. (1993).The scent of orchids. *Elsevier Science Publishers*, Amsterdam.
Kamemoto, H. & Amore, T.D. (1990). Inheritance of semi-alba and alba in Dendrobium. In: *Proc. 13th World Orchid Conference* (Kernohan, J. et al eds.) World Orchid Conf. Proc. Trust, Auckland, pp. 242–244.
Kamemoto, H. (1950). Polyploidy in cattleyas. *Am. Orchid Soc. Bull.* 19: 366–373.
Kamemoto, H., Amore, T.D. and Kuehnle, A.R. (1999). In: *Breeding Dendrobium Orchids in Hawaii.* University of Hawaii press, Honululu, pp. 166.
Keith, A. (2000). *Cymbidium devonianum*-the first danglers. *Orchids*, 108 (1233): 141–143.
Ketsa, S. & Rugkong, A. (1999). Senescence of Dendrobium 'Pompadour' flowers following pollination. *J. Hort. Sci. & Biotechnol.*, 74(5): 608–613.

Klein, T.M., Wolf, E.D., Wu, R. & Senford, J.C. (1987). High velocity micro-projectile for delivering nucleic acids into living cell. *Nature*, 327: 70–73.

Knapp, J.E., Kausch, A.P. & Chandlee, J.M. (2000). Transformation of three genera of orchid using the bar gene as a selectable marker. *Plant Cell Reports*, 19: 893–898.

Kuehnle , A.R., Fujii, T., Mudalige, R. & Alvarez, A (2004). Gene and genome mélange in breeding of *Anthurium* and *Dendrobium* orchid. *Acta Horticulturae*, 651 (2): 115–122.

Kuehnle, A. R. & Sujji, N. (1992). Transformation of Dendrobium orchid using particle bombardment of protocorms. *Plant Cell Reports*, 11: 484–488.

Kuehnle, R.A. (2006). Orchids. In: *Flower Breeding and Genetics* (Anderson, N.O. ed), Springer, Netherlands, 539–560.

Kuehnle,R.A., Lewis, D., Markham, K., Mitchell, K. Davies, K. & Jordan, B. (1997). Floral flavonoids and pH in *Dendrobium* species and hybrids. *Euphytica*, 95: 187–194.

Kumar, K. Madhu & Sheela,V.L. (2007). Status of breeding in orchids-A review. *Journal of Ornamental Horticulture,* 10: 199–208.

Lee, N. & Lin, G.M. (1984). Effect of temperature on growth and flowering of *Phalaenopsis* white hybrid. *J. Chinese Soc. Hort. Sci.*, 30(4): 223–231.

Lekawatana, S. (2010). Thai orchid: Current Situation. In: Proceedings of Taiwan International Orchid Symposium, 5 March, 2010, Taiwan Orchid Plantation, Tainan, Taiwan.

Lenz, L.W. and Wimbler, D.E. (1959). Hybridization and inheritance in Orchids. In: *The Orchids – A Scientific Survey* (Withner, C. ed.), Ronald Press Co.Ltd., New York, 216–314.

Liau, C.H., You, S.J., Prasad, V., Hsiao, H.H., Lu, J.C., Yang, N.S. & Chan, M.T. (2003). *Agrobacterium tumefaciens* mediated transformation of an Oncidium orchid. *Plant Cell Reports*, 21 (10): 993–998.

Lindner, R.C. (1946). Studies on packaging and storage of Vanda (Joaquim) flowers. *Howaii Agric. Exp. Stat. Progress Notes*, 49: 1–5.

Liu, H., Chang, L.F., Luo, Y.B., Chen, B.S., Wang, Z.S. & Gu, H.Y. (2010). Potential challenges of climate change to orchid conservation in a wild orchid hotspot in Southwestern China. *Bot. Rev.*, 76: 174–192.

Lopez, R.G. & Runkle, E.S. (2005). Environmental Physiology of Growth and Flowering of Orchids. *HortScience*, 40 (7): 1969–1973.

Mathews, M.R. (1996). Brassavola hybrids. *Orchids*, 65(9): 944–949.

Mehlquist, G.A.L. (1952). Chromosome numbers in genera. *Cymbidium Soc.* News.7.

Mehlquist, G.A.L. (1958). Genetics of orchid breeding. *Proc. 2nd World Orchid Conf.* pp. 200–209.

Men, S.Z., Ming, X.T., Liu, R.W., Wei, C.H. & Li, Y. (2003b). Agrobacterium mediated genetic transformation of a Dendrobium orchid. *Plant Cell, Tissue and Organ Culture*, 75 : 63–71.

Menninger, Emma, D. (1954). Tetraploid cymbidiums. *Am. Orchid Soc. Bull.* 23: 158–161.

Mitchell, T. & Tanner, T. (2006). Adapting to climate change -Challenges and opportunities for the development community. Tearfund, pp. 40.

Mountford, N.K. (2001). Phalaenopsis: searching for the perfect 'Blue'. *Orchid Review*, 109 (1238): 96–98.

Nagrare, V.S. & Ram Pal (2008). Cultivating potted orchids fetches more. *Indian Horticulture*, March-April issue, 24–26.

Nan, G.L. & Kuehnle, A.R. (1995b). Genetic transformation in Dendrobium. In: Biotechnology in Agriculture and Forestry, Vol. 34, Plant Protoplasts and Genetic Engineering VI (eds. Y.P.S. Bajaj), Springerverlag, New York, United States of America, pp. 149–160.

Nayar, M. P. & Sastry, A. R. K. (1990). *Red Data Book of Indian Plants.* Volume III. Botanical Survey of India.

Nayar, M. P.& Sastry, A. R. K. (1987). *Red Data Book of Indian Plants.* Volume I. Botanical Survey of India.

Nayar, M. P.& Sastry, A. R. K. (1988). *Red Data Book of Indian Plants*. Volume II. Botanical Survey of India.

Nayar, M.P. & Sastry, A.R.K. (1997–1998, 1999). *Red data Book of Indian Plants*. Volume 1–3, Botanical Society of India, Calcutta.

Nayar, M.P. (1996). *Hotspots of endemic plants of India, Nepal and Bhutan*. Tropical Botanic Garden and Research Institute, Trivandrum, India.

Ng, C.K.Y. & Hew, C.S. (1996). Pathway of phloem loading in the C_3 tropical orchid hybrid *Oncidium* Goldiana. *J. Exp. Botany*, 47: 1935–1939.

Northen, R.T. (1949). In: Home Orchid Growing. D. Van Nostrand Co., Inc. New York, pp. 286.

O'Neil, S.D. (1997). Pollination regulation of flower development. *Ann. Rev. Plant Physiol. & Plant Mol. Biol.*, 48: 547–574.

Ong, H.T. & Lim, L.L.(1983). Use of silver nitrate and citric acid to improve shelf life of Oncidium 'Golden Showers' flowers. *Orchid Rev.*, 90: 264–266.

Pritchard, H.W. (1989). Modern Methods in Orchid conservation: The Role of Physiology. Ecology and management, Cambridge University Press, UK.

Rao, A.N. (2004). Medicinal orchid wealth of Arunachal Pradesh. Newsletter of *Envis Node on Indian Medicinal Plants*, 1(2): 1–5.

Rao, A.N., Rajeevan, P.K., Srivastava, Manoj, Dhiman, S.R. Dhiman & De, L.C. (2012). Guidelines for the Conduct of Test for Distinctiveness, Uniformity and Stability on Orchid *Cattleya* and *Phalaenopsis* orchids, Protection of Plant Varieties and Farmers Rights Authority, NASC Complex, New Delhi-110012. *Plant Variety Journal of India,* 6 (11): 7–116.

Rao, A.N., Rajeevan, P.K., Sood, S.K., De, L.C. & Rawat, G.S. (2011). Guidelines for the Conduct of Test for Distinctiveness, Uniformity and Stability on Orchid *Cymbidium, Dendrobium* and *Vanda* orchids, Protection of Plant Varieties and Farmers Rights Authority, NASC Complex, New Delhi-110012. *Plant Variety Journal of India,* 5 (10): 5–83.

Rao, N.A. (2006). Variability Identification of Paphiopedilum, Anthogonium and some vandaceous species. Bulletin of Arunachal Pradesh, 22: 45–51.

Rao, T.A. & Sridhar, S. (2007). *Wild Orchids of Karnataka – A Pictorial Compendium*, INCERT, Bangalore, India.

Rogerson, W.P. (1991). Hybridization of white paphiopedilums, Part-I, *Am. Orchid Soc. Bull*, 60 (7), 673–682.

Rolfe, R.A. & Hurst, C.C. (1909). In: *The Orchid Stud Book*. London, pp. 327.

Rotor, G.B. (1959). The photoperiodic and temperature responses of orchids, pp. 397–416. In: C.L. Withner (Ed.). *The Orchids*. Ronald Press, New York.

Sander, F.K. (1946). In: Sanders Complete List of Orchid Hybrids, London, pp. 308.

Sarkar, Indrajit, Mandal,T. Naveen Kumar, P., Kumar, Rajiv , Misra, Sanyat, Misra, R.L. & Kishan Pal Singh (2009). *Temperate orchids*. AICRP on Floriculture, Technical Bulletin No.28, IARI, Pusa, New Delhi-12, pp. 77.

Sharief, M.U. (2011). Survey and conservation of rare and endemic orchids of Western Ghats. *J.Orchid Soc.India.*, 25(1–2): 89–99.

Sheehan, (1954). Respiration of cut flowers of Cattleya mossiae. *Amer. Orchid Soc. Bull.*, 23: 241–246.

Shibuya, K., Yoshioka, T., Hashiba, T. & Satoh, S. (2000). Role of the gynoecium in natural senescence of carnation (*Dianthus caryophyllus* L.) flowers. *J. Exp. Bot.*, 51: 2067–2073.

Singh, Amritpal & Duggal, Sanjiv (2009). Medicinal orchids: An overview. *Ethnobotanical Leaflets,* 13: 351–63.

Singh, B.M. (2005). Collection and conservation of endangered medicinal and aromatic plants, In: *Course compendium – Winter school on advances in Medicinal Aromatic and Underutilized Plants Research* (eds. Tyagi, C.S., Verma, P.K., Hooda, J.S., Yadav, O.P., and Goyal, R.K.) Sept 29-Oct, 19 , 2005 at CCSHAU, Hisar, Haryana, India, pp. 1–7.

Singh, F. & Prakash, D. (1995). Cytogenetics of Indian orchids. *Advances in Horticulture Vol-12-Ornamental Plants* (Chadha, K.L. and Bhattacharjee, S.K. eds.) Malhotra Publishing House, New Delhi, pp. 391–481.

Singh, Foza (1990). Indian orchids. *Indian Horticulture*, 35(1): 14–15.

Singh, K.P., Phukan, S. & Bujarbarua, P. (2001). Orchidaceae in *Floristic Diversity and Conservation Strategies in India*. Vol. IV (eds. N.P. Singh and D.K. Singh) pp. 1735–1827., Dehradun, India.

Storey, W.B. (1952). Chromosome numbers in some Vanda species and hybrids. *Am. Orchid Soc. Bull.* 21: 801–806.

Strasburger, E. (1888). Ueber Kern-und Zelltheilung im Pflanzenreich, nebst einem anhang uber Befruchtung. *Histol. Beitrage I*. Fischer, Jena.

Sun, WY, Liao, YJ, Hung, SY, Chang JC & Sung, MJ (2011). Development of ITS based SCAR markers for discrimination of *Paphiopedilum armeniacum, Paphiopedilum micranthum, Paphiopedilum delenatii* and their hybrids. *Scientia Horticulturae*, 127: 405–410.

Tandon, P. Abrol, Y.P. & Kumaria, Suman (2007). In: *Biodiversity and its significance*, published by I.K. International Pvt Ltd., pp. 370.

Taylor, S. (2009). How to make an orchid tree. *http: // www. Bellaonline.com/articles*.

Ten Have, A. & Woltering, E.J. (1997). Ethylene biosynthetic genes are differently expressed during carnation (*Dianthus caryophyllus* L.) flower senescence. *Plant Mol. Biol.*, 34: 89–97.

Teoh Eng Soon (1998). A joy forever, *Vanda* 'Miss Joaquim', Singapore's National Flower, Times Edition.

Teoh, Eng-Soon (2005). In: *Orchids of Asia*. Times Edition, Singapore, pp. 367.

Thammasiri, K., Tang, C.S., Yamamoto, H.Y. & Kamemoto, H. (1986). Carotenoids and chlorophylls in yellow flowered *Dendrobium* species. *Lindleyana*, 1: 215–218.

Thomas, F.O.H. (2001). Modern white phalaenopsis: origin and current status. *Orchid Digest*, 65(4): 148–154.

Tim Wing Yam (2001). Vanda 'Miss Joaquim', the first FCC/ RHS *Vanda* hybrid. *Orchid Review*, 109 (1237): 25–27.

Van Doorn, W.G. (1997). Water relation of cut flowers. *Hort. Review*, 18: 1–85.

Wang, Y.T. (1995). *Phalaenopsis* orchid light requirement during the induction of spiking. *HortScience*, 30(1): 59–61.

Wells, E.W. (1956). The elusive tetraploid I. Counting chromosomes in cymbidiums. *Am. Orchid Soc. Bull.* 25: 84–88.

Wimber, D.E. (1954). Note on Cymbidium Flambing "nobilior". Cymbidium Soc. News, 9: 16–17

Wimbler, D.E. and Hernlund, K. (1955). Aneuploid chromosome numbers in cymbidiums. *Am. Orchid Soc. Bull.*, 24: 743–745.

Winter, K., Wallace, B.J., Stocker, G.C. & Roksandic, Z. (1983). Crassulacean acid metabolism in Australian vascular epiphytes and some related species. *Oecologia*, 57: 129–141.

Yang, J., Lee, H.J., Shin, D.H., Oh, S.K. Seon, J.H., Paek, K.Y. & Han, K.H. (1999). Genetic transformation of Cymbidium orchid by particle bombardment. *Plant Cell Reports*, 18: 978–984.

Yong, J.W.H. & Hew, C.S. (1995). Partitioning of ^{14}C assimilates between sources and sinks in the sympodial thin-leaved orchid *Oncidium* Goldiana. *Int. J. Plant Sci.*, 156: 188–196.

Yong, J.W.H. & Hew, C.S. (1996). Which orchid leaf is the key supplier of carbon for flower development. *Malayan Orchid Review*, 30: 25–46.

Yong, J.W.H. & Hew, C.S. (1995). The importance of photoassimilate contribution from the current shoot and connected back shoots to inflorescence size in the thin-leaved sympodial orchid *Oncidium* Goldiana. *Int. J. Plant Sci.*, 156: 450–459.

Yong, J.W.H. & Hew, C.S. (1995). The patterns of photoassimilate partitioning within connected shoots for the thin-leaved sympodial orchid *Oncidium* Goldiana during different growth stages. *Lindleyana*, 10: 92–108.

Yu, Z.H., Chen, M.Y., Nie, L., Lu, H.F. Ming, X.T., Zheng, H.H., Qu, L.J. & Chen, Z.L. (1999). Recovery of transgenic orchid plants with hygromycin selection by particle bombardment of protocorms. *Plant Cell, Tissue and Organ Culture*, 58: 87–92.

Zheng, X.N., Wen, Z.Q. & Hew, C.S. (1992). Responses of *Cymbidium sinense* to drought stress. *J. Horticultural Sci.*, 67: 295–299.

Zimmerman, J.K., (1990). Role of pseudobulb in growth and flowering of *Catasetum viridiflavum*(Orchidaceae). *Am. J. Botany* 77, pp. 533–542.

List of Figures

List of tables

Bei Fragen zur Produktsicherheit wenden Sie sich bitte an:
If you have any questions regarding product safety,
please contact:

Walter de Gruyter GmbH
Genthiner Straße 13
10785 Berlin
productsafety@degruyterbrill.com